AN INTRODUCTION TO

equations of state

THEORY AND APPLICATIONS

SHALOM ELIEZER

Department of Plasma Physics, SOREQ Nuclear Research Centre, Israel

AJOY GHATAK

Department of Electrical Engineering, National University of Singapore

HEINRICH HORA

Department of Theoretical Physics, University of New South Wales, Australia

with a Foreword by

EDWARD TELLER

Lawrence Livermore National Laboratory, USA

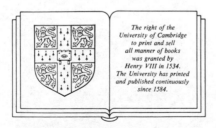

The right of the
University of Cambridge
to print and sell
all manner of books
was granted by
Henry VIII in 1534.
The University has printed
and published continuously
since 1584.

CAMBRIDGE UNIVERSITY PRESS

Cambridge

London   New York   New Rochelle

Melbourne   Sydney

Published by the Press Syndicate of the University of Cambridge
The Pitt Building, Triumpington Street, Cambridge CB2 1RP
32 East 57th Street, New York, NY 10022, USA
10 Stamford Road, Oakleigh, Melbourne 3166, Australia

First published 1986

Printed in Great Britain at the University Press, Cambridge

British Library cataloguing in publication data

Eliezer, Shalom
    An introduction to equations of state: theory
    and applications.
    1. Mathematical physics   2. Equations of state
    I. Title   II. Ghatak, A.K.   III. Hora, Heinrich
    530.1′5   QC20.7.E6/

Library of Congress cataloging in publication data

Eliezer, Shalom.
    An introduction to equations of state.
    Bibliography: p.
    Includes index.
    1. Equations of state.   I. Ghatak, Ajoy K.
II. Hora, Heinrich, 1931–.   III. Title.
QC173.4.E65E45   1986   530.4   85–29094

ISBN   0 521   30389 3

CONTENTS

This book is dedicated to our wives,
Yaffa Eliezer, Gopa Ghatak and Rose Hora
who gave us encouragement and
showed patience and understanding
during the many hours spent
on its preparation.

FOREWORD BY EDWARD TELLER

Is physics completed? At the time when the behavior of atoms was finally understood this seemed to be a very real question.

A few problems remained. These included the relations between elementary particles and the unification of gravitation, electromagnetism and nuclear physics. About the latter not much was known and its importance was underestimated.

Today particles appear less elementary and unification looks more ambitious than ever.

This book on the equation of state gives impressive evidence that a great chapter of science including most of physics, chemistry and a great portion of astronomy is, indeed, completed and unified. Just one idea had to be added to the systematic knowledge of the elementary building blocks. That idea developed into the theory of probability and statistical physics.

The structure of statistical mechanics is deceptively simple. The main point appears to be to separate a bigger system into components whose energies can be added while their probabilities can be multiplied. From this statement it follows that probability depends on energy in an exponential manner. Classical equation of state theories in all of their approaches depend on this one circumstance. The quantities and ideas used in the comprehensive description of the development of two centuries are remarkably homogeneous, even uniform.

Yet, the theory of the equation of state spans an enormous range. Obviously, huge numbers are involved. More importantly, the methods of thinking are quite diverse. On the one end we have reversible processes. On the other irreversibility is the main rule. The distinction between causality and probabilistic arguments is even more basic. Einstein never was reconciled to 'laws' of probability from which there was no appeal to the supreme court of causality.

A discussion of the equation of state gives an impressive demonstration of how great a fraction of the inorganic world is explained by the revolution called quantum mechanics. At the same time the equation of state presents the tools by which our experimental knowledge can be extended into regions of extreme concentrations of energy and matter.

All this leads up to a part of physics which may move faster and farther than any other branch: astrophysics. Neutron stars and black holes are realistic examples of the two frontiers of physics: nuclear forces and general relativity. The authors use the same mathematics, the same concepts to introduce problems of astrophysics as they have used to explain the common properties of matter.

The first man whose ideas about atoms are still remembered is a philosopher from Thessaly: Democritos. Almost two and a half millennia ago he suggested that matter may not be divisible without limitations. A serious revival of his idea started around 1800. Since that time science has been accelerating. It still produces unexpected facts and the completion of physics is not in sight. My preconceived idea is that science is open-ended. In this book the reader will find an exciting review of the rich past and he also will get a glimpse into a future which may be unlimited.

PREFACE

The equation of state is the relation between the pressure, the temperature and the density (specific volume) of a physical system and is related both to fundamental physics and to applications in astrophysics, gases and condensed matter, and nuclear and elementary particle theory.

In the same way that Newtonian mechanics can be regarded as the foundation of physics, so the equation of state of ideal gases can be considered to be the foundation of thermodynamics, hydrodynamics and chemistry. Furthermore, as mechanics was extended to take account of relativity and quantization, so the equation of state had to be developed to describe states of matter in extreme density and temperature domains.

The main aim of this book is to provide the reader with a basic understanding of the development of the equation of state. It should be of value to undergraduate and graduate students with an interest in astrophysics, solid state matter under extreme conditions, plasma physics and shock waves, as well as special aspects of nuclear and elementary particle physics.

The systematic derivation of essential physical theory includes several original expressions. The elements of classical statistics and the Bose–Einstein and Fermi–Dirac equations of state (EOS) are based on partition functions and the Thomas–Fermi model derivation includes the exchange interactions which lead to the Thomas–Fermi–Dirac equation. Special attention is given to the virial theorem. A new treatment of the Grüneisen EOS, based on Einstein's and Debye's models of solids, is given which highlights Einstein's ingenious contributions. Fluid mechanics and the kinetic theory are derived with particular emphasis on shock waves and a special meaning is given to the relation between the equation of state, Grüneisen coefficients, cold pressure and high pressure shock waves in solids.

The book also describes several important applications of the equation of state together with accounts of the most recent research results including original new presentations. The study of inertial confinement fusion, especially laser fusion or particle beam fusion, is one of the fields in which pressures of nearly 10 gigaatmospheres and temperatures of about 100 million degrees are being achieved. These extreme conditions are described in a detailed and novel manner. The application of the equation of state to the extreme conditions of astrophysics is dealt with, special attention being given to the stability of normal stars based on radiation pressure and particle pressure. The equation of state in nuclear matter is described using the Hagedorn theory; this subject can either be related to the theory of strong interaction or to the model of the early universe.

This book is only a stroll through the equations of state in science and many more applications can be considered. In particular the physics of phase transitions could be the basis for a complementary volume on the equations of state. We believe that the book emphasizes the importance of equations of state in science and that further academic study and research is required to bring this subject to the attention of science students.

We would like to thank our secretaries, Mrs Marie Wesson and Ms Noemi Francisco of the University of New South Wales, for their precise and neat work. Our thanks are also extended to Mrs Catherine Faust (UNSW) for her immaculate preparation of the artwork.

Finally, the authors wish to acknowledge the support of the Gordon–Godfrey bequest, the Australian Research Grant Scheme and the Australian Academy of Sciences which was essential for the preparation of this book. These grants enabled the authors to meet in the australian summer of 1983 at the University of New South Wales in Sydney, where they began their collaboration on the book.

March, 1986

Shalom Eliezer
Ajoy Ghatak
Heinrich Hora

1

Introduction

1.1 General remarks

Important branches of physics were developed or originated from the equation of state while (in return) more and more complex formulations of the equations of state were due to the developments of modern physics. The ideal gas law was the first quantitative treatment of chemical kinetics and the beginning of thermodynamics and statistics, resulting in the first complete formulation of the laws of energy conservation and entropy. From the thermodynamics of entropy of radiation, Max Planck discovered the atomistic structure of action (quantization), one of the basic properties of nature. The properties of molecular interaction were described quantitatively for the first time by van der Waal's equation of state.

Nowadays, physics has developed towards research of such extreme conditions of matter, as shock waves in metals at several million atmospheres pressure and strong non-equilibrium states in chemical reactions using laser or plasma technology, while laser produced plasmas now provide matter at pressures of up to a thousand million atmospheres in the laboratory. In astrophysics, we observe objects at much more extreme conditions with pressures and temperatures of many orders of magnitudes beyond the extreme points reached in the laboratory. These astrophysical objects, previously topics of speculation only, are now very serious research fields producing detailed knowledge following the breathtaking development of space technology. The conditions of neutron stars and beyond involve nuclear matter such that the study of the equations of state could become a new route for solving fundamental problems in the physics of elementary particles and nuclei where the combined effort involves the physics of condensed matter and of ordinary high density plasmas.

While the study of equations of state was one of the most important problems in physics in the first of the nineteenth centurt, there is again now a high priority in modern physics to solve the numerous deficiencies in our knowledge of equations of state. The development of equations of state is again sought to meet the needs of the rapidly advancing modern physics of extreme states of matter.

Recognizing the lack of knowledge of the equations of state in modern physics, an introductory summary of knowledge of the development of the equations of state will be presented in the following chapters together with a presentation of the derivations and typical results, with the aim of giving an improved, generalized and shorter presentation of this developing field. Together with these basic concepts, numerous important and exciting applications, in the physics of high pressure in solids, metals and plasmas, for dense nuclear fusion, astrophysics and for nuclear and high energy physics, are given.

1.2   Phenomena at various densities and temperatures

The states of all substances are relatively simply described if the density is very low and the temperature is not too close to 0 K. In this case,

Fig. 1.1. Ranges of density and temperature for which equations of state are to be considered.

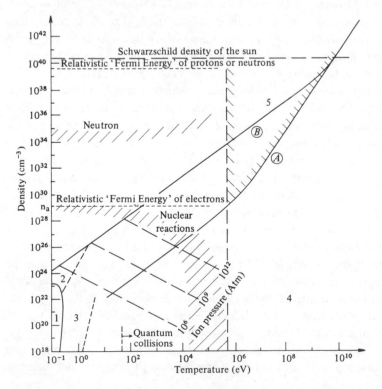

the ideal equation of state for pressure P in a volume V is determined by the temperature T

$$PV = RT \tag{1.1}$$

where the constant R depends on the number of moles in the volume V and the degree of freedom of the molecules. Expressing V as $1/n$, the density of n atoms of dissociated molecules (or ions) in the gas, eqn. (1.1) reads

$$P = nkT \tag{1.2}$$

where $k = 1.38 \times 10^{-16} \, \text{erg deg}^{-1}$ is Boltzmann's constant.

Fig. 1.1 is a diagram of density and temperature of matter which permits a classification of different states without the pressure – and subsequently the equation of state, being given. The density–temperature diagram of matter in this section is therefore a first classification of the problems discussed in the book. The next step will be to explain the Fermi-pressure in Section 1.3, and Section 1.4 will then provide an overview in terms of a preliminary pressure–temperature diagram for the range of states and their limitations to be treated and extended in the subsequent main sections of the book.

In Fig. 1.1, the range of the ideal equation of state is that of nonionized matter at low densities. At higher densities, range 1 covers condensed matter and range 2 may cover the conditions that ions are fixed in space (crystals), limiting their motion to defined centres of oscillation only (Brush 1967).

Range 3 has no sharp limits of high density and temperature and represents mainly a gaseous state with partial ionization. This state is determined by the Saha equation for describing the degree of dissociation by Boltzmann factors (first studied by Schottky (1920)) and covers strong coupling between electrons and ions (Hansen 1973; Golden 1983; Pines and Noziéres 1966) including the metallic state in the upper part of range 1. Because of the high density and low temperature, the electrostatic plasma effects not only decrease the ionization energy but cause the merging of spectral lines into the vacuum levels or the 'drowning of spectral lines' (Inglis and Teller 1939; Traving 1959) (which as well as the polarization shift (Griem 1981) can be explained in a straightforward way by an electrostatic atom model (Hora 1981, p. 28; Henry and Hora 1983; Henry 1983)). The whole physics of range 3 is not yet extensively developed, neither are the highly complicated equations of state in this range.

At higher temperatures and low densities when there is a high degree of ionization of atoms or the state of fully ionized plasma is reached – range 4 – the resulting plasma is gaseous and kinetically rather simple: a classical ionized gas governed by the ideal equation of state. This state, however, has non-classical properties: the classical Coulomb collisions change into quantum collisions at a temperature T^*

$$kT^* = \tfrac{4}{3}Z^2 m_0 c^2 \alpha^2 \tag{1.3}$$

as detected in plasmas as anomalous resistivity which can be explained quantitatively by this process (Hora 1981, p. 37; Hora 1981a). In (1.3), Z is the number of charges (degree of ionization) of the plasma ions, m_0 is the electron rest mass, c is the speed of light, and the fine structure constant is $\alpha = e^2/\hbar c$ where e is the electron charge and $\hbar = h/2\pi$: h is Planck's constant. Further, in range 4, thermonuclear reactions at temperatures around and above 10^4 eV will begin. Another example of non-classical behaviour is that at thermal equilibrium with equilibrium (opaque) black body radiation in the plasma, a strong coupling between electrons and black body radiation will occur, which we discuss in Section (1.5). Above a temperature $mc^2 = 0.52$ MeV, pair production and other inelastic interactions will occur. Nevertheless, the equation of state may well be the ideal one and of a simple nature.

The range 4 is limited towards higher densities by the curve A in Fig. 1.1 where the quantum effect of degeneracy of electrons will occur. The particles will then have a quantum energy E_q which is higher than the thermal energy. This quantum energy $E_q = p^2/2m$ corresponding to a momentum p is determined by the length $x = 1/n^{1/3}$ of the cube into which the electron with its density n is packed, which has to obey quantization

$$xp = h \tag{1.4}$$

resulting in

$$E_q = \frac{h^2}{2m} n^{2/3} \tag{1.5}$$

Taking into account that each of these volumes can be occupied by two electrons due to the spin and including the correct geometric factors for the spherical atoms, the quantum energy E_q is given by the Fermi energy with the electron density n given in cm^{-3}

$$\varepsilon_F = \frac{h^2}{2m_0} n^{2/3} \frac{(3/\pi)^{2/3}}{4} \tag{1.6}$$

$$= 5.82 \times 10^{-27} n^{2/3} \text{erg} = 3.652 \times 10^{-15} n^{2/3} \text{eV}$$

It is worth noting that this quantum energy is not only related to particles with a spin of $\tfrac{1}{2}$ (Fermions) following the Fermi–Dirac statistics (Dirac

1926; Fermi, 1928) but it is a basic quantum energy as described before. Only the density of states can be larger than that of the electrons. It is therefore of interest to note curve B for protons in Fig. 1.1, above which density the quantum energy is higher than the temperature.

The relativistic extension of these limits by using the particle velocity v and the electron rest mass m_0 for the energy $E = m_0 c^2 [(p^2/m_0^2 c^2) + 1]^{1/2}$ where $p = m_0 v/(1 - v^2/c^2)^{1/2}$, one arrives at the relativistically generalized Fermi energy

$$\varepsilon_F = \frac{(3/\pi)^{2/3}}{4} \frac{h^2}{2m_0} n^{2/3} \frac{1}{(\lambda_c/2)[n + 1/(\lambda_c/2)^3]^{1/3}} \tag{1.7}$$

where the Compton wave length of the electron $\lambda_c = h/m_0 c$ was used. Above the density $n_c = \lambda_c^{-3}$, the Fermi energy of the electrons is relativistic

$$\varepsilon_F = hcn^{1/3} \frac{(3/\pi)^{2/3}}{4} \quad (n > \lambda_c^{-3}) \tag{1.8}$$

and does not depend on the particle mass. For lower densities, (1.7) reduces to (1.6). The quantum (Fermi) energies for protons and electrons merge into the same lines in Fig. 1.1 at high energies, where the relativistic particles cannot be distinguished by their mass. The same phenomenon is known from charges oscillating in a laser field or black body radiation: if the oscillation energy becomes relativistic, the particles have the same energy and cannot be distinguished by their mass (Hora 1981). For temperatures above $m_0 c^2$ and at degenerate densities (Range 5), the electrons cannot follow Fermi–Dirac statistics as will be shown in the Section 1.5 because of strong coupling to the black body radiation.

The remaining part of Fig. 1.1 for temperatures below $m_0 c^2$ and above curve A of degeneracy, is characterized by the limit of relativistic Fermi energy $n_c = (1/\lambda_c)^3$, by the range where nuclear reactions are starting – even for low temperatures – which is for example related to picnonuclear reactions (Harrison 1964) as seen also from an increase of fusion cross-sections at high temperatures (Ichimaru 1984; Niu 1981). We refer also to Brush (1967) concerning these low temperature high density nuclear reactions.

At densities above $10^{34} \, \text{cm}^{-3}$, the reaction between protons p and electrons e to produce neutrons n and neutrinos ν

$$p + e = n + \nu \tag{1.9}$$

will start. The line of relativistic 'Fermi energy' of baryons (protons or neutrons) is therefore heuristic only. The cold fusion of protons to neutrons at these densities of $10^{35} \, \text{cm}^{-3}$ is easily understood from the fact that each electron is then compressed to a diameter of about 2×10^{-12} cm. At these densities electrons have sufficient quantum energy of compression, and can add to the protons which have the same radius. Reaction (1.9) should not be considered as the usual collision process. At lower densities than 10^{35} cm^{-3} the electron simply cannot unite with the proton because it would first need quantum energy, to be compressed to the size of the proton. This is the reason why we – contrary to Brush (1967) – drew the shaded area of the neutron generation for increasing density at higher temperatures, as the thermal motion will reduce the fusion reaction. Huge numbers of neutrinos should be emitted at these densities.

This is the range of densities where nuclear matter can be studied by heavy nuclei collisions. Shock waves are then generated and the density can reach five times the density of nuclear matter. Knowledge of the valid equation of state is then essential (Scheid *et al.* 1974; Stöcker *et al.* 1978; Stöcker 1986) and its details can be studied indirectly from the reactions following the 10^{-13} s duration state of the united nuclei (Stock 1984).

If the mass within the Schwarzschild radius of the Sun

$$R = 2M_0\gamma_0/c^2 \tag{1.10}$$

($\gamma_0 \approx$ gravitation constant, $M_0 \approx$ solar mass) is given the resulting density is the Schwarzschild density as shown in Fig. 1.1. One should note that younger stars of larger mass have a lower Schwarzschild density.

1.3   Quantum pressure and compressibility

In order to demonstrate that the quantum (Fermi) pressure really acts as a pressure even if the temperature of the electrons is very much lower than ε_F, we mention here as an example of its action the compressibility of solids by the quantum pressure.

For the ideal equation of state, (1.2), the compressibility (using $V = 1/n$) is

$$\kappa = -\frac{1}{V}\frac{\partial V}{\partial P} = \frac{1}{P} \tag{1.11}$$

For solids there was the theory of Madelung (1918) (see, e.g., Joos 1976) which was modified slightly later. This was based on the ad hoc assumption of a potential where the best possible fit of the exponent for the radial dependence was found to be of the number nine. Contrary to this highly hypothetical relation, we can explain the compressibilities of solids with a better fit and slope of the plot of the experimental values of the compressibility over three orders of magnitude in dependence on the electron density, if we take the quantum pressure of the valence electrons (with one or two electrons in the outermost shell). Taking ε_F as the energy per volume of the electrons in an atom ($P = \varepsilon_F/V$) and calculating

$$\kappa = -\frac{1}{V}\frac{\partial V}{\partial P} = \frac{6m_0}{h^2 n^{5/3}} \tag{1.12}$$

where the balance of the quantum energy with the electrostatic energy (Hora, 1981, p. 28) was included. Without this inclusion, the value of κ is half.

In Fig. 1.2, the results for one and two valence electrons are drawn and compared with the Madelung value. The quantum pressure theory of

Fig. 1.2. Compressibility of solid materials compared with Madelung's theory (Joos 1976) (dashed line) and with the quantum pressure including the counteracting electrostatic energy density (upper line); without the electrostatic energy, the lower line will follow where an interpretation of spin coupling for two valence electrons is possible per quantum state.

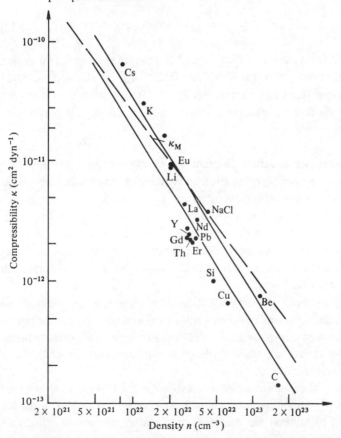

compressibility (Hora and Romatka 1982) reproduces the experimental values to a good approximation. Experimental deviations from the plots are due to the effective mass in the solids (in (1.12) the vacuum electron only was used; for an alternative derivation of the Schrödinger equation for the effective mass, see Appendix 1 of Hora (1981)), and due to additional electrostatic energy densities of attraction and repulsion by electron–electron, electron–ion or ion–ion interaction.

The result is that solids cannot be compressed easily because of the quantum (Fermi) pressure which is nearly 10 eV per atom (corresponding to 100 000 K temperature) but it acts as pressure only, not as temperature of the ideal equation of state. This follows for metals as well as for insulators. It demonstrates that the Fermi pressure is still present even in the unionized condensed material. The pressure of a substantial energy in the quantum (Fermi) energy can also be seen from the shaded line in Fig. 1.1 for nuclear reactions at very low temperatures of 0.1 eV or less: the quantum energy of protons (with one particle per energy level) of mass m_p

$$\varepsilon_F = \frac{h^2}{2m_p} \frac{(3/\pi)^{2/3}}{4} n^{2/3} 2^{2/3} \tag{1.13}$$

is then 17 keV for a density of 10^{29} cm^{-3}, just enough for nuclear fusion reactions with a very low rate. Without following up the sophisticated theory for these reactions (Brush 1967), (1.13) indicates that the ions even at low temperature do have enough quantum energy for performing the nuclear reactions.

1.4   Pressure–temperature diagram

When considering pressure, a knowledge of the equation of state is necessary. For our introductory overview we now consider the various regions in Fig. 1.3 and shall explain the states of degeneracy in more detail then in the preceding sections aiming to give a further introduction to the subsequent sections on Thomas–Fermi, Thomas–Fermi–Dirac and further models. Various domains in the calculation of the equation of state can also be illustrated by Fig. 1.3.

Region 1 represents the normal state of matter and, in general, the equations of state are controlled by ordinary chemical forces. This region includes phase transitions such as solid–liquid transition, critical phenomena (such as ferromagnetism), phase transitions due to changes of symmetry in solids, etc.

We increase the temperature corresponding to region 1, without increasing the pressure and enter region 2. At about $kT \approx 1$ eV ($T \approx 10^4$ K), the condensed or neutral gas matter is dissociated into atoms which are described to a good approximation by the equation of state for an ideal gas (see (1.1))

$$PV \approx NkT \tag{1.14}$$

$$E \approx \tfrac{3}{2}NkT \tag{1.15}$$

where P, V, T, N and E represent respectively the pressure, volume, temperature, total number of particles N and total energy E of the gas and $k(\approx 1.38 \times 10^{-16}\,\mathrm{erg\,K^{-1}})$ represents the Boltzmann constant.

Further increase of temperature (see region 3) leads to electron dissociation (i.e., to ionization) and one may use the Saha equation to describe the equation of state.

Before discussing region 4 we would like to mention that a gas consisting of free electrons is said to be non-degenerate when,

$$kT \gg \varepsilon_{\mathrm{F}} \tag{1.16}$$

where ε_{F} is known as the Fermi energy of the electron gas at $T = 0\,\mathrm{K}$ and is given by (1.6). When this is satisfied, the electron gas behaves as a classical gas and the pressure and total energy will be given by (1.14) and (1.15) respectively. On the other hand, when,

$$kT \ll \varepsilon_{\mathrm{F_0}} \tag{1.17}$$

Fig. 1.3. Scheme of a temperature–pressure diagram for the various states of matter.

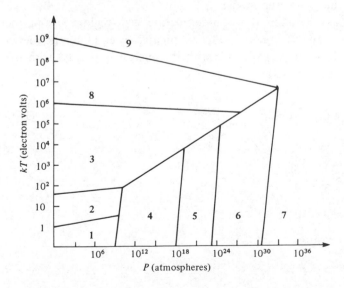

quantum effects dominate and the gas is said to be highly degenerate. At $T = 0\,\mathrm{K}$, the gas is said to be completely degenerate. If we assume the electron gas to be completely degenerate then elementary application of Fermi–Dirac statistics gives

$$
\left.
\begin{aligned}
\varepsilon &\approx \tfrac{3}{5}\varepsilon_{F_0}\\[4pt]
P &\approx \tfrac{2}{3}n\varepsilon_{F_0}
\end{aligned}
\right\}
\tag{1.18}
$$

and

where ε represents the average energy of the electron. Assuming, $n \approx 10^{26}\,\mathrm{cm}^{-3}$, we get,

$$
\left.
\begin{aligned}
\varepsilon_{F_0} &\approx 7.9 \times 10^2\,\mathrm{eV}\\[4pt]
\varepsilon &\approx 4.7 \times 10^2\,\mathrm{eV}\\[4pt]
P &\approx 5 \times 10^{16}\,\mathrm{dyne\,cm^{-2}} \approx 5 \times 10^{10}\,\mathrm{atm}
\end{aligned}
\right\}
\tag{1.19}
$$

Thus the average energy of an electron is much greater than the binding energy of the electrons in atoms and the electrons behave as a degenerate gas. Region 4 corresponds to $P \geqslant 10^9$ atm and $(kT/\varepsilon_{F_0}) \ll 1$ and the equation of state is satisfactorily described by the Thomas–Fermi model. However, the transition regions (between 1 and 4, and 2 and 4) are most difficult to describe satisfactorily and, in general, corrections to the Thomas–Fermi model are necessary. In this intermediate region, which can be as broad as $10^5 \leqslant P \leqslant 10^{12}$ atm, one tries to describe the electrons by a Thomas–Fermi model modified by exchange and quantum corrections. In this region the nuclei can be considered as a classical gas; see (1.14) and (1.15), or can be described by the Grüneisen model.

The boundary between the non-degenerate gas (region 3) and the degenerate electron gas (region 4) can be estimated by equating the energy associated with the ideal gas (1.15) with the energy associated with the degenerate Fermi gas (1.18),

$$
\tfrac{3}{2}kT \approx \tfrac{3}{5}\varepsilon_{F_0} = (3\pi^2)^{2/3}\frac{3h^2}{10m}n^{2/3}
\tag{1.20}
$$

or

$$
n \approx \frac{1}{3\pi^2}\left(\frac{5m}{h^2}\right)^{3/2}(kT)^{3/2}
$$

$$
\approx 9 \times 10^{39}(kT)^{3/2}\,[\mathrm{cm}^{-3}]
\tag{1.21}
$$

where kT is measured in ergs. Thus,

$$
P \approx nkT \approx 9 \times 10^{39}(kT)^{5/2}\,\mathrm{dyne\,cm^{-2}}
\tag{1.22}
$$

For $kT \approx 10^2 \, \text{eV} \approx 1.6 \times 10^{-10} \, \text{erg}$ we get

$$P \approx 3 \times 10^{15} \, \text{dyne cm}^{-2} \approx 3 \times 10^9 \, \text{atm}. \tag{1.23}$$

For $kT \approx 10^4 \, \text{eV}$, $P \approx 3 \times 10^{14} \, \text{atm}$. These estimates indicate the order of the pressures and temperatures at the boundary of regions 3 and 4.

At very high pressures (assuming complete degeneracy) the Thomas–Fermi equation of state (EOS) is given (as detailed later) by

$$P = \frac{h^2}{5m}(3\pi^2)^{2/3}\bar{n}^{5/3}\left| 1 - \frac{1}{2\pi a_0}\left(\frac{4Z^2}{\bar{n}}\right)^{1/3} + \cdots \right| \tag{1.24}$$

where $a_0 (= h^2/me^2)$ is the Bohr radius, Z is the total number of electrons in an atom and

$$\bar{n} = \frac{Z}{V} \tag{1.25}$$

represents the average number of electrons per unit volume; V is the volume associated with each atom. Obviously, at high pressures, $\bar{n}$ becomes very large and (1.24) reduces to (1.18), implying the applicability of the simple Fermi–Dirac model. However, when the pressures become very high (region 5), the electrons become relativistic and we should therefore use the proper relativistic theory for a degenerate electron gas, which has been extensively used in many astrophysics problems. We should mention that in many stars, the temperature T may be as high as 10 million degrees (giving $kT \approx 10^3 \, \text{eV}$); however, the matter may be so compressed that the Fermi energy ε_F may be $10^6 \, \text{eV}$. Thus the condition $kT \ll \varepsilon_F$ will be satisfied and we may apply the theory corresponding to completely degenerate electron gas (i.e., assume $T = 0$).

In region 6, the electrons are degenerate and relativistic while the protons and neutrons of the matter are degenerate but non relativistic, and in region 7 the pressures are so high that the reaction (1.9) becomes possible and we have a neutron star.

1.5 Radiation effects

Considering equilibrium black body radiation under the conditions of Fig. 1.1, problems with the equation of state will appear which we mention here as an example of a yet unsolved problem of the equation of state. This problem relates only to cases where the black body radiation is opaque and in equilibrium. In the laboratory, there are many extreme plasmas e.g. at inertial confinement fusion where the mean free path of the x-ray quanta of a relevant black body radiation is much larger than the plasma dimension. In this case one may neglect the radiation when discussing the equation of state. The following remarks refer primarily therefore to large and extreme astrophysical objects.

These considerations were derived from a question of applied physics about the electron emission current density j from cathodes at a temperature T and the spectrum of the quantum yield of photo cathodes (Görlich *et al.* 1957; Görlich and Hora 1958). The thermionic Richardson equation arrives at $j \approx T^2 \exp(-h v_0/kT)$ for the emission from a metal with a work function $h v_0$. This is simply due to the evaporation (given by a pressure) of the electrons in the interior of the metal with a Fermi–Dirac distribution crossing a surface potential $h v_0$. Richardson's initial assumption that the emission is due to photo effect by the black body radiation and not by evaporation was proved not to be true for cathodes, but would well dominate above 10^7 K because $j \approx T^3 \exp(h v_0/kT)$ as there is a stronger exponent of T (Hora and Müller 1961). At these high temperatures, the

Fig. 1.4. Ranges of different statistics or strong coupling to black body radiation determining the pressure of electrons.

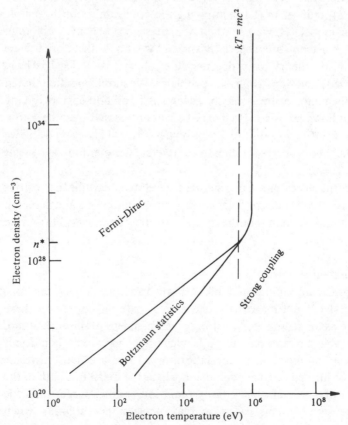

evaporation – and therefore the pressure and the equation of state – was determined by black body radiation.

The final analysis of the problem (Hora 1978) resulted in an equation of the current density j for a body of temperature T with opaque equilibrium black body radiation and a work function $h\nu_0$ (which is equivalent to the order of magnitude of the potential of an ordinary double layer of a plasma surface (Hora *et al.* 1984)) given by

$$j = \begin{cases} (4\pi m_0 k^2 e/h^3)\, T^2 \exp\left[-h\nu_0/kT\right] \\ \quad \text{for } (2m_0 kT)^{3/2} \ll \min\left[n_e h^3; (4\pi m_0{}^2 c^2)^{3/2}\right], \\[2mm] n_e e(k/2\pi m_0)^{1/2}\, T^{1/2} \exp\left[-h\nu_0/kT\right] \\ \quad \text{for } (2m_0 kT)^{3/2} \gg n_e h^3 > \dfrac{(2\pi)^{1/2}}{2m^2 c^2}(2m_0 kT)^{5/2}, \\[4mm] \dfrac{2e}{c^2}\dfrac{k^3}{h^3}\, T^3 \exp\left[-h\nu_0/kT\right] \\ \quad \text{for } \dfrac{(2\pi)^{1/2}}{2m_e{}^2 c^2}(2m_0 kT)^{5/2} > \min\left[n_e h^3; 4\pi^2(m_0{}^2 c^2)^{3/2}\right]. \end{cases} \qquad (1.26)$$

where n_e is the electron density. It turns out that the pressure is determined by the Fermi–Dirac statistics in the first range, by Boltzmann statistics in the second range (obviously as a bridging gap only), and by the strong coupling to the black body radiation in the remaining range (Fig. 1.4). Above the density

$$n^* = \left(\frac{2}{(2\pi)^{1/2}}\right)^{3/2}\left(\frac{1}{\lambda_c}\right) = 1.017 \times 10^{29}\, \text{cm}^{-3} \qquad (1.27)$$

with the Compton wave length λ_c, only Fermi–Dirac statistics is possible for temperatures below $m_0 c^2$, but a change of Fermi–Dirac statistics into strong coupling occurs at temperatures above $m_0 c^2$.

The optical intensity of the black body radiation at $kT = m_0 c^2$ is $7 \times 10^{27}\, \text{W cm}^{-2}$ and the corresponding electric field amplitude results in the threshold of pair production from the vacuum polarization (Heisenberg 1934; Euler 1936). Nevertheless, the phenomenological derivation (Hora 1978) arrives at the general result for any temperature and density, that the electrons are to $\approx 95\%$ behaving classically by oscillating in the radiation field, while the fraction $15/\pi^5 = 4.9\%$ will undergo any kind of inelastic interaction (pair production etc.).

These results may cause a generalization of Pauli's (1955) theorem that particles can appear only as fermions or bosons. The fact that fermions or bosons are simply described by symmetric or antisymmetric functions may require the further inclusion of coupling to radiation. It may be that

Gentile's concept of intermediary statistics (Hora and Müller, 1961) is a possible approach for this generalization. We have no answer at present but we mention this question in the introduction to underline what rich research the problems of the equation of state may lead to.

The fact that the formulation of the appropriate equations of state at these extreme conditions is still under discussion and is not finalized, should be seen not only from this example of black body radiation. While Zeldovich and Novikov (1965) postulate a pressure $p \approx n^2$ at densities above 10^{38} cm^{-3}, because the acoustic velocity is that of the speed of light, Harrison (1965) concludes that $p \approx n^{4/3}$ from nucleon and hyperon interactions. While a $p = nkT$ and $kT = \varepsilon_F \approx n^{1/3}$ (1.8), would support $p \approx n^{4/3}$, the argument in favour of Zeldovich (Bushman and Fortov 1983) or even for a more complicated radiation determined pressure, has to be left open to further research (Wulfe, 1976).

2

A summary of thermodynamics

Though it is assumed that the reader of this monograph is familiar with the theory of thermodynamics from the level of undergraduate studies (see, e.g., Becker 1955; Landau and Lifshitz 1966) a summary of some essential facts, ideas and concepts of thermodynamics is presented in this Section. We shall consider mainly the ideal gas with a low particle density. The inclusion of quantization will follow in Section 3.

2.1   Phenomenology

The study of the phenomenon of heat was introduced into physics essentially in the nineteenth century though some links go back to the preceding century, e.g. the theory of friction by Coulomb and Benjamin Thompson (Graf von Rumford). One problem was the transport of heat first covered by the differential equation of heat transport and the various theories for the thermal conductivity coefficient. This was essentially a question of a mathematical theory of scalar fields for the temperature and was mathematically straightforward. The more difficult quantification was that heat is equivalent to energy (first law of thermodynamics) with the restriction that it can flow only from a body of higher temperature to one of lower temperature and not vice versa (second law of thermodynamics). The structure of such theory was not so straightforward within a mathematical framework but needed a more conceptual consideration as a pheno- menological theory. It is therefore no surprise that the first law was discovered by a medical man (Robert Mayer) watching nature when cruising through the tropics as a ship's doctor.

A conservation law for kinetic and potential energy was known from mechanics, in the preceding century, from a study of mechanical systems of a number of point-like masses (Newton, d'Alembert, Lagrange) and of the equivalent quantities in fluids: kinetic pressure and static pressure

(Bernoulli). Discovering then that the pressure P of a gas and its volume V in an isothermal compression follows the relation

$$PV = \text{const} \tag{2.1}$$

(the Boyle–Mariotte law) and that this can be related to a temperature T (to be defined as the absolute temperature) as found by Gay–Lussac to derive the first equation of state of the ideal gas of volume V

$$PV = RT \tag{2.2}$$

with a constant R, a quantitative relation between pressure and temperature was found. It turned out that the gas constant R is determined by the specific heat at constant pressure c_p and that at constant volume c_v, for molar scaling

$$R = c_p - c_v \tag{2.3}$$

It was evident from mechanics with the definition of pressure as force per unit area or as energy per unit volume, that a change of volume is related to the change of energy in the system of the gas by compression work $dA = -PdV$. But it was not before the discovery of the first law of thermodynamics that the concept came of an amount of energy Q being exchanged with the gas (restricted still to terms of the historical ideal gas):

$$dQ = dU + PdV \tag{2.4}$$

where U is the internal energy represented by RT describes the energy which is exchanged as heat. Equation (2.4) is the expression of the first law of thermodynamics for gases and was formulated in Dulong–Petit's law for solids. It formulates the (measured) equivalence of mechanical energy and heat measured by the warming up of lead samples falling under gravity onto a base-plate.

For the formulation of the second law of thermodynamics it was interesting that a quantity S, the entropy,

$$dS = dQ/T \tag{2.5}$$

could be formulated as a mathematically integrable quantity like a potential, which expressed the degree of irreversibility of a cyclic process. Following a Carnot-process on a P–V diagram (Fig. 2.1) the quantity dS integrated over one cycle is zero, if the process is performed infinitely slowly, only going through near equilibrium processes when transferring energy during the isothermals ($T = \text{const}$) in Fig. 2.1, while changing the volume and pressure. If this thermal exchange is faster, the value of S throughout one cycle is not constant but is increased. The amount of the increase measures the degree of irreversibility, the degree to which heat was flowing quickly from a higher temperature to a lower temperature – a mechanism which cannot be reversed according to Planck's formulation of the second law: Der Prozess der Warmeleitung lässt sich in keinerlei Weise vollständig rückgängig machen (the process of heat conduction is in no way completely reversible).

In gases it was realized immediately (Dalton's law) that chemically reacting gases at the same temperature and pressure reacted in definite multiples of quantities only, e.g. 2 litres of hydrogen and 1 litre of oxygen to form 2 litres of water vapour, or 1 litre of chlorine and 1 litre of hydrogen to form 2 litres of hydrogen chloride. This led to the discovery of molecular structure, and to the atomic weight. The amount of an element, measured in grams, comprising the atomic or molecular weight, was called a mole. Expressing the thermodynamic quantities in multiples of one mole, the 'specific' quantities are then defined in the following equations where v is called the molar volume:

$$\delta q - du + P dv \qquad \text{instead of (2.4) } u = \text{specific energy} \qquad (2.6)$$

$$d\varepsilon = T ds - P dv \qquad (2.7)$$

$$df = -s dT - P dV \qquad f = u - Ts = \text{specific free energy} \qquad (2.8)$$

$$dg = -s dT + v dP \qquad g = f + Pv = u - Ts + Pv = \text{specific potential (Gibbs' potential)} \qquad (2.9)$$

Fig. 2.1. $P-V$ diagram of the Carnot cycle.

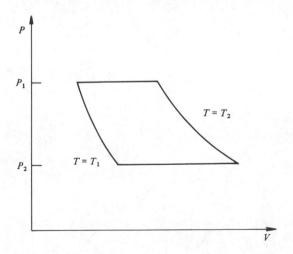

$$d\psi = Tds + vdP \qquad \psi = u + Pv = \text{specific enthalpy} = \text{thermo-}$$
$$\text{dynamic potential} \qquad (2.10)$$

The coefficients of the specific heat for constant volume or pressure are then respectively:

$$c_v = \left.\frac{\partial u}{\partial T}\right|_v = \left.\frac{\delta q}{\delta T}\right|_v \qquad (2.11)$$

$$c_p = \left.\frac{\partial u}{\partial T}\right|_p + p\left.\frac{\partial v}{\partial T}\right|_p = \left.\frac{\delta q}{\delta T}\right|_p \qquad (2.12)$$

$$\alpha = \frac{1}{v_0}\left.\frac{\partial v}{\partial T}\right|_p \qquad \text{thermal expansion coefficient} \qquad (2.13)$$

$$\kappa = -\frac{1}{v_0}\left.\frac{\partial v}{\partial p}\right|_T \qquad \text{compressibility} \qquad (2.14)$$

$$\sigma = \frac{1}{p_0}\left.\frac{\partial p}{\partial T}\right|_v \qquad \text{coefficient of strain} \qquad (2.15)$$

where v_0 and p_0 are constants of normalization. We have the following identities and can express the differential quotients of the quantities T, s, P, and v by the scheme of Table 2.1.

Between the coefficients (2.11) to (2.15), there are the following relations:

$$\sigma p_0 = \alpha/\kappa; \quad c_p - c_v = Tp_0\sigma v_0 = T\alpha^2 v_0/k \qquad (2.16)$$

$$\left.\frac{\partial T}{\partial v}\right|_s = -\left.\frac{\partial p}{\partial s}\right|_v \qquad (2.17)$$

$$\left.\frac{\partial s}{\partial v}\right|_T = \left.\frac{\partial p}{\partial T}\right|_v \qquad (2.18)$$

$$\left.\frac{\partial s}{\partial p}\right|_T = -\left.\frac{\partial v}{\partial T}\right|_p \qquad (2.19)$$

$$\left.\frac{\partial T}{\partial p}\right|_s = \left.\frac{\partial v}{\partial s}\right|_p \qquad (2.20)$$

$$\frac{\partial(s, T)}{\partial(v, P)} = \left.\frac{\partial s}{\partial v}\right|_p\left.\frac{\partial T}{\partial p}\right|_v - \left.\frac{\partial s}{\partial p}\right|_v\left.\frac{\partial T}{\partial v}\right|_p = 1 \qquad (2.21)$$

$$\frac{\partial p}{\partial v}\bigg|_T \frac{\partial v}{\partial T}\bigg|_p \frac{\partial T}{\partial p}\bigg|_v = -1 \tag{2.22}$$

If the number N of particles in the system is not constant, e.g. due to chemical reactions or pair production, the chemical potential μ defines additional terms in the functions in (2.6)–(2.10) for molar quantities. Expressed in non-molar quantities, the definitions and relations are given in Table 2.2.

The relativistic generalization of thermodynamics has to include the fact that any mass $M = M_0(1 - v^2/c^2)^{-1/2}$ has to be expressed by the rest mass M_0 and the velocity v of the system in which the thermodynamics is described. The laws of thermodynamics are then unchanged

$$dQ = dU - dA \tag{2.23}$$

and

$$dS = dQ/T \tag{2.24}$$

Table 2.1. *Definition and relations of thermodynamic quantities*

Quantity (thermodynamic function)	Variables	Relations	
E[energy]	S, V, N	$E = TS - PV + \mu N$	$dE = TdS - PdV + \mu dN$
F[(Helmholtz') free energy]	T, V, N	$F = E - TS$ $F = -PV + \mu N$	$dF = -SdT - PdV + \mu dN$
G[Gibb's potential; free enthalpy]	T, P, N	$G = E - TS + PV$ $G = \mu N$	$dG = SdT + VdP + \mu dN$
Φ[thermodynamic potential]	T, V, μ	$\Phi = F - \mu N$ $\Phi = E - TS - \mu N$	$d\Phi = -SdT - PdV - Nd\mu$ resulting in
		$\Phi = -PV$	$S = -\left(\dfrac{\partial \Phi}{\partial T}\right)_{v,\mu};$ $P = -\left(\dfrac{\partial \Phi}{\partial V}\right)_{\tau,\mu}$ $N = -\left(\dfrac{\partial \Phi}{\partial \mu}\right)_{T,V}$

Table 2.2. *Relation between thermodynamic quantities*

		T	s	p	v
	$\big\|_v$	$\dfrac{k}{\alpha}$	$\dfrac{\kappa c_v}{\alpha T}$	1	0
$\dfrac{\partial}{\partial p}\Big\|_T$	$\big\|_T$	0	$-\alpha v_0$	1	$-kv_0$
	$\big\|_s$	$\dfrac{\alpha T v_0}{c_p}$	0	1	$-\dfrac{c_v}{c_p}v_0 k$
	$\big\|_p$	$\dfrac{1}{\alpha v_0}$	$\dfrac{c_p}{\alpha T v_0}$	0	1
$\dfrac{\partial}{\partial v}\Big\|_T$	$\big\|_T$	0	$\dfrac{\alpha}{k}$	$-\dfrac{1}{k\alpha_0}$	1
	$\big\|_s$	$-\dfrac{\alpha T}{kc_v}$	0	$-\dfrac{c_p}{c_v}\dfrac{1}{kv_0}$	1
	$\big\|_p$	1	$\dfrac{c_p}{T}$	0	αv_0
$\dfrac{\partial}{\partial T}\Big\|_v$	$\big\|_v$	1	$\dfrac{c_v}{T}$	$\dfrac{\alpha}{\kappa}$	0
	$\big\|_s$	1	0	$\dfrac{c_p}{\alpha T v_0}$	$\dfrac{c_v k}{\alpha T}$
	$\big\|_p$	$\dfrac{T}{c_p}$	1	0	$\dfrac{\alpha T v_0}{c_p}$
$\dfrac{\partial}{\partial s}\Big\|_v$	$\big\|_v$	$\dfrac{T}{c_v}$	1	$\dfrac{\alpha T}{kc_v}$	0
	$\big\|_T$	0	1	$-\dfrac{1}{\alpha v_0}$	$\dfrac{k}{\alpha}$

The variables of state are then (subscript 0 indicates the non-relativistic value)

$$P = P_0 \tag{2.25}$$

$$V = V_0\sqrt{1 - v^2/c^2} \tag{2.26}$$

$$T = T_0\sqrt{1 - v^2/c^2} \tag{2.27}$$

$$S = S_0 \tag{2.28}$$

$$U = \frac{1}{(1 - v^2/c^2)^{1/2}}\left(U_0 + \frac{v^2}{c^2}P_0V_0\right) \tag{2.29}$$

$$F = \frac{1}{(1 - v^2/c^2)^{1/2}}\left[F_0 + \frac{v^2}{c^2}(T_0S_0 + P_0V_0)\right] \tag{2.30}$$

$$G = \frac{1}{(1 - v^2/c^2)^{1/2}}\left(G_0 + \frac{v^2}{c^2}T_0S_0\right) \tag{2.31}$$

$$M = M_0/(1 - v^2/c^2)^{1/2} \tag{2.32}$$

In (2.2) a first historic expression of the equation of state for the ideal gas was given. A more general derivation will be given in Chapter 3. It should be mentioned that for real gases a generalization of (2.2) was derived by van der Waals

$$(P + a/V^2)(V - b) = nRT \tag{2.33}$$

where a and b are constants and n represents the numbers of moles, which expresses the scaling factor not used in (2.2).

The isothermal curves of (2.33) in the P–V-diagram of Fig. 2.1 are of higher order and define the so called 'critical point' where $\partial P/\partial V|_T = 0$ and $\partial^2 P/\partial V^2|_T = 0$ corresponding to the critical pressure and the critical temperature. The 'inversion temperature' is the value T_I where $T(\partial v/\partial T)|_P = V$.

2.2 Statistical picture

Though the phenomenological properties of materials derived from thermodynamics are macroscopic entities, some reflect atomistic properties, as was recognized very early, e.g. from the already mentioned Dalton's law. It was then evident from the multiples of the chemical reaction components that the gases hydrogen and oxygen consist of molecules each of two atoms. One could easily determine that the mechanical motion of these molecules has several degrees of freedom, f, of motion, e.g. the three degrees of freedom for translative motion and two degrees for the rotation of the diatomic molecule. Larger molecules had at least one additional degree of freedom if the numbers of degrees of freedom had not then been increased by oscillations between the atoms, and vibrations of the long molecules or ionization processes. A straightforward method to measure these values, f, was from the adiabatic compression of the gases. These are changes based on equation (2.4) where no energy dQ is put into the system or taken out. This is identical with an isentropic process (constancy of entropy) from (2.4).

$$dU = - PdV \tag{2.34}$$

Using (2.11), (2.2) and (2.3) one finds

$$c_v dT = - T(c_p - c_v)dV/V \tag{2.35}$$

and using

$$\gamma = c_p/c_v \tag{2.36}$$

we find

$$dT/T + (\gamma - 1)dV/V = 0$$

and from integration

$$TV^{\gamma-1} = \text{const} \tag{2.37}$$

or after using (2.4)

$$PV^\gamma = \text{const} \tag{2.38}$$

The value of γ is given by the degrees of freedom

$$\gamma = (f + 2)/f \tag{2.39}$$

as could be measured in the gases. Going further it is possible to find the number of particles in a mole, which is given by

$$N = 6.023 \times 10^{23} \text{ molecules per mole}$$

known as the Loschmidt number or the Avogadro number. At thermal equilibrium for low density of the gas, the energy per degree of freedom in any molecule is

$$E = kT/2 \tag{2.40}$$

which result defines the equation of state of the ideal gas (2.4) by using the number n of molecules per cm^3

$$P = fnkT/2 \tag{2.41}$$

where the Boltzmann constant

$$k = 1.38 \times 10^{-16} \, \text{erg deg}^{-1} \tag{2.42}$$

2.3 Maxwell–Boltzmann distribution

Restricting consideration to the thermal equilibrium of an ideal gas at low density, the particles described in the preceding section interact and the final result is a statistical distribution of the energy amongst the particles. This will now be derived from the result described for entropy, as Boltzmann's is an expression for the most probable distribution of the energy amongst the particles.

There is a correlation between the entropy S_{12} and the probability W_{12} of the microscopic structure of the states of two thermodynamic systems with the respective values S_1, S_2, W_1, and W_2 (Joos 1949; Hora 1981)

$$S_{12} = S_1 + S_2 \tag{2.43}$$

$$W_{12} = W_1 W_2 \tag{2.44}$$

from their definition. The function which verified the correlation

$$f(x_1 x_2) = f(x_1) + f(x_2) \tag{2.45}$$

is given by Boltzmann's equation, combining (2.43) and (2.44)

$$S = k \ln W \tag{2.46}$$

using the Boltzmann constant k as the gas constant per particle.

The concept of describing a plasma by the probabilities of the distribution of energy to its individual particles is a limited picture and may not cover all the facts of reality. It presumes, for example, that the forces between the particles are small or in first order negligible, or only during negligible times, while the interactions are necessary on the other hand to achieve equilibrium. The other extreme is the description of phenomena fully deterministic by all details of the interaction by differentiable or by analytic (holomorphic) functions which may run into the insufficiency of super-determinism (Laplace, Cauchy). This can even be a consequence of quantum mechanics (not only in the Schrödinger picture) if the correlation between object and measuring apparatus is considered.

In Boltzmann statistics – in contrast to quantum statistics – it is assumed that one can distinguish between the individual particles of an ensemble. Using six-dimensional volume elements $\Delta\tau_i = \Delta x \Delta y \Delta z \Delta v_x \Delta v_y \Delta v_z$, the number N_i of particles in this element is given by a distribution function $f(i)$

$$N_i = f(i)\Delta\tau_i \tag{2.47}$$

The total number N of particles should be constant

$$N = \sum f(i)\Delta\tau_i \quad \delta N = \sum \delta f(i)\Delta\tau_i = 0 \tag{2.48}$$

where the phase space element $\Delta\tau_i$ is constant at any variation due to the Liouville theorem. The energy $U(\delta)$ of the particles in the ith cell results in an energy $N_i u(i)$ in the cell. The total energy U should be constant

$$U = \sum u(i)f(i)\Delta\tau_i \quad \delta U = \sum \delta f(i)u(i)\Delta\tau_i = 0 \tag{2.49}$$

The probability of the system with a weighting of the states of each cell by

$$G_i = \Delta\tau_i \tag{2.50}$$

is given by the number of combinations of all cases $N! \prod G_i$ which are possible by permutation without repetition if we use distinguishable particles

$$W = \frac{N! \prod G_i^{N_i}}{\prod N_i!} \tag{2.51}$$

Using (2.47), (2.50) and the Stirling formula, Appendix 2, (A2.3) (as approximation for large N)

$$N! = \frac{N^N}{e^N}$$

we find from (2.51)

$$W = \frac{N^N \prod \Delta\tau_i^{f(i)\Delta\tau_i}}{\prod (f(i)\Delta\tau_i)^{f(i)\Delta\tau_i}} \tag{2.52}$$

and the entropy from (2.46)

$$S = KN \ln N - k\sum f(i)\Delta\tau_i \ln f(i) \tag{2.53}$$

Equilibrium corresponds to the value of highest probability W, or $\delta S = 0$, while fulfilling the secondary conditions of constant total particle number

N (2.48) and total energy U (2.49), from (2.50)

$$0 = \sum \delta_i f(i) \Delta \tau_i \ln f(i) + \sum \delta f(i) \Delta \tau_i \tag{2.54}$$

To formulate the secondary conditions, the method of multiplicators is used by adding conditions (2.48) after multiplying with α to (2.54) and by adding (2.49) after multiplying with β in the same way. The result is

$$\ln f(i) + 1 + \alpha + \beta u(i) = 0 \tag{2.55}$$

This leads immediately to the desired energy distribution function

$$f(i) = A \exp(- \beta u(i)) \tag{2.56}$$

where

$$A = \exp[-(1 + \alpha)] \tag{2.57}$$

is given by the constant number N of all particles from (2.48) and (2.56)

$$N = A \sum \exp(- \beta u(i)) \Delta \tau_i \tag{2.58}$$

or

$$f(i) = \frac{N \exp[- \beta u(i)]}{\sum \exp[- \beta u(i)] \Delta \tau_i} \tag{2.59}$$

The denominator is called the (phase space weighted) partition function

$$\sigma = \sum \Delta \tau_i \exp(- \beta u(i)) \tag{2.60}$$

The physical interpretation of the multiplicator β is given from the definition of the entropy. Using (2.59) and (2.60) in (2.53)

$$S = K \ln N - K \sum \frac{N}{\sigma} \Delta \tau_i \exp(- \beta u(i)) [\ln N - \beta u(i) - \ln \sigma] \tag{2.61}$$

and (2.49)

$$\frac{N}{\sigma} \sum \Delta \tau_i \exp(- \beta u(i)) = U \tag{2.62}$$

(2.61) reduces to

$$S = k \beta U + K N \ln \sigma \tag{2.63}$$

Thermodynamics defines the relation between S, U and the temperature T for conditions of constant volume V

$$\frac{1}{T} = \left(\frac{\partial S}{\partial U} \right)_v \tag{2.64}$$

U is a function of β (2.62), therefore

$$\left(\frac{\partial S}{\partial U}\right)_v = \frac{dS}{d\beta}\left(\frac{\partial \beta}{\partial U}\right)_v = \frac{dS}{d\beta}\frac{1}{(\partial U/\partial \beta)_v} \tag{2.65}$$

By differentiating (2.63) we find

$$\frac{dS}{d\beta} = KU + K\beta\frac{\partial U}{\partial \beta} + \frac{KN}{\sigma}\frac{\partial \sigma}{\partial \beta} \tag{2.66}$$

by substitution of the differences of (2.60) and (2.62) by differentials

$$\frac{\partial \sigma}{\partial \beta} = -\sum u(i)\Delta\tau_i \exp[-\beta u(i)] = -\frac{U\sigma}{N} \tag{2.67}$$

and from (2.66)

$$\frac{dS}{d\beta} = k\beta\frac{\partial U}{\partial \beta} \tag{2.68}$$

Taking the differential form and (2.65) we find

$$\left(\frac{\partial S}{\partial U}\right)_v = \beta K = \frac{1}{T} \tag{2.69}$$

and finally

$$\beta = \frac{1}{KT} \tag{2.70}$$

The distribution function $f(i)$, (2.59) turns out to be the Maxwell–Boltzmann distribution

$$f(i) = \frac{N\exp(-u(i)/KT)}{\sum \Delta\tau_i \exp(-u(i)/KT)} \tag{2.71}$$

as used subsequently in (12.21)

It should be noted that the discussion of the fact that the quantity σ in (2.63) is not dimensionless (σ is of the dimension of an action to the cube). led Planck (1900) to the addition of an arbitrary constant h in the entropy of (2.63)

$$S = \frac{U}{T} + KN\ln\frac{\sigma}{h^3} \tag{2.72}$$

arriving at a free energy

$$F = U - TS = -KNT\ln\frac{\sigma}{h^3} \tag{2.73}$$

The correct value of h, Planck's constant, was derived first from the fit of the then valid Planck's radiation law with the measurements. This was the very first observation of the quantization (atomistic structure) of action. One should note this result as a consequence of Boltzmann statistics, and furthermore its equivalence to the need of the stimulated emission of radiation derived by Einstein (1916).

3

Equation of state for an ideal gas

3.1 The partition function

Probably the first step in the understanding of the equation of state is to consider an ideal gas consisting of N non-interacting particles. Although the corresponding equation of state is derived from elementary kinetic theory in almost all books on heat, we will first calculate the partition function and then obtain the equation of state. This will give us a simple example of calculating the partition function.

We first mention that if we solve the Schrödinger equation for a free particle confined in a cubical box of side L then the corresponding energy levels are given by (see Appendix 1)

$$\varepsilon_n = \frac{\pi^2 \hbar^2}{2mL^2}(n_x^2 + n_y^2 + n_z^2) \tag{3.1}$$

where $\hbar = h/2\pi$ (h being Planck's constant), m represents the mass of the particle and $n_x, n_y, n_z = 1, 2, 3, \ldots$; the subscript n on ε represents the three integers n_x, n_y and n_z. The partition function (2.60) of the system is given by

$$Q = \sum_i e^{-E_i/kT} \tag{3.2}$$

where E_i represents the *total* energy of the system and the summation is over *all* the energy states. Let the ith state of the total system correspond to the first particle in the state characterized by n_1, the second particle in the state characterized by n_2, etc.; thus

$$E_i = \varepsilon_{1n_1} + \varepsilon_{2n_2} + \cdots$$
$$= \varepsilon_{n_1} + \varepsilon_{n_2} + \cdots \tag{3.3}$$

where in the last step we have taken account of the fact that all particles are identical so that the possible energy levels of each particle are the same.

Thus

$$Q = \sum_{n_2} \sum_{n_2} \cdots \exp[-\{\varepsilon_{n_1} + \varepsilon_{n_2} + \cdots + \varepsilon_{n_N}\}/kT]$$

$$= \left[\sum_{n_1} \exp[-\varepsilon_{n_1}/kT] \right]^N \tag{3.4}$$

Now, the spacing between the energy levels (3.1) is given approximately by

$$\Delta\varepsilon \approx \frac{\pi^2\hbar^2}{2mL^2} \tag{3.5}$$

For hydrogen atoms $m \approx 1.67 \times 10^{-24}$ g and for $L \approx 10$ cm we have

$$\Delta\varepsilon \approx \frac{(3.14)^2 \times (1.055 \times 10^{-27})^2}{2 \times 1.67 \times 10^{-24} \times 100} \approx 3.3 \times 10^{-32} \text{ ergs}$$

$$\approx 2 \times 10^{-20} \text{ eV} \tag{3.6}$$

which represents an extremely small energy difference; we may recall that the thermal energy ($\approx kT$) corresponding to $T \approx 300$ K is about 0.025 eV, which is about 10^{18} times greater than the energy given by (3.6)! Thus the summation in (3.4) can be replaced by an integration and we may write

$$\sum_n e^{-\varepsilon_n/kT} \approx \int_0^\infty e^{-\varepsilon/kT} g(\varepsilon) \mathrm{d}\varepsilon \tag{3.7}$$

where $g(\varepsilon)\mathrm{d}\varepsilon$ represents the number of energy states in the energy interval $\mathrm{d}\varepsilon$ and is given by (see Appendix 1)

$$g(\varepsilon) = \frac{(2m)^{3/2} V}{4\pi^2\hbar^3} \varepsilon^{1/2} \tag{3.8}$$

$V(= L^3)$ represents the volume of the box. Substituting the above expression for $g(\varepsilon)$ in (3.7) and carrying out the integration we get

$$\sum_n e^{-\varepsilon_n/kT} \approx \left(\frac{mkT}{2\pi\hbar^2} \right)^{3/2} V$$

where we have used the relation

$$\int_0^\infty e^{-\varepsilon/kT} \varepsilon^{1/2} \mathrm{d}\varepsilon = (kT)^{3/2} \int_0^\infty e^{-x} x^{1/2} \mathrm{d}x$$

$$= (kT)^{3/2} \Gamma(\tfrac{3}{2}) = \tfrac{1}{2}\pi^{1/2}(kT)^{3/2}$$

Thus the partition function is given by

$$Q = q^N = \left[\left(\frac{mkT}{2\pi\hbar^2} \right)^{3/2} V \right]^N \tag{3.9}$$

where

$$q = \left(\frac{mkT}{2\pi\hbar^2}\right)^{3/2} V \tag{3.10}$$

represents the single particle partition function.

We further give here the Grand Partition Function Z_G for a Hamiltonian with the number of particles N and the chemical potential μ

$$Z_G \equiv \sum_N \sum_i e^{-\beta(E_i - \mu N)} \quad \beta \equiv \frac{1}{k_B T} \tag{3.11a}$$

$$= \sum_N \sum_i \langle Ni | e^{-\beta(\hat{H} - \mu\hat{N})} | Ni \rangle \tag{3.11b}$$

$$= \text{Tr}(e^{-\beta(\hat{H} - \mu\hat{N})}) \tag{3.11c}$$

where the thermodynamic potential Φ is

$$\Phi(T, V, \mu) = -k_B T \ln Z_G \tag{3.12}$$

3.2   Thermodynamic functions

Once the partition function is known, the free energy F and the pressure P can readily be calculated:

$$F = -kT \ln Q = -NkT \ln\left[\left(\frac{mkT}{2\pi\hbar^2}\right)^{3/2} V\right] \tag{3.13}$$

and

$$P = -\left(\frac{\partial F}{\partial V}\right)_T = +\frac{NkT}{V}$$

giving

$$PV = NkT \tag{3.14}$$

which represents the equation of state for an ideal gas consisting of non-interacting particles in thermodynamic equilibrium. The total energy E and the specific heat c_v are given by

$$E = kT^2 \left(\frac{\partial \ln Q}{\partial T}\right)_V = \tfrac{3}{2} NkT \tag{3.15}$$

and

$$c_v = \left(\frac{\partial E}{\partial T}\right)_V = \tfrac{3}{2} Nk \tag{3.16}$$

Further, the entropy is given by

$$S = -\left(\frac{\partial F}{\partial T}\right)_V = Nk \ln\left[\left(\frac{mkT}{2\pi\hbar^2}\right)^{3/2} V\right] + \tfrac{3}{2} Nk \tag{3.17}$$

3.3 The Gibbs' paradox

We next consider two perfect gases at the same temperature T separated by a partition as shown in Fig. 3.1. Let N_1 and N_2 represent the numbers of particles in compartments 1 and 2 respectively and let V_1 and V_2 be their volumes. The entropy associated with the two gases will be given by

$$S_1 = N_1 k \ln\left[\left(\frac{mkT}{2\pi\hbar^2}\right)^{3/2} V_1\right] + \tfrac{3}{2}N_1 k \tag{3.18a}$$

and

$$S_2 = N_2 k \ln\left[\left(\frac{mkT}{2\pi\hbar^2}\right)^{3/2} V_2\right] + \tfrac{3}{2}N_2 k \tag{3.18b}$$

If the partition is removed and the two gases are allowed to mix, then the total entropy will be given by

$$S_{\text{tot}} = \left\{N_1 k \ln\left[\left(\frac{mkT}{2\pi\hbar^2}\right)^{3/2}(V_1 + V_2)\right] + \tfrac{3}{2}N_1 k\right\}$$
$$+ \left\{N_2 k \ln\left[\left(\frac{mkT}{2\pi\hbar^2}\right)^{3/2}(V_1 + V_2)\right] + \tfrac{3}{2}N_2 k\right\} \tag{3.18c}$$

The increase in entropy will therefore be given by

$$\Delta S = S_{\text{tot}} - (S_1 + S_2)$$
$$= N_1 k \ln\frac{V_1 + V_2}{V_1} + N_2 k \ln\frac{V_1 + V_2}{V_2} \geqslant 0 \tag{3.19}$$

Thus ΔS is always positive, which should indeed by the case when we are mixing two different gases which is an irreversible process. Indeed, (3.19) is the correct expression for the increase in entropy when we are mixing two different gases. However, if we are considering the same gas to be in the two compartments (see Fig. 3.1) having the same density, i.e.

$$\frac{N_1}{V_1} = \frac{N_2}{V_2} = \frac{N_1 + N_2}{V_1 + V_2} \tag{3.20}$$

then removing the partition leads to a reversible process because we can get back to the same state by reinserting the partition. Thus we should expect

Fig. 3.1. Mixing of two gases which are at the same temperature. N_1 and N_2 represent the number of particles occupying volumes V_1 and V_2 respectively.

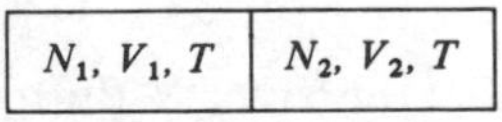

| N_1, V_1, T | N_2, V_2, T |

no change in entropy (i.e., $\Delta S = 0$) however, (3.19) tells us that $\Delta S > 0$. This is known as *Gibbs' paradox* and is due to the fact that the particles have been assumed to be *distinguishable* and there has been overcounting of states. We will have a detailed discussion on the indistinguishability of particles in Chapter 5; we may mention here that Gibbs resolved this paradox by dividing the expression for Q by $N!$:

$$Q = \frac{1}{N!}\left[\left(\frac{mkT}{2\pi\hbar^2}\right)^{3/2} V\right]^N \tag{3.21}$$

Thus

$$F = -kT\ln Q = -NkT\ln\left[\left(\frac{mkT}{2\pi\hbar^2}\right)^{3/2}\frac{V}{N}\right] - NkT \tag{3.22}$$

where we have used Stirling's formula (see Appendix 2)

$$\ln N! \approx N\ln N - N \tag{3.23}$$

It is easy to see that using (3.22) for the free energy one obtains the same expression for the equation of state (3.14) and for the total energy and specific heat (3.15) and (3.16). However, the expression for entropy is given by

$$S = -\left(\frac{\partial F}{\partial T}\right)_V = Nk\ln\left[\left(\frac{mkT}{2\pi\hbar^2}\right)^{3/2}\frac{V}{N}\right] + \tfrac{5}{2}Nk \tag{3.24}$$

Equation (3.24) is known as the *Sackur–Tetrode* equation. Now, if we have two different gases (like helium and neon) in two separate compartments (of volumes V_1 and V_2) and if the two compartments are connected then the increase in entropy would be given by

$$\Delta S = k\left[N_1\ln\frac{V_1 + V_2}{N_1} + N_2\ln\frac{V_1 + V_2}{N_2} - N_1\ln\frac{V_1}{N_1} - N_2\ln\frac{V_2}{N_2}\right]$$

$$= k\left[N_1\ln\frac{V_1 + V_2}{V_1} + N_2\ln\frac{V_1 + V_2}{V_2}\right] \tag{3.25}$$

which is identical to (3.19). However, if the two gases are the same (having the same initial temperature) then the change in entropy will be

$$\Delta S = k\left[\left(N_1 + N_2\right)\ln\frac{V_1 + V_2}{N_1 + N_2} - N_1\ln\frac{V_1}{N_1} - N_2\ln\frac{V_2}{N_2}\right] \tag{3.26}$$

Thus if the initial densities are equal (i.e., (3.20) is satisfied) then

$$\Delta S = 0 \tag{3.27}$$

as it indeed should be. We should point out that the factor $N!$ (introduced in

(3.21)) represents the number of ways in which N particles can be arranged among themselves provided, of course, each state is occupied by not more than one particle. Thus the factor $N!$ does not properly account for the indistinguishability of particles – the proper treatment will be given in Chapter 5. Nevertheless, by putting an (*ad hoc*) factor of $(1/N!)$ the entropy becomes an extensive property of the system and Gibbs' paradox is resolved.

Finally, Gibbs' free energy and the chemical potential are given by

$$G = F + PV = -NkT \ln\left[\left(\frac{mkT}{2\pi\hbar^2}\right)^{3/2} \frac{V}{N} \right] \tag{3.28}$$

and

$$\mu = G/N = -kT \ln\left[\left(\frac{mkT}{2\pi\hbar^2}\right)^{3/2} \frac{V}{N} \right] \tag{3.29}$$

4

Law of equipartition of energy and effects of vibrational and rotational motions

4.1 Classical considerations

Considering an ideal gas in which the molecules are represented by structureless particles, the total energy of a molecule is simply the translational energy and is given by

$$\varepsilon = \frac{1}{2m}(p_x^2 + p_y^2 + p_z^2) \tag{4.1}$$

where p_x, p_y and p_z represent the x, y and z components of the linear momentum of the molecule whose mass is m. However, in general, the molecules are capable of vibrational and rotational motions which give rise to additional terms in the expression for energy. For example, if we consider a diatomic molecule like CO then the molecule can have vibrational motion along the axis and also rotational motion along two mutually perpendicular axes, we are assuming each atom to be represented by a point particle. Now, the vibrational energy (ε_v) of the molecule can be written as

$$\varepsilon_v = \tfrac{1}{2}\mu v_\xi^2 + \tfrac{1}{2}k\xi^2 = \frac{1}{2\mu}p_\xi^2 + \tfrac{1}{2}k\xi^2 \tag{4.2}$$

where ξ represents the displacement of the C and O atoms from their equilibrium separation, $\mu[= m_1 m_2/(m_1 + m_2)]$ is the reduced mass and k the force constant. The rotational kinetic energy can be easily calculated by assuming the centre of mass to be fixed at the origin and each atom moving over the surface of a sphere; thus

$$\varepsilon_r = \tfrac{1}{2}m_1(r_1^2\dot{\theta}^2 + r_1^2\sin^2\theta\dot{\phi}^2) + \tfrac{1}{2}m_2(r_2^2\dot{\theta}^2 + r_2^2\sin^2\theta\dot{\phi}^2)$$
$$= \tfrac{1}{2}I(\dot{\theta}^2 + \sin^2\theta\dot{\phi}^2) \tag{4.3}$$

where $I = (m_1 r_1^2 + m_2 r_2^2)$ represents the moment of inertia of the molecule about an axis passing through the centre of mass and perpendicular to the line joining the atoms, r_1 and r_2 are the distances of the two atoms from the centre of mass and θ and ϕ represent the angular coordinates of one of the atoms (the values of $\dot{\theta}^2$, $\sin^2\theta$ and $\dot{\phi}^2$ would be the same for both the atoms). Equation (4.3) is usually written in the form

$$\varepsilon_r = \frac{1}{2I}\left(p_\theta{}^2 + \frac{p_\phi{}^2}{\sin^2\theta}\right) \tag{4.4}$$

where

$$p_\theta = \frac{\partial \varepsilon_r}{\partial \dot{\theta}} = I\dot{\theta} \quad\text{and}\quad p_\phi = \frac{\partial \varepsilon_r}{\partial \dot{\phi}} = I\sin^2\theta\,\dot{\phi} \tag{4.5}$$

represent the *momenta canonical* to the coordinates θ and ϕ respectively. Thus the total energy of the diatomic molecule (taking into account the rotational and vibrational motions) will be

$$\varepsilon = \frac{1}{2m}(p_x{}^2 + p_y{}^2 + p_z{}^2) + \left(\frac{1}{2\mu}p_\xi{}^2 + \tfrac{1}{2}k\xi^2\right)$$

$$+ \frac{1}{2I}\left(p_\theta{}^2 + \frac{1}{\sin^2\theta}p_\phi{}^2\right) \tag{4.6}$$

where $m = m_1 + m_2$ and p_x, p_y and p_z represent the x, y and z components of the momentum of the center of mass. It may be noted that in (4.6) the canonical coordinates (like ξ) and canonical momenta (like p_x, p_ξ, p_θ, etc.) appear as a quadratic power. Indeed, in any dynamical system if the total energy

$$E = E(q_1, q_2, \ldots q_n; \quad p_1, p_2, \ldots p_n) \tag{4.7}$$

(where qs and ps represent generalized coordinates and momenta) can be written in the form

$$E = \tfrac{1}{2}\alpha_j\zeta_j{}^2 + E' \tag{4.8}$$

where ζ_j represents any one of the position or momentum coordinates and E' does *not* depend on ζ_j (but may depend on other coordinates) then, according to the law of equipartition of energy, the mean value of

$$\varepsilon_j = \tfrac{1}{2}\alpha_j\zeta_j{}^2 \tag{4.9}$$

(at thermal equilibrium) will be $\tfrac{1}{2}kT$. The proof is based on Boltzmann's law according to which the distribution of particles in the volume element $dq_1 \cdots dp_n$ will be given by

$$f = Ce^{-E/kT}dq_1 \cdots dp_n \tag{4.10}$$

where C is a constant. Thus

$$\langle \varepsilon_j \rangle = \frac{\int_{-\infty}^{+\infty} \cdots \int e^{-E/kT} \varepsilon_j \, dq_1 \cdots dp_n}{\int_{-\infty}^{+\infty} \cdots \int e^{-E/kT} \, dq_1 \cdots dp_n} \tag{4.11}$$

If we now use (4.8) and (4.9) we will get

$$\langle \varepsilon_j \rangle = \frac{\int_{\infty}^{\infty} \varepsilon_j e^{-\varepsilon_j/kT} \, d\zeta_j}{\int_{-\infty}^{+\infty} e^{-\varepsilon_j/kT} \, d\zeta_j} \tag{4.12}$$

where the integrals over all other variables cancel out in the numerator and denominator. Now

$$\int_{-\infty}^{+\infty} \varepsilon_j e^{-\varepsilon_j/kT} \, d\zeta_j = \tfrac{1}{2}\alpha \int_{-\infty}^{+\infty} \zeta_j^2 \exp\left(-\tfrac{1}{2}\alpha \frac{\zeta_j^2}{kT} \right) d\zeta_j$$

$$= \tfrac{1}{2}\alpha 2 \int_0^\infty \zeta_j^2 \exp\left(-\tfrac{1}{2}\alpha \frac{\zeta_j^2}{kT} \right) d\zeta_j$$

$$= 2 \int_0^\infty (xkT) e^{-x} \frac{1}{2}\left(\frac{2kT}{\alpha} \right)^{1/2} x^{-1/2} \, dx$$

where $x = \tfrac{1}{2}\alpha\zeta_j^2/(kT)$. Thus

$$\int_{-\infty}^{+\infty} \varepsilon_j e^{-\varepsilon_j/kT} \, d\zeta_j = (kT)\left(\frac{2kT}{\alpha} \right)^{1/2} \tfrac{1}{2}\pi^{1/2} \tag{4.13}$$

where we have used the relation

$$\int_0^\infty x^{1/2} e^{-x} \, dx = \Gamma(\tfrac{3}{2}) = \tfrac{1}{2}\pi^{1/2} \tag{4.14}$$

Similarly,

$$\int_{-\infty}^{+\infty} e^{-\varepsilon_j/kT} \, d\zeta_j = \left(\frac{2kT}{\alpha} \right)^{1/2} \Gamma(\tfrac{1}{2}) = \left(\frac{2kT}{\alpha} \right)^{1/2} \pi^{1/2} \tag{4.15}$$

Substituting in (4.12) we get

$$\langle \varepsilon_j \rangle = \tfrac{1}{2}kT \tag{4.16}$$

Thus if a generalized position or momentum coordinate occurs in the expression for the total energy as a quadratic term then it contributes energy $\tfrac{1}{2}kT$ to the total energy of the system. This is known as the *law of equipartition of energy* and as an obvious consequence of this law, each quadratic term would contribute $\frac{1}{2}k$ to the heat capacity of the system.

Now, the average energy associated with each term of the translational kinetic energy is $\frac{1}{2}kT$, i.e.,

$$\frac{1}{2m}\langle p_x^2 \rangle = \frac{1}{2m}\langle p_y^2 \rangle = \frac{1}{2m}\langle p_z^2 \rangle = \tfrac{1}{2}kT \tag{4.17}$$

Thus, for a monatomic molecule (like helium, argon, etc.) the average energy (per molecule) is $(\frac{3}{2})kT$ and if N represents the total number of molecules we get

$$E = \tfrac{3}{2}NkT \tag{4.18}$$

as the total energy of the system. Thus the specific heat (at constant volume) is given by

$$c_v = \left(\frac{\partial E}{\partial T}\right)_V = \tfrac{3}{2}Nk \tag{4.19}$$

Equations (4.18) and (4.19) are identical to (3.15) and (3.16). Using

$$N \approx 6.022 \times 10^{23} \, \text{mole}^{-1} \text{ (Avogadro's or Loschmidt's number)} \tag{4.20}$$

and

$$k \approx 1.38 \times 10^{-16} \, \text{erg K}^{-1} \text{ (Boltzmann constant)} \tag{4.21}$$

we get

$$c_v = \tfrac{3}{2}R \tag{4.22}$$

where

$$R \approx 8.31 \times 10^7 \, \text{erg mole}^{-1} \text{K}^{-1} \tag{4.23}$$

represents the molar gas constant. Further, for a perfect gas

$$c_p - c_v = R \tag{4.24}$$

giving

$$c_p = \tfrac{5}{2}R \tag{4.25}$$

Thus the ratio of the specific heats, γ, is given by

$$\gamma \equiv \frac{c_p}{c_v} = \tfrac{5}{3} \approx 1.67 \tag{4.26}$$

The value of γ as given by the above equation agrees well with the experimental values for monatomic gases like helium, argon etc. (see Table 4.1).

For diatomic molecules like H_2, O_2, CO,... etc., if we neglect the vibrational motion but take into account the rotational motion then for each molecule we will have two extra degrees of freedom (associated with p_θ and p_ϕ – see (4.6)) and we will get

$$E = \tfrac{5}{2}NkT \tag{4.27}$$

Thus

$$c_v = \tfrac{5}{2}R \tag{4.28}$$

$$c_p = \tfrac{7}{2}R \tag{4.29}$$

and

$$\gamma = \tfrac{7}{5} = 1.4 \tag{4.30}$$

which agrees reasonably well with the experimental data at room temperatures (see Table 4.1). We may mention here that at low temperatures the value of c_p falls to $\tfrac{5}{2}R$ (see Fig. 4.1). This can be explained by assuming that the rotational motions (approximately) disappear below a certain temperature – it is also customary to say that the rotational degrees of freedom are frozen below a certain temperature. We denote this temperature by T_{rot}. For example,

$$T_{\mathrm{rot}}(O_2) \approx 2\,\mathrm{K} \quad \text{and} \quad T_{\mathrm{rot}}(H_2) \approx 85\,\mathrm{K} \tag{4.31}$$

Table 4.1. *Molecular specific heat in calories at 20 °C and atmospheric pressure*

Gas	c_p	c_v	$\gamma = \dfrac{c_p}{c_v}$	Remarks
Argon (A)	4.97	2.98	1.666	Monatomic
Helium (He)	4.97	2.98	1.666	
Hydrogen (H_2)	6.865	4.88	1.408	
Oxygen (O_2)	7.03	5.035	1.396	
Nitrogen (N_2)	6.95	4.955	1.402	
Nitric oxide (NO)	7.10	5.10	1.39	Diatomic
Hydrochloric acid (HCl)	7.04	5.00	1.41	
Carbon monoxide (CO)	6.97	4.98	1.40	
Chlorine (Cl_2)	8.29	6.15	1.35	
Air	6.950	4.955		
Carbon dioxide (CO_2)	8.83	6.80	1.299	
Sulphur dioxide (SO_2)	9.65	7.50	1.29	Triatomic
Hydrogen sulphide (H_2S)	8.3	6.2	1.34	
Ammonia (NH_3)	8.80	6.555	1.315	
Methane (CH_4)	8.50	6.50	1.31	
Ethane	12.355	10.30	1.20	Polyatomic
Acetylene	10.45	8.40	1.24	
Ethylene	10.25	8.20	1.25	

Note: Values taken from Saha and Srivastava (1958), p. 105.

We should point out that the decrease in the value of the specific heat is quite gradual as can be seen from Fig. 4.1. The classical theory and the law of equipartition of energy are inadequate to explain this behaviour which can be understood only through quantum mechanical considerations which are to be discussed in later sections of this chapter.

Similarly, the two terms due to vibrational motion in (4.2) contribute $\frac{1}{2}kT$ each to $\langle \varepsilon \rangle$ and therefore if we also take into account the rotational motions we would get

$$E = \frac{7}{2}NkT \tag{4.32}$$

giving

$$c_v = \frac{7}{2}R \tag{4.33}$$

$$c_p = \frac{9}{2}R \tag{4.34}$$

and

$$\gamma = \frac{9}{7} \approx 1.29 \tag{4.35}$$

As in the case of rotational motion, here also the value of c_p (or c_v) decreases with decrease in temperature (Fig. 4.1) and for each gas there is a temperature (which we denoted by T_{vib}) below which we may (approximately) assume the vibrational degrees of freedom to be frozen. For example,

$$T_{vib}(O_2) \approx 2200 \text{ K} \tag{4.36}$$

Thus, according to Fig. 4.1, at high temperatures $c_p \to \frac{9}{2}R$, implying that both vibrational and rotational motions contribute to the specific heat. As the temperature is decreased, the vibrational motion becomes slowly frozen and c_p attains a value of about $\frac{7}{2}R$. Further decrease in temperature results

Fig. 4.1. The rotational–vibration specific heat, c_p, of the diatomic gases HD, HT and DT. (Figure adapted from Pathria (1972).)

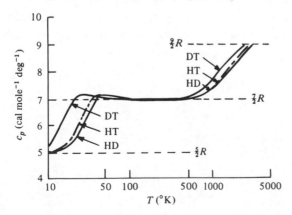

in the gradual disappearance of rotational motions also. Once again, the classical theory (and the law of equipartition of energy) are inadequate to explain this behaviour and one has to resort to quantum mechanical considerations, which is the subject matter of the next few sections.

4.2 The partition function

We consider a gas of N non-interacting molecules. Let q represent the partition function of one molecule which will be given by

$$q = \sum_n e^{-\varepsilon_n/kT} \tag{4.37}$$

where the summation is over all the states of the molecule. The partition function for the gas (consisting of N non-interacting molecules) will be given by (see (3.21))

$$Q = \frac{1}{N!} q^N \tag{4.38}$$

Now, in Chapter 3, we calculated the partition function under the assumption that each particle (or molecule) is capable of having only translational energy. However, as mentioned in the previous section, the molecules can have other energies also (vibrational, rotational etc.) which would alter the form of the partition function. As a first approximation, we may assume the vibrational and rotational motions to be independent of one another so that we may write for the total energy of a molecule

$$\varepsilon = \varepsilon_t + \varepsilon_v + \varepsilon_r + \varepsilon_e + \cdots \tag{4.39}$$

where ε_t, ε_v, ε_r and ε_e represent the translational energy, the vibrational energy, the rotational energy and the energy due to electronic excitations respectively. As mentioned earlier, we assume no coupling between the various forms of energy and since the total energy appears in the exponent (see (4.37)), the partition function for one molecule can be written in the form:

$$q = q_t q_v q_r q_e \cdots \tag{4.40}$$

where

$$q_t = \sum_n \exp\left(-\frac{\varepsilon_{tn}}{kT}\right) \tag{4.41}$$

$$q_v = \sum_n \exp\left(-\frac{\varepsilon_{vn}}{kT}\right) \tag{4.42}$$

$$q_r = \sum_n \exp\left(-\frac{\varepsilon_{rn}}{kT}\right) \tag{4.43}$$

$$q_e = \sum_n \exp\left(-\frac{\varepsilon_{en}}{kT}\right) \tag{4.44}$$

and the summations in (4.41), (4.42), (4.43) and (4.44) are over the translational, vibrational, rotational and electronic states respectively. Since $\ln Q$ appears in the calculation of most thermodynamic quantities, we use (4.38) and (4.40) to obtain

$$\ln Q = N \ln q - \ln N!$$
$$= N \ln q_t + N \ln q_v + N \ln q_r + N \ln q_e + \cdots - (N \ln N - N) \tag{4.45}$$

where use has been made of the Stirling formula (see Appendix 2). Thus the total energy

$$E = kT^2 \left(\frac{\partial \ln Q}{\partial T}\right)_V \tag{4.46}$$

can be written in the form

$$E = E_t + E_v + E_r + E_e + \cdots \tag{4.47}$$

where

$$E_t = NkT^2 \left(\frac{\partial \ln q_t}{\partial T}\right)_V \tag{4.48}$$

$$E_v = NkT^2 \left(\frac{\partial \ln q_v}{\partial T}\right)_V \tag{4.49}$$

$$E_r = NkT^2 \left(\frac{\partial \ln q_t}{\partial T}\right)_V \tag{4.50}$$

and

$$E_e = NkT^2 \left(\frac{\partial \ln q_e}{\partial T}\right)_V \tag{4.51}$$

represent the translational, vibrational, rotational and electronic energy respectively. The expression for q_t was derived in the previous chapter and is given by

$$q_t = \left(\frac{mkT}{2\pi\hbar^2}\right)^{3/2} V \tag{4.52}$$

We will next calculate q_v, q_r and q_e.

4.3   The vibrational partition function

The vibrational energy levels of a diatomic molecule are *approximately* given by[†]

$$\varepsilon_{vn} = (n + \tfrac{1}{2})\hbar\omega \quad n = 0, 1, 2, \ldots \tag{4.53}$$

where $\omega = (k/\mu)^{1/2}$, k being the force constant and $\mu\,(= m_1 m_2/(m_1 + m_2))$ is the reduced mass of the molecule. Thus

$$q_v = \sum_{n=0}^{\infty} \exp\left[-(n + \tfrac{1}{2})\frac{\hbar\omega}{kT} \right]$$

$$= e^{-(1/2)x}[1 + e^{-x} + e^{-2x} + \cdots] \quad x = \frac{\hbar\omega}{kT}$$

[†] In the simplest model describing the vibrational motion of a diatomic molecule the potential energy function is given by

$$V(r) = \tfrac{1}{2}k(r - r_e)^2$$

where k is the force constant and r_e the equilibrium separation of the two atoms. The radial part of the Schrödinger equation is given by (see any text on quantum mechanics)

$$\frac{d^2R}{dr^2} + \frac{1}{r}\frac{dR}{dr} + \frac{2\mu}{\hbar^2}\left[\varepsilon - \frac{j(j+1)\hbar^2}{2\mu r^2} - \tfrac{1}{2}k(r - r_e)^2 \right]R(r) = 0$$

where $\mu\,(= m_1 m_2/(m_1 + m_2))$ is the reduced mass and we have used $j(j+1)$ instead of $l(l+1)$ to be consistent with the literature ($j = 0, 1, 2, \ldots$). $R(r)$ represents the radial part of the wave function; the complete wave function is obviously $R(r)\,Y_{jm}(\theta, \phi)$ where $Y_{jm}(\theta, \phi)$ are the spherical harmonics. We write $u(r) = rR(r)$ to obtain

$$\frac{d^2u}{d\rho^2} + \frac{2\mu}{\hbar^2}\left[\varepsilon - \frac{j(j+1)\hbar^2}{2\mu r_e^2} - \tfrac{1}{2}k\rho^2 \right]u(\rho) = 0$$

where $\rho = r - r_e$ and in the second term inside the square brackets we have replaced $(r_e + \rho)$ by r_e (valid for small vibrations). The above equation has to be solved subject to the boundary condition that $u(r = 0) = 0 = u(r = \infty)$ which gives $u(\rho = -r_e) = 0 = u(\rho = \infty)$. We assume r_e to be large enough so that the first condition can be written as $u(r = -\infty) = 0$; thus the problem becomes similar to the one-dimensional harmonic oscillator problem so that the energy eigenvalues are given by

$$\varepsilon = \varepsilon_r + \varepsilon_v$$

where

$$\varepsilon_r = \frac{j(j+1)\hbar^2}{2\mu r_e^2} \quad j = 0, 1, 2, \ldots$$

and

$$\varepsilon_v = (n + \tfrac{1}{2})\hbar\omega \quad n = 0, 1, 2, \ldots$$

represent the rotational and vibrational energies respectively; $\omega = (k/\mu)^{1/2}$. We should mention that if r is not replaced by r_e in the expression $j(j+1)\hbar^2/(2\mu r^2)$ then the rotational and vibrational energies will get coupled (see, e.g., Ghatak and Lokanathan (1984)).

The geometric series can be easily summed to give

$$q_v = \frac{\exp(-\theta_v/2T)}{1 - \exp(-\theta_v/T)} \tag{4.54}$$

where

$$\theta_v \equiv \frac{\hbar\omega}{k} \tag{4.55}$$

is known as the vibrational characteristic temperature. We now use (4.49) to obtain

$$E_v = NkT^2\left(\frac{\partial \ln q_v}{\partial T}\right)_V = Nk\left[\frac{\theta_v}{2} + \frac{\theta_v}{\exp(\theta_v/T) - 1}\right] \tag{4.56}$$

and the corresponding specific heat will be given by

$$(c_v)_v = \left(\frac{\partial E_v}{\partial T}\right)_N = Nk\left(\frac{\theta_v}{T}\right)^2 \frac{\exp(\theta_v/T)}{[\exp(\theta_v/T) - 1]^2} \tag{4.57}$$

Typical values for θ_v for some diatomic molecules are given in Table 4.2. We may note that for $T \gg \theta_v$, $(\theta_v/T) \to 0$ and we have

$$(c_v)_v \approx Nk \tag{4.58}$$

which represents the classical value obtained from the law of equipartition of energy. For $T \ll \theta_v$, $(\theta_v/T) \to \infty$ and

$$(c_v)_v \to 0 \tag{4.59}$$

Table 4.2. *The vibrational characteristics temperature* θ_v, *rotational characteristics temperature* θ_r, *the equilibrium separation of atoms* r_e *and the dissociation energy* D_0 *for some diatomic molecules*

	$\theta_v(K)$	$\theta_r(K)$	$r_e(\text{Å})$	$D_0(eV)$
H_2	6210	85.4	0.740	4.454
N_2	3340	2.86	1.095	9.76
O_2	2230	2.07	1.204	5.08
CO	3070	2.77	1.128	9.14
NO	2690	2.42	1.150	5.29
HCl	4140	15.2	1.275	4.43
HBr	3700	12.1	1.414	3.60
HI	3200	9.0	1.604	2.75
Cl_2	810	0.346	1.989	2.48
Br_2	470	0.116	2.284	1.97
I_2	310	0.054	2.667	1.54

Note: Values taken from Hill (1960), p. 153.

As can be seen from Table 4.2, for most diatomic molecules, θ_v is usually much greater than the room temperature and therefore the vibrational contribution to specific heat is negligible. We can now understand the justification in the statement that at room temperatures the vibrational degrees of freedom (for most molecules) can be assumed to be frozen. Figure 4.2 shows the temperature dependence of $(c_v)_v$ as given by (4.57).

4.4   The rotational partition function

As discussed in Section 4.3 the rotational energy of a diatomic molecule is given by

$$\varepsilon_r = \frac{1}{2I}\left[p_\theta{}^2 + \frac{p_\phi{}^2}{\sin^2 \theta} \right] \tag{4.60}$$

which can be put in the form

$$\varepsilon_r = \frac{\mathbf{L}^2}{2I} \tag{4.61}$$

where $\mathbf{L}$ represents the angular momentum of the molecule about the center of mass. Since the eigenvalues of $\mathbf{L}^2$ are $j(j+1)\hbar^2$ $(j=0,1,2,\ldots)$ the rotational energy levels are given by

$$\varepsilon_r = \frac{j(j+1)\hbar^2}{2I} \tag{4.62}$$

The above expression also follows from the footnote on page 42. Since each rotational state is $(2j+1)$-fold degenerate, the rotational partition

Fig. 4.2. The temperature variation of specific heat corresponding to vibrational motion $((c_v)_v)$ as given by (4.57). The dashed line corresponds to the classical value. (Curve adapted from Gopal (1974).)

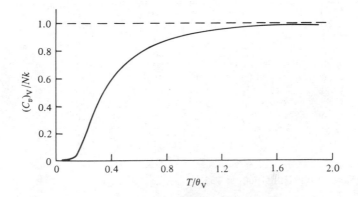

function will be given by

$$q_r = \sum_{j=0}^{\infty} (2j+1) \exp\left[-\frac{j(j+1)\hbar^2}{2IkT}\right]$$

$$= \sum_{j=0}^{\infty} (2j+1)e^{-j(j+1)\theta_r/T} \tag{4.63}$$

where

$$\theta_r \equiv \frac{\hbar^2}{2Ik} \tag{4.64}$$

represents the characteristic temperature for rotation. For $T \ll \theta_r$, the successive terms in the summation decrease very rapidly and one can get a very accurate result by considering the first few terms of the expression

$$q_r = 1 + 3e^{-2\theta_r/T} + 5e^{-6\theta_r/T} + 7e^{-12\theta_r/T} + \cdots \tag{4.65}$$

It is obvious that for $\theta_r/T > 1$, the successive terms decrease with extreme rapidity and q_r can be evaluated very easily. The corresponding expressions for E_r and $(c_v)_r$ can be easily evaluated and one obtains

$$(c_v)_r \approx 12\,Nk\left(\frac{\theta_r}{T}\right)^2 \exp\left(-\frac{2\theta_r}{T}\right) \tag{4.66}$$

in the limit of $T \ll \theta_r$. For $T \gg \theta_r$, the energy levels are very closely spaced and the summation in (4.63) can be replaced by an integral:

$$q_r = \sum_{j=0}^{\infty} (2j+1)e^{-j(j+1)\theta_r/T}$$

$$\approx \int_0^{\infty} (2j+1)e^{-j(j+1)\theta_r/T}\,dj$$

$$= \int_0^{\infty} e^{-x\theta_r/T}\,dx$$

$$= \frac{T}{\theta_r} = \frac{2IkT}{\hbar^2} \tag{4.67}$$

A more accurate expression for q_r (when $T \gg \theta_r$) is given by (see, e.g., Mayer and Mayer (1940)):

$$q_r = \frac{T}{\theta_r}\left(1 + \frac{\theta_r}{3T} + \cdots\right) \tag{4.68}$$

We should mention here that for a diatomic molecule having two identical atoms (like in O_2, N_2 etc.) there has been overcounting of states because there are two configurations which are identical; the situation is similar to the one considered in the previous chapter where we had to divide by $N!$ to take into account of the fact that the molecules are indistinguishable. Thus, for $T \gg \theta_r$, we may write

$$q_r \approx \frac{T}{\sigma\theta_r}\left(1 + \frac{\theta_r}{3T} + \cdots\right) \tag{4.69}$$

where $\sigma = 2$ for symmetric diatomic molecules like O_2, N_2 etc. and $\sigma = 1$ for asymmetric molecules like CO, NO etc. Using only the first term on the right hand side of (4.69) we get for the free energy

$$F_r = -NkT \ln q_r = -NkT \ln\left(\frac{T}{\sigma\theta_r}\right) \tag{4.70}$$

and

$$E_r = NkT^2\left(\frac{\partial \ln q_r}{\partial T}\right)_v = NkT \tag{4.71}$$

$$(c_v)_r = \left(\frac{\partial E_r}{\partial T}\right)_V = Nk \tag{4.72}$$

$$S_r = -\left(\frac{\partial F_r}{\partial T}\right)_V = Nk \ln \frac{T}{\sigma\theta_r} + Nk \tag{4.73}$$

It is of interest to note that (for $T \gg \theta_r$) the expressions for E_r and $(c_v)_r$ are the same as obtained by using the law of equipartition of energy. Since the rotational characteristic temperatures are small compared with the room

Fig. 4.3. The temperature variation of specific heat corresponding to rotational motion ($(c_v)_r$) as obtained by using (4.67) for q_r. (Curve adapted from Gopal (1974).)

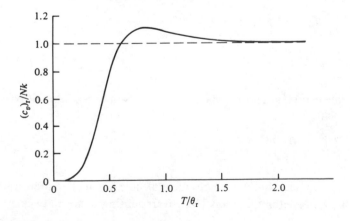

temperature (Table 4.2), it is only at very low temperatures that we find the rotational degrees of freedom to be frozen.

Figure 4.3 gives the variation of $(c_v)_r$ as a a function of temperature obtained by using (4.67) for q_r. Notice that at high temperatures both $(c_v)_v$ and $(c_v)_r$ tend to Nk which is the value predicted from the law of equipartition of energy.

4.5 The electronic partition function

For most gases the first excited electronic state is about a few electron volts separated from the ground state. Thus, at ordinary temperatures[†] the excitation of electronic states can be neglected. However, for gases like O_2 and NO, the electronic excited states lie close to the ground level and as a first approximation we may take into consideration only the ground state and the first excited state. For such a case, the electronic partition function will be given by[‡]

$$
\begin{aligned}
q_e &= g_0 e^{-\varepsilon_0/kT} + g_1 e^{-\varepsilon_1/kT} \\
&= e^{-\varepsilon_0/kT}(g_0 + g_1 e^{-\varepsilon/kT})
\end{aligned}
\tag{4.74}
$$

where g_0 and g_1 represent the degeneracies of the ground state and the first excited state respectively and $\varepsilon\,(=\varepsilon_1 - \varepsilon_0)$ represents the separation of the ground state and the first excited state.[§] Thus

$$
E_e = NkT^2\left(\frac{\partial \ln q_e}{\partial T}\right)_v = N\varepsilon_0 + N\varepsilon\left(1 + \frac{g_0}{g_1}e^{+\varepsilon/kT}\right)^{-1}
\tag{4.75}
$$

and

$$
(c_v)_e = \frac{N\varepsilon^2}{kT}\frac{g_0}{g_1}e^{\varepsilon/kT}\left[1 + \frac{g_0}{g_1}e^{\varepsilon/kT}\right]^{-2}
\tag{4.76}
$$

Notice that $(c_v)_e \to 0$ when $kT \ll \varepsilon$ and also when $kT \gg \varepsilon$. Obviously, a maximum occurs at a particular value of T which depends on the value of g_0/g_1; e.g., for $g_1/g_0 = 1$, the maximum occurs at $kT \approx 0.4\varepsilon$ (see Gopal 1974).

4.6 Summary

We conclude this chapter by noting that if we treat the rotational motion classically (which is justified since θ_r is usually much smaller than

---

[†] We may recall that at $T \approx 300\,\mathrm{K}$, $kT \approx \frac{1}{40}\,\mathrm{eV}$.

[‡] If we assume the zero of the energy to correspond to separated atoms at rest, then $\varepsilon_0 = -D_0 + \frac{1}{2}\hbar\omega$ where D_0 represents the dissociation energy of the molecule at $0\,\mathrm{K}$.

[§] For most gases $g_0 = 1$ and $\varepsilon \gg kT$ (for ordinary temperatures). However NO and O_2 are exceptions: for NO, $g_0 = 2$, $g_1 = 2$, $\varepsilon/k = 178\,\mathrm{K}$ and for O_2, $g_0 = 3$, $g_1 = 2$ and $\varepsilon/k = 11300\,\mathrm{K}$.

normal temperatures), the vibrational motion quantum mechanically and if we neglect the excitation of electronic states then the partition function will be given by the following equation:

$$q = \left[\left(\frac{mkT}{2\pi\hbar^2} \right)^{3/2} V \right]\left[\frac{e^{-\theta_v/2T}}{1 - e^{-\theta_v/T}} \right]\left[\frac{T}{\sigma\theta_r} \right][g_0 e^{-\varepsilon_0/kT}] \tag{4.77}$$

where the successive terms inside the square brackets correspond to translational motion, vibrational motion, rotational motion and electronic excitation respectively. The other thermodynamic function will be given by the following equations:

$$Q = \frac{1}{N!} q^N$$

$$F = -kT\ln Q = -NkT\left\{ \ln\left[\left(\frac{mkT}{2\pi\hbar^2} \right)^{3/2} \frac{V}{N} \right] \right.$$

$$+ 1 - \frac{\theta_v}{2T} - \ln(1 - e^{-\theta_v/T})$$

$$\left. + \ln\left(\frac{T}{\sigma\theta_r} \right) - \frac{\varepsilon_0}{kT} + \ln g_0 \right\} \tag{4.78}$$

$$E = kT^2\left(\frac{\partial}{\partial T}\ln Q \right)_V = NkT\left\{ \frac{3}{2} + \frac{\theta_v}{2T} + \frac{\theta_v/T}{e^{\theta_v/T} - 1} + 1 + \frac{\varepsilon_0}{kT} \right\} \tag{4.79}$$

$$c_v = Nk\left[\frac{3}{2} + 1 + \left(\frac{\theta_v}{T} \right)^2 \frac{e^{\theta_v/T}}{(e^{\theta_v/T} - 1)^2} \right] \tag{4.80}$$

$$S = -\left(\frac{\partial F}{\partial T} \right)_V = Nk\left\{ \ln\left[\left(\frac{mkT}{2\pi\hbar^2} \right)^{3/2} \frac{V}{N} \right] + \frac{7}{2} \right.$$

$$\left. + \frac{\theta_v/T}{e^{\theta_v/T} - 1} - \ln(1 - e^{-\theta_v/T}) + \ln g_0 \right\} \tag{4.81}$$

$$P = -\left(\frac{\partial F}{\partial V} \right)_T = \frac{NkT}{V} \tag{4.82}$$

5

Bose–Einstein equation of state

5.1 Introduction

We begin this chapter by considering the simple problem of calculating the number of ways three identical particles can be distributed in two states such that there are two particles in the first state and one particle in the second state. In classical statistics, the particles are distinguishable and can be labelled as A, B and C and as can be seen from Fig. 5.1 we have three ways for such a distribution. In quantum mechanics, identical particles are indistinguishable and there is therefore only *one* way in which we can have two particles in ε_1 and one in ε_2. This indistinguishability leads to fundamental difference in the calculation of thermodynamic functions.[†] Even for identical indistinguishable particles, there are two types of distribution possible: one in which a state can be occupied by any number of particles – this leads to Bose–Einstein statistics which is the subject matter of this chapter, and the second in which a state can be occupied by not more than one particle – this leads to Fermi–Dirac statistics which is the subject matter of the next chapter. Particles obeying Bose–Einstein statistics are

Fig. 5.1. We have three distinguishable particles which we label *A*, *B* and *C*; there will be three ways in which we can have two particles in ε_1 and one particle in ε_2.

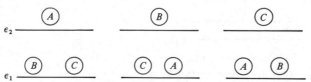

[†] We should mention that there may be several independent states with the same value of energy; this leads to the concept of degeneracy in quantum mechanics. Each linearly independent state has to be considered separately.

known as *Bosons* and those obeying Fermi–Dirac statistics are known as *Fermions*.

Before we discuss the details of Bose–Einstein (or Fermi–Dirac) statistics, we will calculate the partition function for identical particles obeying classical statistics where the particles are assumed to be distinguishable.[†] We should mention that in Chapter 3, we calculated the partition function corresponding to classical statistics, but the method that we will use here will be slightly different, which will enable us to appreciate the difference in the calculation of the partition function for particles obeying classical and quantum statistics.

5.2 Classical statistics

We consider a gas of N identical particles confined in a box of volume V at temperature T. Let the possible energy levels (for each particle) be denoted by $\varepsilon_1, \varepsilon_2, \ldots$ etc. (see (3.1)). Let the number of particles in the energy states $\varepsilon_1, \varepsilon_2, \varepsilon_3, \ldots$ be denoted by $n_1, n_2, n_3, \ldots$ respectively (see Fig. 5.2). Thus the total energy of the system will be given by

$$E = n_1\varepsilon_1 + n_2\varepsilon_2 + \cdots \tag{5.1}$$

and the partition function of the system will be given by

$$Q = \sum e^{-(n_1\varepsilon_1 + n_2\varepsilon_2 + \cdots)/kT} \tag{5.2}$$

where the summation is over all possible states of the gas. We have to sum the above series subject to the condition that

$$\sum_j n_j = N \tag{5.3}$$

which represents the total number of particles. Now, in classical statistics, the particles are *distinguishable* and the number of ways in which we can

Fig. 5.2. The energy level ε_1 is occupied by n_1 particles, ε_2 by n_2 particles etc.

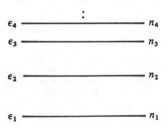

[†] Classical statistics is often referred to as Maxwell–Boltzmann statistics.

have n_1 particles in state 1, n_2 particles in state 2,... is

$$\frac{N!}{n_1! n_2! n_3!} \cdots \tag{5.4}$$

The above equation is consistent with Fig. 5.1 which corresponds to $N = 3$, $n_1 = 2$ and $n_2 = 1$. Thus the partition function will be

$$Q = \sum_{n_1} \sum_{n_2 \cdots} \frac{N!}{n_1! n_2! \cdots} e^{-(n_1 \varepsilon_1 + n_2 \varepsilon_2 + \cdots)/kT} \tag{5.5}$$

where the values of $n_1, n_2, \ldots$ should be such that (5.3) is satisfied. The above expression is simply the binomial expansion of

$$(e^{-\varepsilon_1/kT} + e^{-\varepsilon_2/kT} + \cdots)^N \tag{5.6}$$

and therefore

$$Q = \left[\sum_n e^{-\varepsilon_n/kT} \right]^N \tag{5.7}$$

We may recall that in Chapter 3, using a slightly different method, we derived an identical expression for the partition function – obviously, there also the particles were assumed to be distinguishable. Starting from (5.7) we can obtain the equations of state and other thermodynamic functions (see Chapters 3 and 4).

We next consider Bose–Einstein statistics where the particles are indistinguishable and each state can be occupied by any number of particles. However, two different cases are possible one in which there is no restriction on the total number of particles and the second in which the total number of particles is always conserved. We will consider the two cases separately.

5.3 Bose–Einstein statistics without restriction on the total number of particles: photons

Since the particles are indistinguishable, the partition function will be given by

$$Q = \sum_{n_1} \sum_{n_2 \cdots} e^{-(n_1 \varepsilon_1 + n_2 \varepsilon_2 + \cdots)/kT} \tag{5.8}$$

Notice that the factor given by (5.4) is not present. Since there is no restriction on the total number of particles and hence on $n_1, n_2, \ldots$ etc., the summations in (5.8) can be written in the form

$$Q = \sum_{n_1} (e^{-\varepsilon_1/kT})^{n_1} \sum_{n_2} (e^{-\varepsilon_2/kT})^{n_2} \sum_{n_3 \cdots} \tag{5.9}$$

Carrying out the sum of geometric series, we get

$$Q = \frac{1}{1 - e^{-\varepsilon_1/kT}} \frac{1}{1 - e^{-\varepsilon_2/kT}} \cdots \tag{5.10}$$

or

$$\ln Q = -\sum_j \ln(1 - e^{-\varepsilon_j/kT}) \tag{5.11}$$

This is indeed the case when we consider photon statistics where there is no limit on the number of photons occupying a particular state. The average number of particles in a particular state will be given by

$$\langle n_j \rangle = -\frac{1}{kT} \frac{\partial}{\partial \varepsilon_j} (\ln Q) = \frac{e^{-\varepsilon_j/kT}}{1 - e^{-\varepsilon_j/kT}} \tag{5.12}$$

or

$$\langle n_j \rangle = \frac{1}{e^{\varepsilon_j/kT} - 1} \tag{5.13}$$

which is nothing but the Planck distribution.[†]

Now, if we consider the radiation field inside a box of volume V then using a method similar to that used in Appendix 1 we find that there will be (see (A1.27))[‡]

$$2 \times \frac{V}{8\pi^3\hbar^3} \times 4\pi p^2 \mathrm{d}p$$

[†] Equation (5.12) follows from the fact that the probability of finding n_1 photons in ε_1, n_2 photons in ε_2,... is proportional to the Boltzmann factor $\exp[-(n_1\varepsilon_1 + n_2\varepsilon_2 + \cdots)/kT]$; thus

$$\langle n_j \rangle = \frac{\displaystyle\sum_{n_1 \, n_2 \cdots} n_j \exp[-(n_1\varepsilon_1 + n_2\varepsilon_2 + \cdots)/kT]}{\displaystyle\sum_{n_1 \, n_2 \cdots} \exp[-(n_1\varepsilon_1 + n_2\varepsilon_2 + \cdots)/kT]}$$

$$= \frac{\displaystyle\sum_{n_j = 0,1\cdots}^{\infty} n_j \exp[-n_j\varepsilon_j/kT]}{\displaystyle\sum_{n_j = 0,1\cdots}^{\infty} \exp[-n_j\varepsilon_j/kT]}$$

$$= [1 - e^{-\varepsilon_j/kT}]\left[-kT \frac{\mathrm{d}}{\mathrm{d}\varepsilon_j} \sum_{n_j} e^{-n_j\varepsilon_j/kT}\right]$$

$$= [1 - e^{-\varepsilon_j/kT}]\left[-kT \frac{\mathrm{d}}{\mathrm{d}\varepsilon_j} \frac{1}{1 - e^{-\varepsilon_j/kT}}\right] = \frac{1}{e^{\varepsilon_j/kT} - 1}$$

[‡] If we solve the wave equation (corresponding to the radiation field) inside a cubical box of volume V and use the periodic boundary conditions then the allowed values of k_x, k_y and k_z will be given by (A1.23). The equation $\mathbf{p} = \hbar\mathbf{k}$ and (A1.25), (A1.26) and (A1.27) will also remain valid. The relationship between energy and momentum will however be given by (5.14). (See also Section 6.2 and Appendix D of Thyagarajan and Ghatak (1981).)

states with the photon momentum lying between p and $p + dp$; the additional factor of 2 arises from the fact that for each state there are two independent modes of polarization. While discussing photon statistics, it is more convenient to use the frequency ω as the variable which is related to p through the following equation,

$$\varepsilon = \hbar\omega = pc \tag{5.14}$$

Thus the number of states with frequency lying between ω and $\omega + d\omega$ will be given by

$$g(\omega)d\omega = \frac{V}{\pi^2 c^3}\omega^2 d\omega \tag{5.15}$$

Now, as in Chapter 3, we replace the summation in (5.11) by an integration to obtain

$$\ln Q = -\int_0^\infty \ln[1 - e^{-\hbar\omega/kT}]g(\omega)d\omega \tag{5.16}$$

Using (5.15) and integrating by parts we obtain

$$\ln Q = \frac{V}{\pi^2 c^3}\left[\int_0^\infty \frac{(\hbar/kT)\exp(-\hbar\omega/kT)\,\omega^3}{[1 - \exp(-\hbar\omega/kT)]}\frac{d\omega}{3}\right] = \frac{\pi^2}{45}V\left(\frac{kT}{\hbar c}\right)^3 \tag{5.17}$$

where we have used the following result

$$\int_0^\infty \frac{x^3 dx}{e^x - 1} = \frac{\pi^4}{15}$$

Thus the total energy E and the pressure P will be given by

$$\left.\begin{aligned} E &= kT^2\left(\frac{\partial \ln Q}{\partial T}\right)_V = \frac{\pi^2}{15}V\frac{(kT)^4}{(\hbar c)^3} \\ \\ P &= kT\left(\frac{\partial \ln Q}{\partial V}\right)_T = \frac{\pi^2}{45}\frac{(kT)^4}{(\hbar c)^3} \end{aligned}\right\} \tag{5.18}$$

and

For the photon gas we therefore have

$$P = \tfrac{1}{3}u = \tfrac{1}{3}aT^4 \tag{5.19}$$

where $u(= E/V)$ represents the energy density and

$$a = \frac{1}{15}\frac{k^4}{\hbar^3 c^3} \approx 7.564 \times 10^{-15}\,\mathrm{erg\,cm^{-3}\,K^{-4}}$$

is known as the radiation pressure constant. In order to have a numerical appreciation we note that at $T \approx 10^6$ K, the radiation pressure will be given by

$$P \approx \tfrac{1}{3} \times 7.564 \times 10^{-15} \times 10^{24} \approx 2.5 \times 10^9 \text{ dyne cm}^{-2}$$
$$\approx 2.5 \times 10^3 \text{ atm}$$
$$(1 \text{ atm} \approx 1.01 \times 10^6 \text{ dyne cm}^{-2})$$

It is therefore of interest to mention that if we use the equation corresponding to the first law of thermodynamics

$$dQ = dU + PdV$$

where dQ represents the heat added to the system, dU the increase in internal energy of the system and PdV the mechanical work done by the system, then for the photon gas we have

$$dQ = d(uV) + \tfrac{1}{3}udV = Vdu + \tfrac{4}{3}udV$$

For an adiabatic change $dQ = 0$ and therefore

$$Vdu + \tfrac{4}{3}udV = 0$$

or

$$uV^{4/3} = \text{constant}$$

Since $u = aT^4$, we have

$$TV^{1/3} = \text{constant} \tag{5.20}$$

If we now use the relation $P = (a/3)T^4$, we would get

$$PV^{4/3} = \text{constant} \tag{5.21}$$

Thus the photon gas behaves like a perfect gas with the ratio of specific heat $\gamma = \tfrac{4}{3}$.

Now, if $u(\omega)d\omega$ represents the energy of the radiation field (per unit volume) in the frequency interval $d\omega$, then

$$u(\omega)d\omega = (\text{energy associated with each state}) \times (\text{number of states per unit volume in the frequency interval } d\omega)$$

$$= (\langle n \rangle \hbar\omega) \times \left(\frac{1}{\pi^2 c^3} \omega^2 d\omega \right)$$

or

$$u(\omega) = \frac{\hbar\omega^3}{\pi^2 c^3} \frac{1}{\exp(\hbar\omega/kT) - 1} \tag{5.22}$$

which is the famous Planck's law. Another quantity of interest is the emissivity $e_\omega(T)$ which is defined such that $e_\omega(T)d\omega$ represents the amount of energy coming out of an unit area per unit time. Experimentally, it is the emissivity (rather than the total energy density E/V which is measured). Now it is well known in the kinetic theory of gases that the number of particles coming out of a unit area per unit time is $n\bar{c}/4$ where $\bar{c}$ represents the average velocity (see, e.g., Pathria 1972, p. 149). Thus

$$e_\omega(T)\mathrm{d}\omega = \frac{c}{4}u(\omega)\mathrm{d}\omega$$

or $\qquad\qquad\qquad\qquad\qquad\qquad\qquad\qquad\qquad\qquad$ (5.23)

$$e_\omega(T) = \frac{\hbar\omega^3}{4\pi^2 c^2}\frac{1}{\exp(\hbar\omega/kT) - 1}$$

The total energy radiated will be given by

$$e(T) = \int_0^\infty e_\omega(T)\mathrm{d}\omega = \frac{\hbar}{4\pi^2 c^2}\left(\frac{kT}{\hbar}\right)^4 \int_0^\infty \frac{x^3}{e^x - 1}\mathrm{d}x$$

or

$$e(T) = \sigma T^4 \qquad\qquad\qquad\qquad\qquad\qquad (5.24)$$

where

$$\sigma = \frac{\pi^2}{60}\frac{k^4}{\hbar^3 c^2} \approx 5.67 \times 10^{-5}\,\mathrm{erg\,s^{-1}\,cm^{-2}\,K^{-4}} \qquad (5.25)$$

represents the Stefan's constant. Equation (5.24) represents the famous Stefan–Boltzmann law.

5.4 Bose–Einstein statistics for a constant number of particles

In the previous section we calculated the partition function Q (5.8) without the restriction given by (5.3). Such a consideration will be valid for a photon gas, however; when the total number of particles is conserved (as in helium gas) then the restriction imposed by (5.3) should be taken into account and the calculation becomes very difficult. It is, however, possible to circumvent this restriction by considering the grand canonical ensemble which consists of a large number of systems each of volume V and through the walls of which the particles can diffuse to other systems. The whole ensemble is in thermodynamic equilibrium at temperature T. Due to the fact that the walls of each system are permeable to molecules, the number of molecules in a particular system will not remain fixed and indeed the probability that a system contains N particles and has a *total* energy $E_i (= n_{1i}\varepsilon_1 + n_{2i}\varepsilon_2 + \cdots) \left[\sum_j n_{ji} = N\right]$ is given by[†]

[†] The grand canonical ensemble and equations (5.26)–(5.36) are discussed in most texts on statistical mechanics; see, e.g., Hill (1960), Landau and Lifshitz (1958), Pathria (1972) and Ter Haar (1962).

$$\mathscr{P}_i(N, V, T) = \frac{e^{N\mu/kT}e^{-E_i/kT}}{Z} \tag{5.26}$$

where μ is known as the chemical potential (it is introduced as an undetermined multiplier and is determined by (5.31); $n_{1i}, n_{2i}, \ldots$ respectively represent the number of particles in states $\varepsilon_1, \varepsilon_2, \ldots$ when the total energy is E_i and

$$Z = \sum_N e^{\mu N/kT} \sum_i e^{-E_i/kT} \tag{5.27}$$

is known as the *grand partition function*. If we use (5.2) we get

$$Z = \sum_N e^{\mu N/kT} Q(N, V, T) \tag{5.28}$$

Now, the probability of the system having N particles irrespective of its value of total energy will be given by

$$\mathscr{P}(N, V, T) = \sum_i \mathscr{P}_i(N, V, T) = \frac{1}{Z} e^{N\mu/kT} \sum_i e^{-E_i/kT}$$

$$= \frac{1}{Z} e^{\mu N/kT} Q(N, V, T) \tag{5.29}$$

Thus the average value of N (over all the systems) will be

$$\bar{N} = \sum_N N \mathscr{P}(N, V, T)$$

$$= \frac{1}{Z} \sum_N N e^{\mu N/kT} Q(N, V, T)$$

$$= \frac{1}{Z} kT \left(\frac{\partial Z}{\partial \mu} \right)_{V,T}$$

$$= kT \left(\frac{\partial q}{\partial \mu} \right)_{V,T}$$

where

$$q = \ln Z \tag{5.30}$$

is known as the grand potential. Since each system will, on average, consist of an extremely large number of particles we must have $\bar{N} = N$ and therefore

$$N = kT \left(\frac{\partial \ln Z}{\partial \mu} \right)_{V,T} = kT \left(\frac{\partial q}{\partial \mu} \right)_{V,T} \tag{5.31}$$

The above equation is used for determining μ. The mean occupation number of the level E_j will be given by

$$\langle n_j \rangle = \sum_N \sum_i n_{ji} \mathscr{P}_i(N, V, T)$$

$$= \frac{1}{Z} \sum_N e^{\mu N/kT} \sum_i n_{ji} e^{-(n_{1i}\varepsilon_1 + n_{2i}\varepsilon_2 + \cdots)/kT}$$

or

$$\langle n_j \rangle = -kT \frac{1}{Z} \left(\frac{\partial Z}{\partial \varepsilon_j} \right)_{\mu, T, \text{ all other } \varepsilon s}$$

$$= -kT \left(\frac{\partial q}{\partial \varepsilon_j} \right)_{\mu, T, \text{ all other } \varepsilon s} \tag{5.32}$$

The total energy will be given by

$$E = \sum_N \sum_i \varepsilon_i \mathscr{P}_i(N, V, T)$$

$$= +kT^2 \left(\frac{1}{Z} \frac{\partial Z}{\partial T} \right)_{\mu, V} = +kT^2 \left(\frac{\partial q}{\partial T} \right)_{\mu, V} \tag{5.33}$$

Other thermodynamic functions like the pressure P, the Helmholtz free energy F and the entropy S are respectively given by (see, e.g., Hill (1960), Pathria (1972)):

$$P = kT \left(\frac{\partial \ln Z}{\partial V} \right)_{\mu, T} = kT \frac{\ln Z}{V} = \frac{kT}{V} q \tag{5.34}$$

$$F = N\mu - PV \tag{5.35}$$

and

$$S = \frac{E - F}{T} = \frac{E - N\mu + PV}{T} \tag{5.36}$$

We next calculate the grand partition function for a Bose–Einstein gas. Since

$$E_i = n_{1i}\varepsilon_1 + n_{2i}\varepsilon_2 + \cdots \tag{5.37}$$

(5.27) can be written in the form

$$Z = \sum_N \sum_i e^{\mu(n_{1i} + n_{2i} + \cdots)/kT} e^{-(n_{1i}\varepsilon_1 + n_{2i}\varepsilon_2 + \cdots)/kT}$$

$$= \sum_{N=0}^{\infty} \left\{ \sum_i [e^{(\mu - \varepsilon_1)/kT}]^{n_{1i}} [e^{(\mu - \varepsilon_2)/kT}]^{n_{2i}} \cdots \right\} \tag{5.38}$$

Notice that the sum over the curly brackets is over the numbers $n_{1i}, n_{2i}, \ldots$ such that the total value is a particular number N and then we have a sum over *all* possible values of N. This double sum is the same as the sum over all possible values of $n_1, n_2, \ldots$ independent of one another.[†] Thus

$$Z = \left\{ \sum_{n_1 = 0, 1, \ldots}^{\infty} [e^{(\mu - \varepsilon_1)/kT}]^{n_1} \right\} \left\{ \sum_{n_2 = 0, 1, \ldots} [e^{(\mu - \varepsilon_2)/kT}]^{n_2} \right\} \cdots \qquad (5.39)$$

Since each quantity inside the curly brackets is a geometric series we have

$$Z = \left[\frac{1}{1 - e^{(\mu - \varepsilon_1)/kT}} \right] \left[\frac{1}{1 - e^{(\mu - \varepsilon_2)/kT}} \right] \cdots \qquad (5.40)$$

where we have presupposed that μ is a negative quantity. Thus

$$q = \ln Z = - \sum_j \ln[1 - e^{(\mu - \varepsilon_j)/kT}] \qquad (5.41)$$

$$N = kT \left(\frac{\partial q}{\partial \mu} \right)_{V, T} = \sum_j \frac{1}{e^{(\varepsilon_j - \mu)/kT} - 1} \qquad (5.42)$$

$$E = kT^2 \left(\frac{\partial q}{\partial T} \right)_{\mu, V} = \sum_j \frac{\varepsilon_j}{e^{(\varepsilon_j - \mu)/kT} - 1} \qquad (5.43)$$

$$P = \frac{kT}{V} q = - \frac{kT}{V} \sum_j \ln[1 - e^{(\mu - \varepsilon_j)/kT}] \qquad (5.44)$$

and the mean occupation number of the level ε_j will be (5.32)

$$\langle n_j \rangle = - kT \left(\frac{\partial q}{\partial \varepsilon_j} \right)_{\mu, T, \text{ all other } \varepsilon s}$$

$$= \frac{1}{e^{(\varepsilon_j - \mu)/kT} - 1} \qquad (5.45)$$

which is usually referred to as the Bose–Einstein distribution. Equations (5.42), (5.43) and (5.45) imply

$$N = \sum_j \langle n_j \rangle \quad \text{and} \quad E = \sum_j \langle n_j \rangle \varepsilon_j \qquad (5.46)$$

which should indeed be the case.

[†] The argument can be understood if we consider just 2 states E_1 and E_2; we consider the sum

$$\sum_{N=0}^{\infty} \left[\sum_i A^{n_{1i}} B^{n_{2i}} \right] = \{A^0 B^0\} + \{A^0 B^1 + A^1 B^0\} + \{A^0 B^2 + A^1 B^1 + A^2 B^0\}$$

$$+ \{A^0 B^3 + A^1 B^2 + A^2 B^1 + A^3 B^0\} + \cdots$$

where the successive terms inside the curly brackets correspond to $N = 0$ $N = 1$, $N = 2$ and $N = 3$ respectively. Obviously, the right hand side is equal to $(A^0 + A^1 + A^2 + \cdots)(B^0 + B^1 + B^2 + \cdots)$
The analysis can be easily extended to $3, 4 \cdots$ states.

If we assume an almost continuous distribution of states then each of the summations in (5.41)–(5.44) can be replaced by an integral:

$$\sum_j \cdots \approx \int_0^\infty \cdots g(\varepsilon)d\varepsilon \tag{5.47}$$

where

$$g(\varepsilon) = G\frac{(2m)^{3/2}V}{4\pi^2\hbar^3}\varepsilon^{1/2} \tag{5.48}$$

represents the density of states (see (A1.19)), G being the degeneracy parameter. We first evaluate the sum in (5.42)

$$N \approx G\frac{(2m)^{3/2}V}{4\pi^2\hbar^3}\int_0^\infty \frac{\varepsilon^{1/2}d\varepsilon}{e^{(\varepsilon-\mu)/kT}-1} \tag{5.49}$$

If we assume that

$$e^{-\mu/kT} \gg 1 \tag{5.50}$$

then the unity in the denominator of the integrand in (5.49) can be neglected and we will have

$$N \approx G\frac{(2m)^{3/2}V}{4\pi^2}e^{\mu/kT}\int_0^\infty e^{-\varepsilon/kT}\varepsilon^{1/2}d\varepsilon$$

or

$$e^{-\mu/kT} \approx \frac{G}{(N/V)}\left(\frac{mkT}{2\pi\hbar^2}\right)^{3/2} \tag{5.51}$$

where we have used the relation

$$\int_0^\infty \varepsilon^{1/2}e^{-\varepsilon/kT}dE = (kT)^{3/2}\int_0^\infty x^{1/2}e^{-x}dx = (kT)^{3/2}\tfrac{1}{2}\pi^{1/2} \tag{5.52}$$

Equation (5.51) determines μ – however, only when (5.50) is satisfied which is also the condition for validity of classical statistics. This can also be seen from the evaluation of E and P by using (5.43), (5.44), (5.47), (5.48) and (5.50):

$$E \approx G\frac{(2m)^{3/2}V}{4\pi^2\hbar^3}e^{\mu/kT}\int_0^\infty \varepsilon^{3/2}e^{-\varepsilon/kT}dE = \tfrac{3}{2}NkT \tag{5.53}$$

$$P \approx -\frac{kT}{V}\int_0^\infty \ln[1-e^{(\mu-\varepsilon)/kT}]g(\varepsilon)d\varepsilon$$

$$\approx \frac{kT}{V}e^{\mu/kT}G\frac{(2m)^{3/2}V}{4\pi^2\hbar^3}\int_0^\infty \varepsilon^{1/2}e^{-\varepsilon/kT}dE = \frac{NkT}{V} \tag{5.54}$$

where we have used (5.51) and have approximated $\ln(1-x)$ by $-x$.

Equations (5.53) and (5.54) are the same as those derived from classical statistics (see (3.14) and (3.15)). Using (5.50) and (5.51) we get the following condition for the applicability of classical statistics

$$e^{-\mu/kT} \approx \frac{G}{(N/V)}\left(\frac{mkT}{2\pi\hbar^2}\right)^{3/2} \gg 1 \tag{5.55}$$

Conversely, quantum effects will dominate when

$$\frac{G}{(N/V)}\left(\frac{mkT}{2\pi\hbar^2}\right)^{3/2} \lesssim 1 \tag{5.56}$$

which can be due to (a) small particle mass, or (b) low temperatures, or (c) high particle density. Assuming $N/V \approx 10^{22}$ particles/cm^3, (5.56) gives

$$T \lesssim \frac{1}{G^{2/3}}\frac{2\pi\hbar^2}{mk}\left(\frac{N}{V}\right)^{2/3}$$

$$= \frac{1}{G^{2/3}}\frac{2 \times 3.14 \times (1.054 \times 10^{-27})^2}{m \times (1.38 \times 10^{-16})}(10^{22})^{2/3}$$

or

$$T \lesssim \frac{2 \times 10^{-23}}{G^{2/3}m}(\text{K}) \tag{5.57}$$

where m is measured in grams. For an electron gas,[†] $G = 2$ (because of spin degeneracy), $m = 9.1 \times 10^{-28}$ g and thus (5.57) becomes

$$T \lesssim 1.4 \times 10^4 \, \text{K} \tag{5.58}$$

implying that an electron gas will exhibit quantum effects even at room temperatures (see next chapter for more details). On the other hand, if we consider the lightest molecule (viz., H_2) then m is about 3700 times the electron mass and quantum effects will be observed when

$$T \ll 10 \, \text{K} \tag{5.59}$$

where we have assumed $G = 1$. Thus any quantum effect in a molecular gas can only be observed at extremely low temperatures (unless the particle densities are very high).

Returning to (5.43) and (5.44) and replacing the sums by integrals (see (5.47) and (5.48)) we get

$$E = G\frac{(2m)^{3/2}V}{4\pi^2\hbar^3}\int_0^\infty \frac{\varepsilon^{3/2}}{e^{(\varepsilon-\mu)/kT} - 1}\,d\varepsilon \tag{5.60}$$

[†] We should mention here that although electrons obey Fermi–Dirac statistics, the condition for applicability of classical statistics (5.55) and the condition for quantum effects to dominate (5.56) remain the same.

and

$$P = -\frac{kT}{V}\left[G\frac{(2m)^{3/2}V}{4\pi^2\hbar^3}\int_0^\infty \ln(1 - e^{(\mu - \varepsilon)/kT})\varepsilon^{1/2}d\varepsilon \right] \tag{5.61}$$

where we have *not* assumed the condition given by (5.50). In the above expression, if we integrate by parts, we would readily obtain

$$PV = \tfrac{2}{3}E \tag{5.62}$$

The above relation is valid for particles obeying classical statistics, Bose–Einstein statistics as well as Fermi–Dirac statistics (but not for a photon gas – see (5.21)).

When the condition given by (5.50) is not satisfied then the evaluation of the integrals in (5.49), (5.60) and (5.61) is quite difficult. However, if we use the expansion[†]

$$f_n(y) = \frac{1}{\Gamma(n)}\int_0^\infty \frac{x^{n-1}}{(1/y)e^x - 1}dx = y + \frac{y^2}{2^n} + \frac{y^3}{3^n} + \cdots \tag{5.63}$$

where $y \equiv c^{\mu/kT}$, then (5.49) expresses N as a power series in $e^{\mu/kT}$; if we invert the series then μ is expressed in terms of N. Using (5.63) for $n = \tfrac{3}{2}$ and (5.34)–(5.36) we can calculate all thermodynamic quantities of interest.[‡] The final results are (for more details, see, e.g., Hill (1960), Pathria (1972)):

$$P = \frac{NkT}{V}\left[1 \mp \frac{1}{2^{5/2}}\Lambda + \cdots \right] \tag{5.64}$$

$$E = \tfrac{3}{2}PV = \tfrac{3}{2}NkT\left[1 \mp \frac{1}{2^{5/2}}\Lambda + \cdots \right] \tag{5.65}$$

$$c_v = \left(\frac{\partial E}{\partial T}\right)_V = \tfrac{3}{2}Nk\left[1 + \frac{1}{2^{7/2}}\Lambda + \cdots \right] \tag{5.66}$$

$$\mu = kT\left[\ln\Lambda \mp \frac{1}{2^{3/2}}\Lambda + \cdots \right] \tag{5.67}$$

[†] $\dfrac{1}{\Gamma(n)}\displaystyle\int_0^\infty \dfrac{x^{n-1}}{(1/y)e^x - 1}dx = \dfrac{1}{\Gamma(n)}\displaystyle\int_0^\infty y e^{-x}(1 - y e^{-x})^{-1}x^{n-1}dx$

$\qquad = \dfrac{1}{\Gamma(n)}\displaystyle\int_0^\infty y e^{-x}[1 + y e^{-x} + y^2 e^{-2x} + \cdots]x^{n-1}dx$

$\qquad = y + \dfrac{y^2}{2^n} + \cdots$

[‡] Once E and μ are known, P, q, F and S can be determined from (5.62), (5.34), (5.35) and (5.36) respectively.

$$S = Nk\left[\frac{5}{2} - \ln \Lambda \mp \frac{1}{2^{7/2}}\Lambda + \cdots\right] \tag{5.68}$$

where

$$\Lambda \equiv \frac{1}{G}\left(\frac{2\pi\hbar^2}{mkT}\right)^{3/2}\left(\frac{N}{V}\right) \tag{5.69}$$

The above expressions represent the equations of state and for the sake of completeness we have included the results corresponding to Fermi–Dirac statistics which will be discussed in the next chapter; the upper and lower signs in (5.64)–(5.68) correspond to Bose–Einstein and Fermi–Dirac statistics respectively. Obviously, all the expressions would involve quadratic and higher powers of Λ and therefore they will be useful only when $\Lambda \ll 1$, i.e., high temperatures or low density or high values of the mass of each particle. It may be noted that as $\Lambda \to 0$ all the expressions go over to their classical values (as $\Lambda \to 0$, $\mu \to -\infty$ so that (5.50) and all the discussion after (5.50) becomes valid).

5.4.1   *Bose–Einstein condensation*

At very low temperatures an extremely interesting phenomenon occurs for a Bose–Einstein gas. This phenomenon, known as the *Bose–Einstein condensation*, occurs due to the accumulation of a large number of particles in the ground state $(E = 0)$.[†] Obviously, when this happens, it will not be appropriate to replace the sum in (5.42) by an integral (see (5.49)). We overcome this problem by writing

$$N = N_0 + N' \tag{5.70}$$

where

$$N_0 = \frac{1}{e^{-\mu/kT} - 1} \tag{5.71}$$

represents the number of particles in the ground state $E = 0$ and N' represents the number of particles in all the excited states. As a first approximation we may write (cf. (5.49))

$$N' = G\frac{(2m)^{3/2}V}{4\pi^2\hbar^3}\int_0^\infty \frac{\varepsilon^{1/2}d\varepsilon}{e^{(\varepsilon-\mu)/kT} - 1} \tag{5.72}$$

which is justified because (5.48) gives a zero weight to the ground state. Equation (5.71) gives

$$e^{\mu/kT} = \frac{N_0}{N_0 + 1} \approx 1 - \frac{1}{N_0} \tag{5.73}$$

[†] Such a condensation is not possible for a gas obeying Fermi–Dirac statistics where a state cannot be occupied by more than one particle.

Thus when the Bose–Einstein condensation occurs, N_0 is very large compared to unity and μ would be very close to zero (but it will always be negative). Now if we put $\mu = 0$ in (5.72) we would get an upper limit on N':

$$N' < G\frac{(2m)^{3/2}V}{4\pi^2\hbar^3}\int_0^\infty \frac{\varepsilon^{1/2}}{e^{\varepsilon/kT}-1}\,dE \tag{5.74}$$

Using (5.63) we get

$$N' < G\left(\frac{mkT}{2\pi\hbar^2}\right)^{3/2}Vf_{3/2}(1) \tag{5.75}$$

where

$$f_{3/2}(1) = 1 + \frac{1}{2^{3/2}} + \frac{1}{3^{3/2}} + \cdots \approx 2.612 \tag{5.76}$$

Thus for the Bose–Einstein condensation to occur, the temperature should be less than a particular value which we denote by T_c:

$$T < T_c = \frac{1}{G^{2/3}}\frac{2\pi\hbar^2}{mk}\left(\frac{N}{2.612\,V}\right)^{2/3} \tag{5.77}$$

Indeed liquid He[4] undergoes a peculiar phase transition at about 2.19 K which is due to Bose–Einstein condensation. The experimental specific heat variation is shown in Fig. 5.3. The temperature variation near $T = T_c$ is in agreement with the result derived from the above analysis.[†] We may

Fig. 5.3. The experimental variation of specific heat of liquid He[4] under its own vapour pressure (adapted from Pathria 1972; the experimental data is due to Keesom and co-workers).

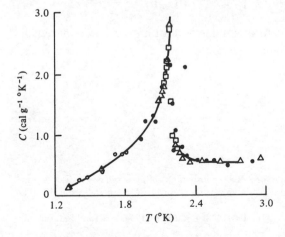

[†] See, e.g., Pathria (1972).

mention that for liquid He^4, $m \approx 6.65 \times 10^{-24}$ g, $(N/V) \approx (6.02 \times 10^{23}/27.6)$ cm^{-3} and $G = 1$ so that

$$T_c \approx 1.96 \text{ K}$$

which is in good agreement with the experimental data. We should mention that the accumulation of a large number of particles in the ground state (which *is* the Bose–Einstein condensation) is a purely quantum mechanical phenomenon and occurs even in the absence of intermolecular forces.[†]

The equations of state around $T \approx T_c$ can be determined from the equations discussed earlier, we give the final results.[‡] (We are assuming $G = 1$):

Case 1: $0 < T < T_c$

$$y = e^{-\mu/kT} \approx 1 \tag{5.78}$$

$$P(T) = \left(\frac{m}{2\pi\hbar^2}\right)^{3/2}(kT)^{5/2}f_{5/2}(1) = 1.341\left(\frac{m}{2\pi\hbar^2}\right)^{3/2}(kT)^{5/2} \tag{5.79}$$

$$E = \tfrac{3}{2}PV = 2.011\left(\frac{m}{2\pi\hbar^2}\right)^{3/2}(kT)^{5/2}V \tag{5.80}$$

$$c_v = \left(\frac{\partial E}{\partial T}\right)_V \approx Nk\left[5.028\left(\frac{mkT}{2\pi\hbar^2}\right)^{3/2}\left(\frac{V}{N}\right)\right]$$

$$= c_v(T_c)\left(\frac{T}{T_c}\right)^{3/2} \tag{5.81}$$

$$S = Nk\left[\tfrac{5}{2}f_{5/2}(1)\frac{1}{\Lambda}\right] \approx 3.352\,Nk\left(\frac{2\pi\hbar^2}{mkT}\right)^{3/2}\left(\frac{N}{V}\right) \tag{5.82}$$

Notice that P is independent of volume and for an adiabatic (for which S and therefore $T^{3/2}V$ remains constant)

$$PV^{5/3} = \text{constant} \tag{5.83}$$

Further

$$c_v(T_c) \approx 1.925\,Nk \tag{5.84}$$

[†] The equations of state for a Bose–Einstein gas of hard spheres is discussed in Ter Haar (1962); more details can be found in Lee, Huang and Yang (1957) and references therein.

[‡] For more details see, e.g., Gopal (1974), Landsberg (1954, 1961) and Pathria (1972).

Case 2: $T \gtrsim T_c$

$y(T)$ and hence $\mu(T)$ is determined from the relationship

$$f_{3/2}(y)\left[= y + \frac{y^2}{2^{3/2}} + \frac{y^3}{3^{3/2}} + \cdots \right] = \Lambda = \left(\frac{2\pi\hbar^2}{mkT} \right)^{3/2} \frac{N}{V}$$

$$= 2.612 \left(\frac{T_c}{T} \right)^{3/2} \tag{5.85}$$

If we plot y as a function of T/T_c, we will get a monotonically decreasing function. The other thermodynamic functions are given by[†]

$$P = \frac{NkT}{V} \frac{f_{5/2}(y)}{f_{3/2}(y)} \tag{5.86}$$

$$E = \tfrac{3}{2} PV = \tfrac{3}{2} NkT \frac{f_{5/2}(y)}{f_{3/2}(y)} \tag{5.87}$$

$$c_v = Nk \left[\frac{15}{4} \frac{f_{5/2}(y)}{f_{3/2}(y)} - \frac{9}{4} \frac{f_{3/2}(y)}{f_{1/2}(y)} \right] \tag{5.88}$$

$$S = \frac{5}{2} \frac{f_{5/2}(y)}{f_{3/2}(y)} - \ln y \tag{5.89}$$

If we plot c_v as a function of temperature using (5.84) and (5.88) we will find that whereas c_v is continuous at $T = T_c$. Its slope is discontinuous at $T = T_c$ which is consistent with the experimental data (see Fig. 5.3).

[†] In deriving (5.88) we have to use the relation $(\partial y/\partial T)_V = (-3y/2T)$ $(f_{3/2}(y)/f_{1/2}(y))$ which is readily obtained from the relation $\partial f_{3/2}(y)/\partial T = (-3/2T)f_{3/2}(y)$ (see (5.85)) and $\partial f_{3/2}(y)/\partial y = f_{1/2}(y)/y$ (see (5.63)).

6

Fermi–Dirac equation of state

6.1 Overview

In the previous chapter we mentioned that according to quantum mechanics there are two types of distributions possible for distinguishable particles: one in which a state can be occupied by any number of particles – this leads to Bose–Einstein statistics which was the subject matter of the previous chapter, and the other in which a state can not be occupied by more than one particle – this leads to Fermi–Dirac statistics which is the subject matter of the present chapter. Once again, an energy state can be degenerate in the sense that there may be more than one wave function associated with it – in that case, each linearly independent wave function can be associated with a state. Elementary particles like electrons, protons, neutrons etc. obey Fermi–Dirac statistics and are referred to as *Fermions*. In this chapter we will first derive the grand partition function for a Fermi–Dirac gas and then calculate the thermodynamic functions.

6.2 The grand partition function and other thermodynamic functions

In Section 4 of the previous chapter we gave the procedure for calculating the grand partition function of a Bose–Einstein gas; the procedure up to (5.39) remains the same. We rewrite this equation

$$Z = \left\{ \sum_{n_1} [e^{(\mu - \varepsilon_1)/kT}]^{n_1} \right\} \left\{ \sum_{n_2} [e^{(\mu - \varepsilon_2)/kT}]^{n_2} \right\} \cdots \tag{6.1}$$

where $n_1, n_2, \ldots$ represent the number of particles in states $1, 2, \ldots$ respectively; the notation is the same as used in the previous chapter: μ represents the chemical potential and $\varepsilon_1, \varepsilon_2, \ldots$ represent the energies of the first, second, $\ldots$ states respectively. For a Bose gas there was no limit on the number of particles that can occupy a state and therefore each summation was from 0 to ∞. However, for a Fermi gas n_1 can be either 0 or 1 and therefore we have

$$Z = (1 + e^{(\mu - \varepsilon_1)/kT})(1 + e^{(\mu - \varepsilon_2)/kT}) \cdots \qquad (6.2)$$

Thus

$$q = \ln Z = \sum_j \ln [1 + e^{(\mu - \varepsilon_j)/kT}] \qquad (6.3)$$

$$N = kT\left(\frac{\partial q}{\partial \mu}\right)_{V,T} = \sum_j \frac{1}{e^{(\varepsilon_j - \mu)/kT} + 1} \qquad (6.4)$$

$$E = kT^2\left(\frac{\partial q}{\partial T}\right)_{\mu,V} = \sum_j \frac{\varepsilon_j}{e^{(\varepsilon_j - \mu)/kT} + 1} \qquad (6.5)$$

$$P = \frac{kT}{V}q = -\frac{kT}{V}\sum_j \ln [1 + e^{(\mu - \varepsilon_j)/kT}] \qquad (6.6)$$

and the mean occupation number of the level ε_j will be

$$\left.\begin{aligned} \langle n_j \rangle &= -kT\left(\frac{\partial q}{\partial \varepsilon_j}\right)_{\mu,T,\text{all other } \varepsilon s} \\ &= \frac{1}{e^{(\varepsilon_j - \mu)/kT} + 1} \end{aligned}\right\} \qquad (6.7)$$

which is usually referred to as the Fermi–Dirac distribution. Equations (6.4), (6.5) and (6.7) imply

$$N = \sum_j \langle n_j \rangle \quad \text{and} \quad E = \sum_j \langle n_j \rangle \varepsilon_j \qquad (6.8)$$

which should indeed be the case. As in the previous chapter, if we assume an almost continuous distribution of states then the summations in (6.3)–(6.6) can be replaced by integrals:

$$\sum_j \cdots \approx \int_0^\infty \cdots g(\varepsilon)d\varepsilon \qquad (6.9)$$

where

$$g(\varepsilon) = G\frac{(2m)^{3/2}V}{4\pi^2\hbar^3}\varepsilon^{1/2} \qquad (6.10)$$

represents the density of states (see (A1.18)). For an electron gas the degeneracy parameter G is equal to 2 to take into account the spin degeneracy of the electrons. Thus (6.4) becomes

$$N \approx G\frac{(2m)^{3/2}V}{4\pi^2\hbar^3}\int_0^\infty \frac{\varepsilon^{1/2}d\varepsilon}{e^{(\varepsilon - \mu)/kT} + 1} \qquad (6.11)$$

If we assume that

$$e^{-\mu/kT} \gg 1 \tag{6.12}$$

then the unity in the denominator of the integrand in (6.11) can be neglected and we will have

$$N \approx G\frac{(2m)^{3/2}V}{4\pi^2\hbar^3}e^{\mu/kT}\int_0^\infty e^{-\varepsilon/kT}\varepsilon^{1/2}dE$$

or

$$e^{-\mu/kT} \approx \frac{G}{(N/V)}\left(\frac{mkT}{2\pi\hbar^2}\right)^{3/2} \tag{6.13}$$

which determines μ as a function of temperature when (6.12) is satisfied. Equation (6.13) is identical to the corresponding equation for a Bose–Einstein gas (5.51) and therefore in the classical limit (i.e., when (6.12) is valid) we have (see (5.53) and (5.54))

$$PV \approx NkT \tag{6.14}$$

and

$$E \approx \tfrac{3}{2}NkT \tag{6.15}$$

which correspond to classical statistics. Using (6.12) and (6.13) we get the following condition for the validity of classical statistics

$$T \gg \frac{2\pi\hbar^2}{mk}\left(\frac{N/V}{G}\right)^{2/3} \tag{6.16}$$

which is the same as for the Bose gas. For electrons in a metal, $G = 2$ and we may assume $(N/V) \approx 10^{22}\ \mathrm{cm}^{-3}$ and therefore (6.16) becomes

$$T \gg 2 \times 10^4\ \mathrm{K} \tag{6.17}$$

Thus, at ordinary temperatures the electron gas would behave very differently from a classical gas and quantum effects will dominate.

Returning to (6.5) and (6.6) and replacing the sums by integrals (see (6.9) and (6.10)) we get

$$E = G\frac{(2m)^{3/2}V}{4\pi^2\hbar^3}\int_0^\infty \frac{\varepsilon^{3/2}d\varepsilon}{e^{(\varepsilon-\mu)/kT}+1} \tag{6.18}$$

and

$$P = -\frac{kT}{V}\left[G\frac{(2m)^{3/2}V}{4\pi^2\hbar^3}\int_0^\infty \ln\left[1+e^{(\mu-\varepsilon)/kT}\right]\varepsilon^{1/2}d\varepsilon\right] \tag{6.19}$$

or

$$PV = -kT\left[G\frac{(2m)^{3/2}V}{4\pi^2\hbar^3}\left(\ln\{1+e^{(\mu-\varepsilon)/kT}\}\tfrac{2}{3}\varepsilon^{3/2}\Big|_0^\infty \right.\right.$$
$$\left.\left. -\frac{2}{3}\int_0^\infty \frac{\varepsilon^{3/2}d\varepsilon}{e^{(\varepsilon-\mu)/kT}+1} \right)\right]$$

Thus

$$PV = \tfrac{2}{3}E \qquad\qquad (6.20)$$

Since we have not assumed (6.12), the above equation is valid for *all* temperatures – provided, of course, (6.10) is valid.

6.2.1 *The Fermi–Dirac distribution function*

In order to have a greater appreciation of the quantum effects, we study the energy variation of the Fermi–Dirac distribution function (usually referred to as the Fermi function) which we denote by $F(\varepsilon)$:

$$F(\varepsilon) = \frac{1}{e^{(\varepsilon-\varepsilon_F)/kT}+1} \qquad\qquad (6.21)$$

where, to be consistent with the literature, we have used the symbol ε_F for μ (cf. (6.7)). The quantity ε_F is referred to as the Fermi energy. The variation of the Fermi function with energy when $\varepsilon_F/kT \gg 1$ (the quantum limit) and when $\varepsilon_F/kT \ll -1$ (the classical limit) are shown in Fig. 6.1. At absolute zero $(T = 0)$, we have:

$$F(\varepsilon) = \begin{cases} 1 & \varepsilon < \varepsilon_F \\ 0 & \varepsilon > \varepsilon_F \end{cases} \qquad\qquad (6.22)$$

Since $F(\varepsilon)$ represents the mean occupation number of the energy level ε,

Fig. 6.1. The variation of the Fermi function with energy.
$\Lambda = \tfrac{1}{2}(2\pi\hbar^2/mkT)^{3/2}n$ (see (5.69)).

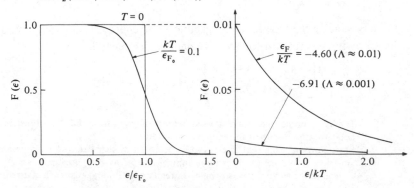

(6.22) tells us that (at $T = 0$) all states below the Fermi level are filled and all states above the Fermi level are empty;[†] obviously, a state cannot be occupied by more than one electron.[‡] We may mention that the gas is said to be *completely degenerate* when all the lowest quantum states are occupied and thus at $T = 0$, the electron gas is completely degenerate. Now, at $T = 0$, (6.11) becomes

$$N \approx G \frac{(2m)^{3/2} V}{4\pi^2 \hbar^3} \int_0^{\varepsilon_{F_0}} \varepsilon^{1/2} d\varepsilon \qquad (6.23)$$

where

$$\varepsilon_{F_0} \equiv \varepsilon_F (T = 0) \qquad (6.24)$$

represents the Fermi level at $T = 0$. Equation (6.23) gives

$$\varepsilon_{F_0} = \left(\frac{6\pi^2}{G} \right)^{2/3} \frac{\hbar^2}{2m} n^{2/3} \qquad (6.25)$$

where $n = (N/V)$ represents the number of free electrons per unit volume. Substituting $G = 2$ and the values for the Planck's constant and the electron mass, we get

$$\varepsilon_{F_0} \approx 5.842 \times 10^{-27} n^{2/3} \text{ erg}$$
$$\approx 3.652 \times 10^{-15} n^{2/3} \text{ eV} \qquad (6.26)$$

where n is measured in cm^{-3}. Using the data given in Table 6.1 and assuming the valence electrons to be free we can, using (6.25), calculate ε_{F_0} and $T_{F_0} (= \varepsilon_{F_0}/k)$ which are also given in Table 6.1.

The total energy at absolute zero will be (see (6.18))

$$E_0 = G \frac{(2m)^{3/2} V}{4\pi^2 \hbar^3} \int_0^{\varepsilon_{F_0}} \varepsilon^{3/2} d\varepsilon$$
$$= G \frac{(2m)^{3/2} V}{4\pi^2 \hbar^3} \tfrac{2}{5} \varepsilon_{F_0}^{5/2} \qquad (6.27)$$

which is usually referred to as the ground state energy. Using (6.25) we get the following expression for average energy per particle

$$\frac{E_0}{N} = \frac{3}{5} \left[\left(\frac{6\pi^2}{G} \right)^{2/3} \frac{\hbar^2}{2m} n^{2/3} \right] = \tfrac{3}{5} \varepsilon_{F_0} \qquad (6.28)$$

[†] It may be noted that this behaviour is quite different from the Bose–Einstein condensation discussed in the previous chapter.

[‡] Some authors write that a state can be occupied by two electrons (spin up and spin down). We feel that it is more appropriate to say that, because of spin degeneracy, there are twice as many states (see (6.10)) and that each state can be occupied by not more than one electron.

The pressure at absolute zero will be given by

$$P_0 = \frac{2E_0}{3V} = \tfrac{2}{5}n\varepsilon_{F_0} = (3\pi^2)^{2/3}\frac{\hbar^2}{5m}n^{5/3} \tag{6.29}$$

For an electron gas in a metal, ε_{F_0} is a few eV (see Table 6.1) and therefore even at absolute zero there is considerable energy and pressure associated with the electron gas. This is entirely a quantum effect. We should also mention that the low temperature behaviour of a Fermi gas is very different from that of a Bose gas because Fermi particles cannot accumulate in the lowest energy state.

For $T > 0$, we consider two limiting cases: (a) $kT \ll \varepsilon_F$ (when quantum effects will dominate) and (b) $kT \gg \varepsilon_F$ (when the system will start behaving as a classical gas).

Case 1: $kT \ll \varepsilon_F$ (strongly degenerate gas)
We rewrite (6.11) in the form

$$N = G\frac{(2m)^{3/2}V}{4\pi^2\hbar^3}\int_0^\infty \varepsilon^{1/2}F(\varepsilon)\,\mathrm{d}\varepsilon \tag{6.30}$$

where $F(\varepsilon)$ is the Fermi function (see (6.21)). We integrate by parts to obtain

$$N = G\frac{(2m)^{3/2}V}{4\pi^2\hbar^3}\left[\tfrac{2}{3}\varepsilon^{3/2}F(\varepsilon)\Big|_0^\infty - \frac{2}{3}\int_0^\infty \varepsilon^{3/2}\frac{\mathrm{d}F}{\mathrm{d}\varepsilon}\,\mathrm{d}\varepsilon\right] \tag{6.31}$$

The first term inside the square brackets vanishes at both limits. In order to evaluate the integral we note that for $kT \ll \varepsilon$, the function $F(\varepsilon)$ is very 'flat' if we are slightly away from $\varepsilon = \varepsilon_F$ (see Fig. 6.1); thus $\mathrm{d}F/\mathrm{d}\varepsilon$ will be a very

Table 6.1 *Fermi energy and Fermi temperatures (at absolute zero) for some metals*

	Density at 20 °C (g cm^{-3})	Atomic volume (cm^3 mole^{-1})	Valence	ε_{F_0} (eV)	T_{F_0} (K)
Lithium	0.53	13	1	4.72	5.5×10^4
Sodium	0.97	24	1	3.12	3.7×10^4
Potassium	0.86	45	1	2.14	2.4×10^4
Copper	8.96	7.09	1	7.04	8.2×10^4
Silver	10.49	10.28	1	5.51	6.4×10^4
Aluminium	2.70	9.99	3	11.7	1.4×10^5

Note: Adapted from Kittel (1956).

sharply peaked function around $\varepsilon = \varepsilon_F$ (see Fig. 6.2).[†] We therefore introduce the variable

$$x = \frac{\varepsilon - \varepsilon_F}{kT} \qquad (6.32)$$

to obtain

$$N = -G\frac{(2m)^{3/2}V\varepsilon_F{}^{3/2}}{6\pi^2\hbar^3}\int_{-\varepsilon_F/kT}^{\infty}\left(1 + \frac{xkT}{\varepsilon_F}\right)^{3/2}\frac{\mathrm{d}F}{\mathrm{d}x}\mathrm{d}x \qquad (6.33)$$

where

$$F(x) = \frac{1}{e^x + 1} \qquad (6.34)$$

Since we have assumed $\varepsilon_F/kT \gg 1$, $\mathrm{d}F/\mathrm{d}x$ would be very sharply peaked around $x = 0$ and very little error will be involved in replacing the lower limit by $-\infty$; we also expand $[1 + (xkT/\varepsilon_F)]^{3/2}$ in a binomial series and

Fig. 6.2. At low temperatures, the function $\mathrm{d}F/\mathrm{d}\varepsilon$ is very sharply peaked at $\varepsilon = \varepsilon_F$.

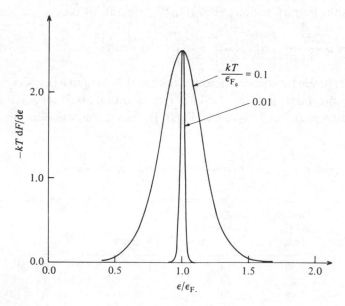

[†] Indeed at $T = 0$, $\mathrm{d}F/\mathrm{d}\varepsilon$ is a delta function:

$$\frac{\mathrm{d}F}{\mathrm{d}\varepsilon} = \delta(\varepsilon - \varepsilon_F) \quad (T = 0)$$

integrate term by term to obtain

$$N \approx -G\frac{(2m)^{3/2}V\varepsilon_F^{3/2}}{6\pi^2\hbar^3}\left[F(x)\bigg|_{-\infty}^{+\infty} - \frac{3}{2}\frac{kT}{\varepsilon_F}\frac{xe^x}{(e^x+1)^2}\,dx\right.$$
$$\left. - \frac{3}{8}\left(\frac{kT}{\varepsilon_F}\right)^2\int_{-\infty}^{+\infty}\frac{x^2e^x}{(e^x+1)^2}\,dx\right] \tag{6.35}$$

Now, $F(\infty)=0$ and $F(-\infty)=1$, thus the first term inside the square brackets is -1. The integrand of the second term is an odd function of x and hence the integral vanishes. Further

$$\int_{-\infty}^{\infty}\frac{x^2e^x}{(e^x+1)^2}\,dx = 2\int_0^{\infty}x^2e^{-x}(1+e^{-x})^{-2}\,dx$$
$$= 2\left[\int_0^{\infty}x^2(e^{-x} - 2e^{-2x} + \cdots)\,dx\right]$$
$$= 4\left[1 - \frac{1}{2^2} + \frac{1}{3^2} - \cdots\right] = \frac{\pi^2}{3} \tag{6.36}$$

Thus

$$N \approx G\frac{(2m)^{3/2}V\varepsilon_F^{3/2}}{6\pi^2\hbar^3}\left[1 + \frac{\pi^2}{8}\left(\frac{kT}{\varepsilon_F}\right)^2 + \cdots\right] \tag{6.37}$$

Inverting the series we get

$$\varepsilon_F(T) \approx \varepsilon_{F_0}\left[1 - \frac{\pi^2}{12}\left(\frac{kT}{\varepsilon_{F_0}}\right)^2 + \cdots\right] \tag{6.38}$$

where ε_{F_0} is given by (6.25). Similarly, the total energy will be given by (6.18):

$$E = \int_0^{\infty}\varepsilon g(\varepsilon)F(\varepsilon)\,d\varepsilon$$
$$\approx -G\frac{(2m)^{3/2}V\varepsilon_F^{5/2}}{6\pi^2\hbar^3}\int_{-\infty}^{\infty}\left(1 + \frac{xkT}{\varepsilon_F}\right)^{5/2}\frac{dF}{dx}(x)\,dx$$

Once again, we make a binomial expansion, integrate term by term and use (6.38) to obtain

$$E \approx \tfrac{3}{2}N\varepsilon_{F_0}\left[1 + \frac{5\pi^2}{12}\left(\frac{kT}{\varepsilon_{F_0}}\right)^2 + \cdots\right] \tag{6.39}$$

Thus

$$P = \frac{2}{3}\frac{E}{V} = \frac{2}{5}\left(\frac{N}{V}\right)\varepsilon_{F_0}\left[1 + \frac{5\pi^2}{12}\left(\frac{kT}{\varepsilon_{F_0}}\right)^2 + \cdots\right] \tag{6.40}$$

The specific heat for the electron gas will therefore be

$$c_v = \left(\frac{\partial E}{\partial T}\right) \approx Nk\left[\frac{\pi^2}{2}\frac{kT}{\varepsilon_{F_0}} + \cdots\right] \tag{6.41}$$

which can be written in the form

$$c_v \approx \tfrac{3}{2}Nk\left(\frac{\pi^2 kT}{3\varepsilon_{F_0}}\right) \tag{6.42}$$

where $\tfrac{3}{2}Nk$ represents the classical expression for the specific heat of the electron gas. For sodium metal $\varepsilon_{F_0} \approx 3\,\text{eV}$ (Table 6.1), and for $T \approx 300\,\text{K}$, $kT \approx \tfrac{1}{40}\,\text{eV}$ and the quantity inside the square brackets is about $\tfrac{1}{40}$. Physically, this is due to the fact that at low temperatures the energy levels much below the Fermi level are completely filled and therefore thermal excitation to unoccupied states would require considerable amounts of energy. Thus only those electrons which are in the vicinity of the Fermi level contribute to the specific heat and therefore at low temperatures the specific heat of an ideal Fermi gas is much smaller than that predicted by classical theory.

Since the lattice specific heat at low temperatures varies as T^3 (see, e.g., Ghatak and Kothari 1972) we expect that the specific heat variation at low temperatures should be given accurately by

$$c_v = AT + BT^3 \tag{6.43}$$

where the first term (on the right hand side) is due to electrons and the

Fig. 6.3. The experimental data on the specific heat for copper in the temperature range $1 - 4\,\text{K}$. The plot is of c_v/T as a function of T^2. Since the experimental data falls on a straight line, (6.43) is very nearly obeyed. The intercept on the vertical axis gives the value of A. (Figure adapted from Pathria 1972; The experimental data is due to Corak *et al.* (1955))

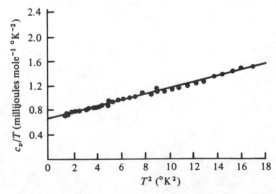

second term is due to lattice vibrations. Equation (6.43) is qualitatively in agreement with experimental data (see Fig. 6.3). Now according to (6.41)

$$A = \tfrac{1}{2}\pi^2 \frac{Nk}{T_F} = \tfrac{1}{2}\pi^2 z \frac{R}{T_F} \tag{6.44}$$

where z represents the number of conduction electrons/atom. For example, for sodium, $z = 1$, $T_F \approx 5.5 \times 10^4$ K and using $R \approx 2$ cal mole^{-1} K^{-1} we get

$$A \approx 2.7 \times 10^{-4} \text{ cal mole}^{-1} \text{ K}^{-2}$$

and the corresponding experimental value is 4.3×10^{-4} cal mole^{-1} K^{-2}. Even for other metals, the experimental value of A is considerably different from the value predicted by (6.44) – see, e.g., Kittel (1956), Table 10.4. This anomaly is due to our oversimplified model of the electron gas inside the metal – a better model would be to assume an 'effective mass' for the electron. For further details see, e.g., Kittel (1956).

Other thermodynamic functions like the Helmholtz free energy, entropy etc. can easily be calculated by using (5.34)–(5.36); we must remember that we are denoting μ by ε_F. We summarize the results when $kT \ll \varepsilon_F$:

$$P = \frac{2}{5}\left(\frac{N}{V}\right)\varepsilon_{F_0}\left[1 + \frac{5\pi^2}{12}\left(\frac{kT}{\varepsilon_{F_0}}\right)^2 + \cdots\right] \tag{6.45}$$

$$E = \tfrac{3}{5}\varepsilon_{F_0}\left[1 + \frac{5\pi^2}{12}\left(\frac{kT}{\varepsilon_{F_0}}\right)^2 + \cdots\right] \tag{6.46}$$

$$\mu = \varepsilon_F = \varepsilon_{F_0}\left[1 - \frac{\pi^2}{12}\left(\frac{kT}{\varepsilon_{F_0}}\right)^2 + \cdots\right] \tag{6.47}$$

$$c_v = \tfrac{3}{2}NK\left[\frac{\pi^2}{3}\left(\frac{kT}{\varepsilon_{F_0}}\right) + \cdots\right] \tag{6.48}$$

$$F = N\mu - PV = \tfrac{3}{2}N\varepsilon_{F_0}\left[1 - \frac{5\pi^2}{12}\left(\frac{kT}{\varepsilon_{F_0}}\right)^2 + \cdots\right] \tag{6.49}$$

$$S = \frac{E - F}{T} = \frac{\pi^2}{2}\left(\frac{kT}{\varepsilon_{F_0}}\right) + \cdots \tag{6.50}$$

with ε_{F_0} given by (6.25).

Case 2: $kT \gg \varepsilon_F$ (non-degenerate gas).
When $kT \gg \varepsilon_F$, the considerations will be very similar to the corresponding case discussed in the previous chapter for a Bose–Einstein gas except that we would have to use the following expansion (cf. (5.63)):

$$g_n(y) \equiv \frac{1}{\Gamma(n)} \int_0^\infty \frac{x^{n-1}dx}{(1/y)e^x + 1} = \frac{1}{\Gamma(n)} \int_0^\infty ye^{-x}(1 + ye^{-x})^{-1}x^{n-1}dx$$

$$= \frac{1}{\Gamma(n)} \int_0^\infty ye^{-x}[1 - ye^{-x} + y^2e^{-2x} + \cdots]x^{n-1}dx$$

$$= y - \frac{y^2}{2^n} + \frac{y^3}{3^n} - \cdots \tag{6.51}$$

The final results for the various thermodynamic functions have already been given in the previous chapter (see (5.64)–(5.68); the lower signs correspond to Fermi–Dirac statistics and for an electron gas $G = 2$).

6.3 Relativistic considerations[†]

The results of the previous section are not valid relativistically because in calculating the density of states (Appendix 1) we had assumed the following (non-relativistic) relation between energy and momentum:

$$\varepsilon = \frac{p^2}{2m} \tag{6.52}$$

However, if we solve Dirac's relativistic equation (instead of the non-relativistic Schrödinger equation) then the density of states in the momentum space happens to be the same, viz:

$$g(p)dp = \frac{V}{\pi^2\hbar^3}p^2dp \tag{6.53}$$

which is the same as (A1.27) (except for the additional factor of 2 which is already incorporated in (6.53)). The derivation of (6.53) from Dirac's relativistic equation can be found in Chandrasekhar (1939), Brush (1967) etc. Since the relativistic relation between energy and momentum is given by

$$\varepsilon^2 = p^2c^2 + m^2c^4 \tag{6.54}$$

(where m represents the rest mass of electron) we have

$$\varepsilon d\varepsilon = pc^2 dp \tag{6.55}$$

giving

$$g(\varepsilon)d\varepsilon = g(p)dp = \frac{V}{\pi^2\hbar^3}p^2dp$$

[†] One of the most important applications of the theoretical results presented in this section is in the study of thermodynamic equilibrium of white dwarf stars which will be discussed in Chapter 16.

or

$$g(\varepsilon) = \begin{cases} \dfrac{V}{\pi^2\hbar^2 c^3}\varepsilon(\varepsilon^2 - m^2 c^4)^{1/2} & \varepsilon \geqslant mc^2 \\[2mm] 0 & \varepsilon < mc^2 \end{cases} \tag{6.56}$$

Using the above expression for the density of states and (6.9), we get the following expressions for various thermodynamic functions ((6.3)–(6.6)):

$$q = \frac{V}{\pi^2\hbar^3 c^3}\int_{mc^2}^{\infty} \varepsilon(\varepsilon^2 - m^2 c^4)^{1/2}\ln\left[1 + e^{(\mu-\varepsilon)/kT}\right]d\varepsilon \tag{6.57}$$

$$N = \frac{V}{\pi^2\hbar^3 c^3}\int_{mc^2}^{\infty}\frac{\varepsilon(\varepsilon^2 - m^2 c^4)^{1/2}}{e^{(\varepsilon-\mu)/kT}+1}d\varepsilon = \frac{V}{\pi^2\hbar^3}\int_0^{\infty}\frac{p^2\,dp}{e^{(\varepsilon-\mu)/kT}+1} \tag{6.58}$$

$$E = \frac{V}{\pi^2\hbar^3 c^3}\int_{mc^2}^{\infty}\frac{\varepsilon^2(E^2 - m^2 c^4)^{1/2}}{e^{(\varepsilon-\mu)/kT}+1}d\varepsilon = \frac{V}{\pi^2\hbar^3}\int_0^{\infty}\frac{\varepsilon p^2\,dp}{e^{(\varepsilon-\mu)/kT}+1} \tag{6.59}$$

$$P = \frac{kT}{V}q = \frac{kT}{\pi^2\hbar^3}\int_0^{\infty}\ln\left[1 + e^{(\mu-\varepsilon)/kT}\right]p^2\,dp$$

$$= \frac{1}{\pi^2\hbar^3}\int_0^{\infty}\frac{1}{e^{(\varepsilon-\mu)/kT}+1}\frac{d\varepsilon}{dp}\frac{p^3}{3}dp$$

where we have integrated by parts, the first term vanishing at both limits. Using (6.54) and (6.55) we get

$$P = \frac{kT}{V}q = \frac{1}{3\pi^2\hbar^3 m}\int_0^{\infty}\frac{1}{e^{(\varepsilon-\mu)/kT}+1}\frac{p^4}{[1+(p/mc)^2]^{1/2}}dp \tag{6.60}$$

We first evaluate the above integrals at absolute zero $(T = 0)$

At absolute zero the Fermi function is a step function, (Fig. 6.1) so that (6.58) becomes

$$N = \frac{V}{\pi^2\hbar^3}\int_0^{p_{F_0}}p^2\,dp = \frac{V}{3\pi^2\hbar^3}p_{F_0}^3 \tag{6.61}$$

where

$$p_{F_0} = \left[\frac{\mu^2(T=0)-m^2 c^4}{c^2}\right]^{1/2} = \left[\frac{\varepsilon_{F_0}^2 - m^2 c^4}{c^2}\right]^{1/2} \tag{6.62}$$

is known as the Fermi momentum (at absolute zero). Equation (6.61) enables us to determine p_{F_0}:

$$p_{F_0} = \left(3\pi^2\frac{N}{V}\right)^{1/3}\hbar \tag{6.63}$$

Once we know p_{F_0}, we can determine E_0 and P_0 where the subscript zero implies that the values correspond to $T = 0$. Thus, using (6.59) and (6.60) we get

$$E_0 = \frac{V}{\pi^2 h^3} \int_0^{p_{F_0}} mc^2 \left(1 + \frac{p^2}{m^2 c^2} \right)^{1/2} p^2 \, dp \qquad (6.64)$$

and

$$P_0 = \frac{1}{3\pi^2 h^3 m} \int_0^{p_{F_0}} \frac{p^4 \, dp}{[1 + (p/mc)^2]^{1/2}} \qquad (6.65)$$

In order to evaluate the integrals we introduce the variable θ such that

$$\sinh \theta = \frac{p}{mc} \qquad (6.66)$$

and therefore

$$\varepsilon = mc^2 \left[1 + \left(\frac{p}{mc} \right)^2 \right]^{1/2} = mc^2 \cosh \theta \qquad (6.67)$$

Thus[†]

$$P_0 = \frac{m^4 c^5}{3\pi^2 h^3} \int_0^\alpha \sinh^4 \theta \, d\theta$$

$$= \frac{m^4 c^5}{24\pi^2 h^2} (2 \sinh^3 \alpha \cosh \alpha - 3 \sinh \alpha \cosh \alpha + 3\alpha) \qquad (6.68)$$

where $\sinh \alpha \equiv (p_{F_0}/mc)$. We may rewrite (6.68) as

$$P_0 = \frac{m^4 c^5}{24\pi^2 h^3} f(x) \qquad (6.69)$$

where

$$f(x) \equiv (2x^3 - 3x)(1 + x^2)^{1/2} + 3 \sinh^{-1} x \qquad (6.70)$$

$$x = \sinh \alpha = p_{F_0}/mc \qquad (6.71)$$

Similarly

$$E_0 = \frac{V m^4 c^5}{\pi^2 h^3} \int_0^\alpha \sinh^2 \theta \cosh^2 \theta \, d\theta = \frac{V m^4 c^5}{24\pi^2 h^3} [8x^3 + g(x)] \qquad (6.72)$$

where

$$g(x) \equiv 8x^3 (1 + x^2)^{1/2} - 8x^3 - f(x) \qquad (6.73)$$

[†] In evaluating the integrals we have used the following relations:

$$\left. \begin{array}{l} \displaystyle \int \sinh^m \theta \cosh^n \theta \, d\theta = \frac{\sinh^{m+1} \theta \cosh h^{n-1} \theta}{m + n} + \frac{n - 1}{m + n} \int \sinh^m \theta \cosh^{n-2} \theta \, d\theta \\[4mm] \displaystyle \qquad = \frac{\sinh^{m-1} \theta \cosh^{n+1} \theta}{m + n} - \frac{m - 1}{m + n} \int \sinh^{m-2} \theta \cosh^n \theta \, d\theta \end{array} \right\} m + n \neq 0$$

The term $8x^3$ in (6.72) gives rise to the rest mass energy Nmc^2. Thus we have

$$E_0 = Nmc^2 + \frac{Vm^4c^5}{24\pi^2\hbar^3}g(x) \tag{6.74}$$

Further

$$\frac{E_0 - Nmc^2}{PV} = \frac{g(x)}{f(x)} \tag{6.75}$$

The functions $f(x)$ and $g(x)$ are monotonically increasing functions of x and have been tabulated by Chandrasekhar (1939), Chapter X. The ratio $g(x)/f(x)$ increases from 1.5 (at $x = 0$) to 3.0 (at $x = \infty$) which may be compared with the value of 1.5 obtained in the non-relativistic expression (6.20).

Now, in the non-relativistic limit, $x \ll 1$ and we may use the following series

$$\left.\begin{aligned} f(x) &= \tfrac{8}{5}x^5 - \tfrac{4}{7}x^7 + \tfrac{1}{3}x^9 - \cdots \\ g(x) &= \tfrac{12}{5}x^5 - \tfrac{3}{7}x^7 + \tfrac{1}{6}x^9 - \cdots \end{aligned}\right\} x \ll 1 \tag{6.76} \tag{6.77}$$

to obtain

$$P_0 = (3\pi^2)^{2/3}\frac{\hbar^2}{5m}\left(\frac{N}{V}\right)^{5/3}\left[1 - \frac{5}{14}\left(3\pi^2\frac{N}{V}\right)^{2/3}\frac{\hbar^2}{m^2c^2} + \cdots\right] \tag{6.78}$$

$$E_0 = Nmc^2 + \frac{3N}{5}(3\pi^2)^{2/3}\frac{\hbar^2}{2m}\left(\frac{N}{V}\right)^{2/3}$$

$$\times\left[1 - \frac{15}{84}\left(3\pi^2\frac{N}{V}\right)^{2/3}\frac{\hbar^2}{m^2c^2} + \cdots\right] \tag{6.79}$$

The quantities outside the square brackets in (6.78) and (6.79) are the expressions in the non-relativistic limit and are the same as derived earlier ((6.28) and (6.29)). The quantities inside the square brackets represent relativistic corrections.

On the other hand, in the extreme relativistic limit we have $x \gg 1$ for which the following expansions are valid[†]

$$\left.\begin{aligned} f(x) &= 2x^4 - 2x^2 + \cdots \\ g(x) &= 6x^4 - 8x^3 + \cdots \end{aligned}\right\} x \gg 1 \tag{6.80} \tag{6.81}$$

[†] All asymptotic expressions are very easy to derive; we have to use (See, e.g., Abramowitz and Stegun (1965))

$$\sinh^{-1}x = x - \tfrac{1}{6}x^3 + \tfrac{3}{40}x^5 - \cdots \qquad |x| < 1$$

$$= \ln 2x + \tfrac{1}{4}x^2 - \frac{3}{32x^4} + \cdots |x| > 1$$

Using (6.74) and (6.75) we obtain

$$P_0 = \frac{ch}{12\pi^2}\left(3\pi^2\frac{N}{V}\right)^{4/3}\left[1 - \frac{3m^2c^2}{2}\frac{1}{h^2}\frac{1}{[3\pi^2(N/V)]^{2/3}} + \cdots\right] \quad (6.82)$$

$$E_0 = Nmc^2 + \frac{3Nhc}{4}\left(3\pi^2\frac{N}{V}\right)^{1/3}\left[1 - \frac{4mc}{3}\frac{1}{h}\frac{1}{[3\pi^2(N/V)]^{1/3}} + \cdots\right] \quad (6.83)$$

The above analysis is valid at absolute zero. For $T > 0$, the evaluation of the integrals in (6.58), (6.59) and (6.60) become rather difficult. We give below some asymptotic expressions together with their domain of validity.[†]

Case 1: $\mu/kT \gg 1$ (almost completely degenerate)

$$N = \frac{Vm^3c^3}{3\pi^2\hbar^3}x^3\left[1 + \pi^2\frac{2x^2+1}{x^4}\left(\frac{kT}{mc^2}\right)^2 + \frac{7\pi^4}{40}\frac{1}{x^8}\left(\frac{kT}{mc^2}\right)^4 + \cdots\right] \quad (6.84)$$

$$E = \frac{Vm^4c^5}{24\pi^2\hbar^3}\left[8x^3 + g(x)\right.$$
$$\left. \times\left(1 + 4\pi^2\frac{(3x^2+1)(x^2+1)^{1/2}-(2x^2+1)}{xg(x)}\left(\frac{kT}{mc^2}\right)^2 + \cdots\right)\right] \quad (6.85)$$

$$P = \frac{m^4c^5}{24\pi^2\hbar^3}f(x)\left[1 + 4\pi^2\frac{x(x^2+1)^{1/2}}{f(x)}\left(\frac{kT}{mc^2}\right)^2\right.$$
$$\left. + \frac{7\pi^4(x^2+1)^{1/2}(2x^2-1)}{15}\frac{1}{x^3f(x)}\left(\frac{kT}{mc^2}\right)^4 + \cdots\right] \quad (6.86)$$

where x and μ are related through the following equation[‡]

$$\mu = mc^2(1+x^2)^{1/2} \quad (6.87)$$

The functions $f(x)$ and $g(x)$ are the same as given by (6.70) and (6.73)

[†] Most of the formulae are quoted from Chandrasekhar (1939) where their derivations have also been given. For more details one may refer to Brush (1967), Chandrasekhar (1939) and references therein.

[‡] Actually in carrying out the integrations an additional parameter θ_0 is also introduced:

$$x = \sinh\theta_0$$

Thus

$$\frac{\mu}{kT} = \frac{mc^2}{kT}\cosh\theta_0$$

respectively. Equation (6.82) determines x, and then (knowing the value of x) one can determine E and P. Chandrasekhar (1939) has shown that the necessary and sufficient condition for the onset of degeneracy is

$$4\pi^2 \frac{x(x^2+1)^{1/2}}{f(x)} \left(\frac{kT}{mc^2}\right)^2 \ll 1 \tag{6.88}$$

Using the asymptotic forms of $f(x)$ as given by (6.74) and (6.78), one can readily see that the gas would become degenerate when (a) either the temperature is very low or (b) the densities are very high. Using (6.83) one can calculate the specific heat which will be given by

$$c_v = N \frac{\pi^2 k^2}{mc^2} \frac{(x^2+1)^{1/2}}{x^2} T \tag{6.89}$$

Case 2: $\mu/kT \ll -1$ (non-degenerate case)
When $\mu/kT \ll -1$, the unity in the denominator of the integral ((6.57)–(6.59)) can be neglected so that we may write

$$N \approx \frac{V}{\pi^2 \hbar^3} e^{+\mu/kT} \int_0^\infty e^{-E/kT} p^2 \, dp \tag{6.90}$$

and similar expressions for E and P. We again introduce the variable

$$\sinh \theta \equiv \frac{p}{mc} \tag{6.91}$$

so that

$$E = mc^2 [1 + (p/mc)^2] = mc^2 \cosh \theta \tag{6.92}$$

Thus

$$N \approx V \frac{m^3 c^3}{\pi^2 \hbar^3} e^{\mu/kT} \int_0^\infty e^{-\Gamma \cosh \theta} \sinh^2 \theta \cosh \theta \, d\theta \tag{6.93}$$

Similarly

$$E \approx V \frac{m^4 c^5}{\pi^2 \hbar^3} e^{\mu/kT} \int_0^\infty e^{-\Gamma \cosh \theta} \sinh^2 \theta \cosh^2 \theta \, d\theta \tag{6.94}$$

and

$$P \approx \frac{m^4 c^5}{3\pi^2 \hbar^3} e^{\mu/kT} \int_0^\infty e^{-\Gamma \cosh \theta} \sinh^4 \theta \, d\theta \tag{6.95}$$

where

$$\Gamma \equiv mc^2/kT \tag{6.96}$$

If we integrate (6.95) by parts, we would obtain

$$P \approx \frac{m^4 c^5}{3\pi^2 \hbar^3} e^{\mu/kT} \frac{1}{\Gamma} \int_0^\infty e^{-\Gamma \cosh \theta} 3 \sinh^2 \theta \cosh \theta \, d\theta$$

$$\approx \frac{m^3 c^3}{\pi^2 \hbar^3} e^{\mu/kT} kT \int_0^\infty e^{-\Gamma \cosh \theta} \sinh^2 \theta \cosh \theta \, d\theta \tag{6.97}$$

Comparing (6.93) and (6.97) we get

$$PV = NkT \tag{6.98}$$

which reproduces the Boyle–Mariott law in the non-degenerate limit.

The integrals in (6.93) and (6.94) can be evaluated in terms of Bessel functions. We give below the final results (the details have been given by Chandrasekhar 1939):

$$N = \frac{V m^3 c^3}{\pi^2 \hbar^3} e^{\mu/kT} \frac{1}{\Gamma} K_2(\Gamma) \tag{6.99}$$

$$E = N \left[\frac{3 K_3(\Gamma) + K_1(\Gamma)}{4 K_2(\Gamma)} \right] \tag{6.100}$$

and

$$\frac{E}{PV} = \Gamma \left[\frac{3 K_3(\Gamma) + K_1(\Gamma)}{4 K_2(\Gamma)} \right] \tag{6.101}$$

where $K_\nu(z)$ is the modified Bessel function having the following asymptotic forms (see, e.g., Abramowitz and Stegun (1964) Section 9.6):

$$K_\nu(z) \approx \begin{cases} \left(\frac{\pi}{2z}\right)^{1/2} e^{-z} \left[1 + \frac{4\nu^2 - 1^2}{1! \, 8z} + \frac{(4\nu^2 - 1^2)(4\nu^2 - 3^2)}{2! \, (8z)^2} + \cdots \right] & z \to \infty \quad (6.102) \\ \dfrac{1}{2} \dfrac{(\nu - 1)!}{(\frac{1}{2}z)^\nu} & z \to 0 \quad\quad (6.103) \end{cases}$$

Using the above asymptotic forms we readily get

$$\frac{E}{PV} = \begin{cases} \frac{3}{2} & \Gamma \to \infty \\ 3 & \Gamma \to 0 \end{cases} \tag{6.104}$$

The variation of $E/(PV)$ with Γ has been tabulated by Chandrasekhar (1939). We may mention that for an electron gas, the conditions $\Gamma \gg 1$ and $\Gamma \ll 1$ imply

$$T \ll 5.9 \times 10^9 \, \text{K} \quad \text{and} \quad T \gg 5.9 \times 10^9 \, \text{K} \tag{6.105}$$

respectively. However, the condition $\mu/kT \ll -1$ must *also* be satisfied.

The expression for the specific heat is given by

$$c_v = \frac{\Gamma^2}{8} Nk \left[\frac{3(K_4 + K_2) + (K_2 + K_0)}{K_2} - \frac{(3K_3 + K_1)(K_1 + K_3)}{K_2^2} \right]$$
(6.106)

Using the asymptotic forms we get

$$c_v \rightarrow \begin{cases} \frac{3}{2} Nk & \Gamma \rightarrow \infty \\ 3 Nk & \Gamma \rightarrow 0 \end{cases}$$
(6.107)

6.4 Adiabatic processes

6.4.1 Non-relativistic case

If we use (6.18) and (6.20), we will obtain the following expression for the thermodynamic potential Ω:

$$\Omega \equiv -PV = -\frac{(2m)^{3/2} V}{3\pi^2 \hbar^3} \int_0^\infty \frac{\varepsilon^{3/2} \, d\varepsilon}{e^{(\varepsilon - \mu)/kT} + 1}$$
(6.108)

where we have substituted $G = 2$. The above equation can be written in the form

$$\Omega = -PV = -VT^{5/2} f_1 \left(\frac{\mu}{T} \right)$$
(6.109)

where

$$f_1 \left(\frac{\mu}{T} \right) = \frac{(2m)^{3/2}}{3\pi^2 \hbar^3 k^{5/2}} \int_0^\infty \frac{z^{3/2} \, dz}{e^{z - (\mu/kT)} + 1}$$
(6.110)

and

$$z = \frac{\varepsilon}{kT}$$
(6.111)

Thus f_1 is function of only one variable, viz. μ/T. Equation (6.109) gives the following expression for entropy

$$\frac{S}{V} = -\frac{1}{V} \left(\frac{\partial \Omega}{\partial T} \right)_{V,\mu} = -\frac{5}{2} T^{3/2} f_1 \left(\frac{\mu}{T} \right) + \left(\frac{\mu}{T} \right) T^{3/2} f_1' \left(\frac{\mu}{T} \right)$$
(6.112)

where primes denote differentiation with respect to the argument. Further

$$\frac{N}{V} = -\frac{1}{V} \left(\frac{\partial \Omega}{\partial \mu} \right)_{T,V} = -T^{3/2} f_1' \left(\frac{\mu}{T} \right)$$
(6.113)

Equations (6.112) and (6.113) yield

$$\frac{S}{N} = \frac{5}{2} \frac{f_1 (\mu/T)}{f_1' (\mu/T)} - \frac{\mu}{T} \equiv f_2 \left(\frac{\mu}{T} \right)$$
(6.114)

For an adiabatic process, $S = $ constant and with a constant number of particles (N), (6.114) implies that

$$\frac{\mu}{T} = \text{const} \tag{6.115}$$

Using the above relation in (6.112) we get (for an adiabatic process)

$$VT^{3/2} = \text{const} \tag{6.116}$$

Equations (6.109) and (6.116) also give the following relations

$$PV^{5/3} = \text{const} \tag{6.117}$$

$$\frac{T^{5/2}}{P} = \text{const} \tag{6.118}$$

valid obviously for an adiabatic process. Equations (6.116), (6.117) and (6.118) coincide with the adiabatic equations for a monatomic gas; however, the factor $\frac{5}{3}$ in (6.117) is *no longer* equal to the specific heat ratio c_p/c_v!; also $c_p - c_v$ is *not* equal to R in this case.

6.4.2   *Extreme relativistic case*

In the extreme relativistic limit, $mc^2 \ll p^2c^4$ so that we may write

$$\varepsilon \approx pc \tag{6.119}$$

Thus (6.59) becomes

$$E \approx \frac{V}{\pi^2 \hbar^3 c^3} \int_0^\infty \frac{\varepsilon^3 \, d\varepsilon}{e^{(\varepsilon - \mu)/kT} + 1} \tag{6.120}$$

In this approximation, (6.60) takes the form

$$P = \frac{1}{3\pi^2 \hbar^3 c^3} \int_0^\infty \frac{\varepsilon^3 \, d\varepsilon}{e^{(\varepsilon - \mu)/kT} + 1} \tag{6.121}$$

Equations (6.120) and (6.121) give the equation of state

$$PV = \tfrac{1}{3}E \tag{6.122}$$

as was obtained in (6.104). Thus

$$\Omega \equiv -PV \approx -\frac{V(kT)^4}{3\pi^2 \hbar^3 c^3} \int_0^\infty \frac{z^3 \, dz}{e^{z - (\mu/kT)} + 1} \tag{6.123}$$

where $z \equiv \varepsilon/kT$. We may therefore write

$$\Omega = VT^4 f_3\left(\frac{\mu}{T}\right) \tag{6.124}$$

Following a procedure exactly similar to the one used in Section 4.1 we get the following equations of state for an adiabatic process ($S = $ constant) with a constant number of particles (N):

$$PV^{4/3} = \text{const} \tag{6.125}$$

$$VT^3 = \text{const} \tag{6.126}$$

$$\frac{T^4}{P} = \text{const} \tag{6.127}$$

These equations correspond to $\gamma = \frac{4}{3}$ but once again, γ is *not* the ratio of specific heats.

7

Ionization equilibrium and the Saha equation

7.1 Introduction

In this chapter we will consider the ionization equilibrium in gases and derive formulae from which we can determine, at a particular temperature and pressure, the percentage ionization of gases (single as well as multiple). The formulae were first derived by Megh Nad Saha in 1919 and although many modifications have been made since then, the original formulae are still extensively used in many diverse areas. In particular, the ionization formulae have played a very important role in the understanding of the spectra of stars.

In Section 7.2 we will derive the necessary thermodynamic formulae and in Section 7.3 we will derive Saha's ionization formulae and discuss their applications.

7.2 The thermodynamic formulation

The partition function for a gas consisting of N non-interacting molecules is given by (see Section 4.2)

$$Q = \frac{1}{N!} q^N \tag{7.1}$$

where q represents the partition function of one molecule and is given by

$$q = \sum_n \exp\left[-\frac{\varepsilon_n}{kT} \right] \tag{7.2}$$

the summation being over all states of the molecule. As discussed in Section 4.2 we may write q as

$$q = q_t q' \tag{7.3}$$

where

$$q_t = \sum_n \exp\left[-\frac{\varepsilon_{tn}}{kT}\right] \tag{7.4}$$

$$q' = \sum_n \exp\left[-\frac{\varepsilon_n'}{kT}\right] \tag{7.5}$$

the sum in (7.4) is over the translational states of the molecule and the summation in (7.5) is over internal states of the molecule. In Chapter 3 we derived the expression for q_t which is given by

$$q_t = \left[\frac{mkT}{2\pi\hbar^2}\right]^{3/2} V \tag{7.6}$$

Thus

$$q = \left(\frac{mkT}{2\pi\hbar^2}\right)^{3/2} V \sum_n \exp\left(-\frac{\varepsilon_n'}{kT}\right) \tag{7.7}$$

and

$$Q = \frac{1}{N!}\left\{\left(\frac{mkT}{2\pi\hbar^2}\right)^{3/2} V \sum_n \exp\left(-\frac{\varepsilon_n'}{kT}\right)\right\} \tag{7.8}$$

Now, the free energy is given by

$$F = -kT \ln Q \tag{7.9}$$

or,

$$F = -NkT \ln\left\{\left(\frac{mkT}{2\pi\hbar^2}\right)^{3/2} V \sum_n \exp\left(-\frac{\varepsilon_n'}{kT}\right)\right\} + NkT \ln(N/e) \tag{7.10}$$

where we have used the Stirling formula for $\ln N!$ (Appendix 2). Following the analysis given by Landau and Lifshitz (1969) we separate the term containing the volume to obtain

$$F = -NkT \ln\left[\frac{eV}{N}\right] + Nf(T) \tag{7.11}$$

where

$$f(T) = -kT \ln\left\{\left(\frac{mkT}{2\pi\hbar^2}\right)^{3/2} \sum_n \exp\left(-\frac{\varepsilon_n'}{kT}\right)\right\} \tag{7.12}$$

is a function only of temperature. Since

$$P = -\left(\frac{\partial F}{\partial V}\right)_T \tag{7.13}$$

we get

$$P = \frac{NkT}{V} \tag{7.14}$$

or

$$PV = NkT \tag{7.15}$$

which is the equation of state for an ideal gas.

We will be assuming the reactions to take place at constant temperature and pressure and therefore it is more convenient to work with the Gibbs free energy which is given by

$$G = F + PV \tag{7.16}$$

Using (7.11) and (7.15) we get

$$G = F + NkT$$

$$= -NkT - NkT \ln\left(\frac{V}{N}\right) + Nf(T) + NkT \tag{7.17}$$

Since[†]

$$dG = -SdT + VdP \tag{7.18}$$

we would like to express G in terms of T and P; we therefore use (7.15) to write (7.17) in the form

$$G = -NkT \ln\left(\frac{kT}{P}\right) + Nf(T)$$

$$= NkT \ln P + N\chi(T) \tag{7.19}$$

where

$$\chi(T) \equiv f(T) - kT \ln kT \tag{7.20}$$

or

$$\chi(T) = -kT \ln\left\{\left(\frac{m}{2\pi\hbar^2}\right)^{3/2} (kT)^{5/2} \sum_n \exp\left(-\frac{\varepsilon_n'}{kT}\right)\right\} \tag{7.21}$$

Comparing the expression for G (7.16) with (5.35) we readily get

$$G = \mu N \tag{7.22}$$

where μ is the chemical potential.

[†] $G = E - TS + PV$

Therefore

$dG = dE - TdS - SdT + PdV + VdP$

But

$TdS = dQ = dE + PdV$

Hence

$dG = -SdT + VdP$

From the above equation we have

$$S = -\left(\frac{\partial G}{\partial T}\right)_P, \quad V = \left(\frac{\partial G}{\partial P}\right)_T$$

We next consider a multicomponent system consisting of different types of particles. Each type is represented by a particular value of the parameter $i(=1, 2, \ldots)$. For such a multicomponent system we will have

$$F = \sum_i F_i \tag{7.23}$$

where

$$F_i = -N_i kT \ln\left[\frac{eV}{N_i}\right] + N_i f_i(T) \tag{7.24}$$

and

$$f_i(T) = kT \ln\left[\left(\frac{m_i kT}{2\pi\hbar^2}\right)^{3/2} \sum_n \exp\left(-\frac{\varepsilon_{in}'}{kT}\right)\right] \tag{7.25}$$

N_i representing the number of particles of type i, m_i and ε_{in}' being the corresponding mass and internal energy states of the molecule. The partial pressure corresponding to the component i is given by

$$P_i = -\left(\frac{\partial F_i}{\partial V}\right)_T = \frac{N_i kT}{V} \tag{7.26}$$

The Gibbs' free energy, G, can be written in the form

$$G = \sum_i G_i \tag{7.27}$$

where

$$G_i = F_i + P_i V \tag{7.28}$$

or

$$G_i = N_i kT \ln P_i + N_i \chi_i(T) \tag{7.29}$$

with

$$\chi_i(T) = f_i(T) - kT \ln kT$$
$$= -kT \ln\left\{\left(\frac{m_i}{2\pi\hbar^2}\right)^{3/2} (kT)^{5/2} \sum_n \exp\left[-\frac{\varepsilon_{in}'}{kT}\right]\right\} \tag{7.30}$$

Since

$$G_i = \mu_i N_i \tag{7.31}$$

we have

$$\mu_i = kT \ln P_i + \chi_i(T) \tag{7.32}$$

7.3 The Saha ionization formula

We follow the analysis given by Brush (1967) and consider a chemical (or ionization) reaction of the form

$$\sum_i \nu_i A_i = 0 \tag{7.33}$$

For example, for the reaction

$$2H_2 + O_2 \leftrightarrows 2H_2O \tag{7.34}$$

$A_1 = H_2$, $A_2 = O_2$, $A_3 = H_2O$, $v_1 = 2$, $v_2 = 1$ and $v_3 = -2$. The double arrow indicates that if we bring together H_2 and O_2 they will combine to form H_2O, however, there will always be some free H_2 and O_2, the amounts of which will depend on temperature and pressure (for a good account of the thermodynamics of chemical equilibria, see, e.g., Saha and Srivastava (1965) Chapter XVIII).

Another type of reaction of great importance is the ionization reaction, the simplest of which is given by

$$H \rightleftarrows H^+ + e^- \tag{7.35}$$

where H represents the neutral hydrogen atom and H^+ the ionized hydrogen atom, which is simply a proton. For the above reaction $A_1 = H$, $A_2 = H^+$, $A_3 = e^-$, $v_1 = 1$, $v_2 = -1$ and $v_3 = -1$.

We assume the chemical (or ionization) reaction to take place at constant temperature and pressure; thus, the Gibbs' free energy must be a minimum with respect to small concentration changes of different elements constituting the chemical reaction. Now,

$$\frac{dG}{dN_1} = \frac{\partial G}{\partial N_1} + \frac{\partial G}{\partial N_2}\frac{\partial N_2}{\partial N_1} + \cdots \tag{7.36}$$

where $\partial N_2/\partial N_1$ represents the change in N_2 when N_1 is varied and other Ns are kept constant. It is obvious from (7.33) that when N_1 is changed by v_1 then N_2 would change by v_2, implying

$$\frac{\partial N_2}{\partial N_1} = \frac{v_2}{v_1} \tag{7.37}$$

Thus for dG/dN_1 to be zero we must have

$$\frac{\partial G}{\partial N_1} + \sum_{\substack{i \\ i \neq 1}} \frac{\partial G}{\partial N_i}\frac{v_i}{v_1} = 0$$

or

$$\sum_i \frac{\partial G}{\partial N_i}\frac{v_i}{v_1} = 0 \tag{7.38}$$

since

$$\frac{\partial G}{\partial N_i} = \mu_i \tag{7.39}$$

(7.38) can be written in the form

$$\sum_i \mu_i \nu_i = 0 \tag{7.40}$$

The above equation represents the general condition for chemical equilibrium. Substituting for μ_i from (7.32) we get

$$\sum_i \nu_i [kT \ln P_i + \chi_i(T)] = 0$$

or

$$\sum_i \nu_i \ln P_i = -\frac{1}{kT} \sum_i \nu_i \chi_i(T) \tag{7.41}$$

But

$$\sum_i \nu_i \ln P_i = \sum_i \ln P_i^{\nu_i} = \ln \prod_i P_i^{\nu_i}$$

Thus

$$\prod_i P_i^{\nu_i} = K_p(T) \tag{7.42}$$

where $K_p(T)$ is known as the equilibrium constant and is given by

$$K_p(T) = \exp\left[-\frac{1}{kT} \sum_i \nu_i \chi_i(T) \right] \tag{7.43}$$

We next consider the ionization reaction

$$M_{n-1} \rightleftarrows M_n + e \tag{7.44}$$

where M_n represents the atom which is n times ionized (i.e., it has $Z - n$ bound electrons, Z representing the atomic number of the atom). Comparing (7.44) with (7.33) we have

$$A_1 = M_{n-1} \quad A_2 = M_n \quad A_3 = e$$

with

$$\nu_1 = 1 \quad \nu_2 = -1 \quad \nu_3 = -1$$

Thus

$$\prod_i P_i^{\nu_i} = \frac{P_{M_{n-1}}}{P_{M_n} P_e} \tag{7.45}$$

Substituting in (7.42) and taking the logarithm we get

$$\ln\left[\frac{P_{M_{n-1}}}{P_{M_n} P_e} \right] = -\frac{1}{kT}[\chi_1(T) - \chi_2(T) - \chi_3(T)]$$

$$= \ln\left[\left(\frac{m_n - 1}{m_n m_e} \right)^{3/2} \frac{(2\pi\hbar^2)^{3/2}}{(kT)^{5/2}} \frac{Q_{n-1}}{Q_n Q_e} e^{\psi_n/kT} \right] \tag{7.46}$$

where

$$\psi_n = \varepsilon_{0,n} - \varepsilon_{0,n-1} \tag{7.47}$$

represents the ionization constant and

$$Q_i = \sum_n \exp\left[-\frac{\varepsilon_{in}'}{kT} \right]; \quad i = n-1, n, e \tag{7.48}$$

represent the internal partition functions. We rewrite (7.46) in the form

$$\ln\left[\frac{P_{M_n} P_e}{P_{M_{n-1}}} \right] \approx -\frac{\psi_n}{kT} + \tfrac{5}{2}\ln T + \ln\left[\left(\frac{m_e}{2\pi h^2} \right)^{3/2} k^{5/2} \right]$$
$$+ \ln\left[2\frac{Q_n}{Q_{n-1}} \right] \tag{7.49}$$

where we have set $Q_e = 2$ (because of the spin degeneracy of the electron) and have assumed $m_{n-1} \approx m_n$.

As a first approximation we neglect the last term on the right hand side of the above equation. As an example, let us consider the heating of a gas of N neutral atoms (in a fixed volume V) to a temperature T. We consider the single ionization of the atom (i.e., $n = 1$), thus we may write (7.44) in the form

$$M \rightleftarrows M^+ + e \tag{7.50}$$

where M and M^+ represent the neutral and singly ionized atom respectively. If ε represents the fraction of atoms ionized then

$$P_e = P_{M^+} = \frac{N}{V}\varepsilon kT \tag{7.51}$$

$$P_M = \frac{N}{V}(1 - \varepsilon)kT \tag{7.52}$$

If P represents the total pressure then

$$P = P + P_{M^+} + P_M$$
$$= \frac{N}{V}(1 + \varepsilon)kT \tag{7.53}$$

Thus, in terms of the total pressure, the partial pressures are given by

$$P_e = P_{M^+} = \frac{\varepsilon}{1 + \varepsilon}P \tag{7.54}$$

and

$$P_M = \frac{1 - \varepsilon}{1 + \varepsilon}P \tag{7.55}$$

Thus

$$\ln\left[\frac{P_{M^+}P_e}{P_M}\right] = \ln\left[\frac{\varepsilon^2}{1-\varepsilon^2}P\right] \tag{7.56}$$

If we substitute the above expression in (7.49), work consistently in c.g.s. units and convert the natural logarithms to common logarithms we would obtain:

$$\log\left[\frac{\varepsilon^2}{1-\varepsilon^2}P\right] \approx \frac{(\log e)\psi_1}{(1.38\ 10^{-16})T} + \tfrac{5}{2}\log T - 0.477 \tag{7.57}$$

where P is measured in dynes cm^{-2}, ψ_n in ergs and T in K and we have neglected the last term on the right hand side. If we measure P in atmospheres and ψ_n in electron volts we will obtain

$$\log\left[\frac{\varepsilon^2}{1-\varepsilon^2}P\,(\text{atm})\right] \approx -\frac{5035\psi_1\,(\text{eV})}{T\,(\text{K})} + \tfrac{5}{2}\log T - 6.48 \tag{7.58}$$

The above equation represents the form of the famous ionization formula

Table 7.1. *Typical values of various parameters appearing in Saha's ionization formula*

Element	$\psi_n(\text{eV})$	g_a	g_i
Na	5.12	2	1
Cs	3.87	2	1
Ca	6.09	1	2
Cd	8.96	1	2
Zn	9.36	1	2
Tl	6.07	2	1

Note: Adapted from Zemansky (1957), Chapter 18.

Table 7.2. *Percentage ionization of calcium as obtained by using equation (7.58)*

Temperature T	Pressure P		
	1 atm	10^{-2} atm	10^{-4} atm
2000 K	–	–	1.7×10^{-3}
4000 K	0.26%	2.6%	26%
6000 K	8.4%	65%	99%
10000 K	86%	99.8%	100%

Note: Table partly adapted from Saha and Srivastava (1965), Chapter 18.

put forward by Saha in 1919. As an example, we consider Ca for which $\psi \approx 6.09 \, \text{eV}$ (see Table 7.1). For given values of T and P we can use the above formula for calculating ε. The result of such a calculation is given in Table 7.2. Thus at $T \approx 6000 \, \text{K}$, the gas is almost completely ionized at $P \approx 10^{-4}$ atmospheres and is only 8.4% ionized at $P \approx 1$ atmospheres. Returning to (7.46), we note that the partition function for the n times ionized atom Q_n can be written in the form

$$Q_n = g_{n,0} + \sum_r g_{n,r} \exp\left[-\frac{\chi_{n,r}}{kT} \right] \tag{7.59}$$

where $g_{n,0}$ represents the statistical weight of the ground state, $g_{n,r}$ the statistical weight of the rth excited state and $\chi_{n,r}$ the excitation energy of the rth state for the n times ionized atom; the summation in (7.59) is carried out over all bound states of the ionized atom.[†] For most cases, not much error is involved in replacing the partition functions by the first term on the right hand side of (7.59). Thus

$$\frac{Q_{n-1}}{Q_n Q_e} \approx \frac{g_{n-1,0}}{2g_{n,0}} \tag{7.60}$$

where we have used the fact that $Q_e = 2$. Using the above equation, (7.49) becomes

$$\ln\left[\frac{P_{M_n} P_e}{P_{M_{n-1}}} \right] \approx \frac{\psi_n}{kT} + \tfrac{5}{2}\ln T + \ln\left[\left(\frac{m_e}{2\pi\hbar^2} \right)^{3/2} k^{5/2} \right] + \ln\left[\frac{2g_{n,0}}{g_{n-1,0}} \right] \tag{7.61}$$

and the single ionization formula (7.58) is modified to

$$\log\left[\frac{\varepsilon^2}{1-\varepsilon^2} P(\text{atm}) \right] \approx -\frac{5035\psi \, (\text{eV})}{T \, (\text{K})} + \tfrac{5}{2}\log T$$
$$- 6.48 + \log\left[\frac{2g_a}{g_i} \right] \tag{7.62}$$

where g_a and g_i represent the statistical weights of the ground state for the neutral atom and the single ionized atom respectively. The values of the parameters g_a and g_i for some typical elements are given in Table 7.1.

The derivation of (7.62) assumes that electrons arise only from the

---

[†] We may mention here that for large values of r, χ_{nr} tends to a constant value and therefore the summation in (7.59) diverges. In practice, this difficulty is avoided by summing over a *finite* number of excited states. This is justified because the presence of 'nearby' particles will obliterate the states which are near the continuum.

ionization of atoms (see (7.50)). Actually, electrons may arise from other sources also so it is usually taken as an independent constituent. If we write n and n^+ as the number density corresponding to neutral and singly ionized atoms respectively, then

$$P_M = nkT \tag{7.63}$$

and

$$P_{M^+} = n^+kT \tag{7.64}$$

so that (7.61) can be written in the form

$$\ln\left(\frac{n^+}{n}P_e\right) \approx -\frac{\psi_1}{kT} + \tfrac{5}{2}\ln T + \ln\left[\left(\frac{m_e}{2\pi\hbar^2}\right)^{3/2}\frac{5}{2}\right] + \ln\left[\frac{2g_a}{g_i}\right] \tag{7.65}$$

In general, if we consider the reaction given by (7.44) then we would get

$$\ln\left(\frac{n^{(r+1)}}{n^{(r)}}P_e\right) \approx -\frac{\psi_r}{kT} + \tfrac{5}{2}\ln T + \ln\left[\left(\frac{m_e}{2\pi\hbar^2}\right)^{3/2}\frac{5}{2}\right]$$
$$+ \ln\left[2\frac{g_{n,0}}{g_{n-1,0}}\right] \tag{7.66}$$

where $n^{(r)}$ represent the density of 'r' times ionized atoms. Equations (7.65) and (7.66) have been applied with great success in astrophysics. For example, while studying the spectra of stars we have lines which originate from atoms (arc lines) and also those which originate from ions (spark lines). By comparing the intensities of characteristic lines emitted from neutral and ionized atoms it is possible to determine the degree of ionization. Thus knowing the pressure of the stellar atmosphere we can use the ionization formulae to determine the temperature of the stellar atmosphere. We refer to the classic treatise by Saha and Srivastava (1965) where details regarding the application of the ionization formula have been given. The book also discusses experimental verification of the ionization formula. For further work on the ionization equilibrium we refer the reader to the references given in the review article of Brush (1967).

8

Debye–Hückel equation of state

8.1 Introduction

Debye–Hückel theory was originally developed (Debye & Hückel 1923a, 1923b) in order to explain quantitatively the osmotic pressure and other thermodynamic quantities of dilute electrolyte solutions. In particular the strong electrolyte (e.g. NaCl) is assumed to behave like a system of charged particles in a dielectric medium (e.g. water) with constant permittivity ε. In this model the medium and the molecules have the same ε. Since the model describing the strong electrolytes assumes a system of charged particles, the results of this theory are also applicable to plasmas (i.e. ionized gases). This model was possible since it was confirmed (Milner 1913) that the solution of electrolytes contains fully isolated ions and no agglomerates, which had been derived from the effect of interionic forces on the osmotic pressure (Milner 1913a).

In this model we consider a medium consisting of charged particles, each having a charge $z_j e$, where $z_j = -1$ for electrons and is a positive or a negative integer, according to the ion under consideration. We denote by n_{j0} the density of charged particles of type j, i.e. their number in a unit volume. The charge neutrality of the system implies the condition

$$\sum_j z_j n_{j0} = 0 \tag{8.1}$$

The departure of the equation of state from an ideal gas is a result of the Coulomb interactions between the charged particles. In order to describe the Debye–Hückel model it is necessary to assume that the energy of the Coulomb interactions between neighbouring particles is small in comparison with the thermal energy of these charged particles, i.e.

$$\frac{(z_j e)^2}{r_0} \ll k_B T \tag{8.2}$$

where k_B is the Boltzmann constant, T is the temperature and r_0 is the average distance between neighbouring particles. r_0 can be estimated from

$$r_0 \approx n_0^{-1/3} \tag{8.3}$$

where n_0 is the average total density (number of particles/cm^3). Equations (8.2) and (8.3) can be combined in order to obtain a condition for an approximate ideal gas,

$$n_0 \ll \left(\frac{k_B T}{z^2 e^2}\right)^3 \approx 2\cdot 2 \times 10^8 \left(\frac{T}{z^2}\right)^3 \text{cm}^{-3} \tag{8.4}$$

where T is measured in degrees Kelvin and z is the average charge of the particles under consideration. For example, if the degree of ionization is of the order one, i.e. $z \approx 1$, and $T \approx 10\,000\,\text{K}$ ($\approx 1\,\text{eV}$), one gets $n_0 \ll 2.2 \times 10^{20}\,\text{cm}^{-3}$, while for $T \approx 300\,\text{K}$ (room temperature) one has $n_0 \ll 6.2 \times 10^{15}\,\text{cm}^{-3}$. For comparison we recall that the Avogadro number is about 6×10^{23} and the number of molecules in air at standard conditions is about $2.7 \times 10^{19}\,\text{cm}^{-3}$.

8.2 Charged particle description

A more precise and a stricter condition for the applicability of the Debye–Hückel theory is given by the requirement of a statistical treatment using Poisson's equation to describe the electrostatic interactions,

$$\mathbf{V}^2 \phi = -4\pi e \sum_j z_j n_j \tag{8.5}$$

where ϕ is the electric potential, n_j is the density of ions of type j (including $j = e$ for electrons) at a particular point in space. The potential energy of a particular ion is

$$E_j = z_j e \phi \tag{8.6}$$

where ϕ in (8.6) is the potential at the location of the ion j. Assuming that the spatial distribution is determined by Boltzmann's formula,

$$n_j = n_{j0} \exp(-z_j e\phi/k_B T) \tag{8.7}$$

one can visualize this distribution surrounding a particular ion, so that for great distance from this central ion

$$\phi \to 0 \quad \text{for} \quad r \to \infty \tag{8.8}$$

implying

$$n_j \to n_{j0} \quad \text{for} \quad r \to \infty \tag{8.9}$$

Using the approximation (8.2), the exponential in (8.7) can be expanded

$$n_j \cong n_{j0} - n_{j0}ez_j\phi/k_BT \tag{8.10}$$

Substituting (8.10) into Poisson's equation (8.5) and using the global neutrality condition given by (8.1), one has

$$\nabla^2\phi = \kappa^2\phi \tag{8.11}$$

where

$$\kappa^2 \equiv \lambda_D^{-2} = \frac{4\pi e^2}{k_BT}\sum_j n_{j0}z_j^2 \tag{8.12}$$

λ_D is usually called the 'Debye length' and it plays an important role in Debye–Hückel theory. The spherical symmetric solution of (8.11) which vanishes at $r \rightarrow \infty$

$$\phi = ez_j\exp(-\kappa r)/r \tag{8.13}$$

where ez_j is the charge of the central ion under consideration. Using (8.11) and (8.13) one gets

$$-\frac{1}{4\pi}\int\nabla^2\phi dV = -\frac{1}{4\pi}\int\kappa^2\phi dV = -\frac{ez_j\kappa^2}{4\pi}\int_0^\infty$$
$$\times\frac{\exp(-\kappa r)}{r}4\pi r^2 dr = -ez_j \tag{8.14}$$

Equation (8.14) shows that the total charge surrounding a central ion of type j is equal and opposite to the charge ez_j of the j ion. From (8.14) we can define

$$q(r)dr \equiv -z_j e\kappa^2\exp(-\kappa r)\cdot rdr \tag{8.15}$$

as the charge between the spherical shells of radius r and $r + dr$. The quantity $q(r)$, which vanishes at $r = 0$ and $r = \infty$, shows a charge opposite to the central ion with a maximum charge distribution at $r = \kappa^{-1} = \lambda_D$,

$$|q| = q_{max} = z_j e\kappa^2\exp(-1)\cdot\lambda_D \quad \text{at} \quad r = \lambda_D \tag{8.16}$$

Moreover, most of the 'charge atmosphere' around the central ion is contained in a few Debye lengths, and therefore the values of $r \approx \lambda_D$ are important in determining the thermodynamic properties of the ionized gas or the electrolyte solution.

The statistical treatment which assumes the Boltzmann distributions (8.7) is justified if the atmospheric cloud contains many particles, i.e. the Debye radius is larger than the average distance between the particles. This condition can be written as

$$\lambda_D \gg r_0 \approx n_0^{-1/3} \tag{8.17}$$

implying

$$n_0 \ll \left(\frac{k_B T}{4\pi e^2 z^2}\right)^3 \approx 10^5 \left(\frac{T}{z^2}\right)^3 \text{cm}^{-3} \tag{8.18}$$

where use of (8.12) has been made together with the definition $n_0 z^2 = \sum_j n_{j0} z_j^2$. The condition given by (8.18) for the Debye–Hückel theory is stricter than the condition of (8.6) for an approximate ideal gas. For example, if $z \approx 1$ and $T \approx 10\,000\,\text{K}(\approx 1\,\text{eV})$ one gets $n_0 \ll 10^{17}\,\text{cm}^{-3}$ in comparison with the value $n_0 \ll 2.2 \times 10^{20}\,\text{cm}^{-3}$ obtained from (8.4) with similar conditions.

8.3 Electrostatic energy

The electrostatic energy of the system is equal to

$$E_{\text{Coub}} = \frac{V}{2}\sum_j e z_j n_{j0} \phi_j \tag{8.19}$$

where V is the volume of the system, $V n_{j0} = N_j$ is the number of particles of type j, and ϕ_j is the potential at ion j due to all other ions. Expanding the potential (8.13) near the central ion under consideration,

$$\phi \approx e z_j / r - e z_j \kappa + \cdots \tag{8.20}$$

one can see that the first term of (8.20) is the potential caused by the central ion, so that the second term can be understood as the potential due to the surrounding charges, i.e. the 'atmosphere' around the central ion. Since all the other terms in (8.20) vanish at $r = 0$, one has to substitute

$$\phi_j = -e z_j \kappa \tag{8.21}$$

into (8.19). Thus the electrostatic energy of the system can be written as

$$E_{\text{Coub}} = -e^3 \left(\frac{\pi}{k_B T V}\right)^{1/2} \sum_j (N_j z_j^2)^{3/2} \tag{8.22}$$

where use has been made of (8.12) and the total number of particles of type j is defined by

$$N_j = n_{j0} V \tag{8.23}$$

Using the thermodynamic relation

$$\frac{E}{T^2} = -\frac{\partial (F/T)}{\partial T} \tag{8.24}$$

one can obtain the electrostatic free energy F_{Coub} by integrating (8.24) from a state with no electrostatic energy given by $T = \infty$ to a state with the temperature T of the system:

$$F_{\text{Coub}} = - T \int_{\infty}^{T} \frac{E_{\text{Coub}}}{T^2} dT \tag{8.25}$$

Substituting (8.22) in (8.25) gives

$$F_{\text{Coub}} = - \frac{2e^3}{3} \left(\frac{\pi}{k_B T V} \right)^{1/2} \left(\sum_j N_j z_j^2 \right)^{3/2} \tag{8.26}$$

The total free energy is obtained by adding this term of the free energy to the ideal gas contribution (3.13). The pressure in the Debye–Hückel model is obtained from

$$P = - \left(\frac{\partial F}{\partial V} \right)_T \tag{8.27}$$

implying

$$P = \sum_j n_j k_B T - \frac{k_B T}{24 \pi \lambda_D^3} \qquad (\text{dyne cm}^{-2}) \tag{8.28}$$

The energy of the Debye–Hückel theory is given by adding the ideal gas energy

$$\tfrac{3}{2} \sum_j N_j k_B T \tag{8.29}$$

to the electrostatic energy given in (8.22),

$$E = \tfrac{3}{2} \sum_j N_j k_B T - \frac{V k_B T}{8 \pi \lambda_D^3} \tag{8.30}$$

The electrostatic pressure and energy are directly related (compare (8.28) and (8.30)),

$$P_{\text{Coub}} = \frac{E_{\text{Coub}}}{3V} \tag{8.31}$$

The electrostatic contribution to the entropy is obtained from the relation

$$S_{\text{Coub}} = - \left(\frac{\partial F_{\text{Coub}}}{\partial T} \right)_V \tag{8.32}$$

and (8.26),

$$S_{\text{Coub}} = \frac{E_{\text{Coub}}}{3T} \tag{8.33}$$

In (8.31) and (8.33) P_{Coub}, E_{Coub} and S_{Coub} are negative quantities. This fact can be understood since the forces between the charged particles are attractive as each ion is surrounded with charges of opposite sign (see (8.14) and (8.15)). Therefore the total energy, the total pressure and the total entropy in the Debye–Hückel model are smaller than the appropriate quantities of the ideal gas. The decrease in these thermodynamic quantities is due to the electrostatic interaction. Note however that the Coulomb corrections are small in comparison with the ideal gas terms since this fact was the starting point for the Debye–Hückel model (8.2). Therefore the energy, the pressure and the entropy cannot become negative in this model.

8.4   Total free energy and equation of state

The equations of state (8.28) and (8.30) of the Debye–Hückel model were obtained assuming that the degree of ionization, z_j, does not change with the temperature and the pressure. However, for a process where z_j changes one has to obtain the equilibrium equations using a technique analogous to that of the Saha equation for an ideal gas (see Chapter 7). The total free energy of the system is

$$F = F_{IG} + F_{Coub} \tag{8.34}$$

where F_{Coub} is given in (8.26) and the ideal gas free energy F_{IG} is

$$F_{IG} = -\sum N_j k_B T \ln\left(\frac{Q_j e}{N_j}\right) - N_e k_B T \ln\left(\frac{Q_e e}{N_e}\right) \tag{8.35}$$

where Q_j and Q_e are the partition functions for an ion ionized j times and an electron respectively. Assuming the ionization process to be in equilibrium for the following reactions

$$X_j \rightleftarrows X_{j+1} + e^- \quad j = 0, 1, 2, \ldots \tag{8.36}$$

where X_j denotes an ion ionized j times. In this case F of (8.34) is minimized with respect to the number of particles keeping T and V constant and using the constraints

$$\delta N_j = -\delta N_{j+1} = -\delta N_e \quad j = 0, 1, 2, \ldots \tag{8.37}$$

The variation of (8.34)

$$\delta F = \sum \left(\frac{\partial F}{\partial N_j}\right)_{T,V} \delta N_j + \left(\frac{\partial F}{\partial N_e}\right)_{T,V} \delta N_e = 0 \tag{8.38}$$

together with (8.37) yield the following Saha equation

$$\frac{N_{j+1} N_e}{N_j} = \frac{Q_{j+1} Q_e}{Q_j} \exp\left(\frac{\Delta I_{j+1}}{k_B T}\right) \tag{8.39}$$

where ΔI_{j+1} is defined by

$$\Delta I_{j+1} = \mu_{j,\text{Coub}} - \mu_{j+1,\text{Coub}} - \mu_{e,\text{Coub}} \tag{8.40}$$

and the Coulomb contributions to the chemical potentials in (8.40) are

$$\mu_{j,\text{Coub}} = \left(\frac{\partial F_{\text{Coub}}}{\partial N_j} \right)_{T,V} \tag{8.41}$$

A direct calculation using (8.40), (8.41) and (8.26) gives,

$$\Delta I_{j+1} = 2(z_j + 1)e^3 \left(\frac{\pi}{k_B T} \right)^{1/2} \left(\sum n_j \cdot z_j^2 \right)^{1/2} = \frac{z_{j+1} e^2}{\lambda_D} \tag{8.42}$$

where λ_D is defined in (8.12). The term ΔI_{j+1} describes a decrease in the ionization potential of an ion ionized j times since one has (see Saha equation, Chapter 7)

$$\frac{Q_{j+1} Q_e}{Q_j} \propto \exp(-I_{j+1}/k_B T) \tag{8.43}$$

From (8.39) and (8.43),

$$\frac{N_{j+1} N_e}{N_j} \propto \exp[-(I_{j+1} - \Delta I_{j+1})/k_B T] \tag{8.44}$$

where ΔI_{j+1} is given in (8.42). It is interesting to point out that the term $z_{j+1} e^2/\lambda_D$ is the ionization energy of an electron located at a Debye radius from the centre of the ion under consideration which is ionized $j + 1$ times. In this model $\lambda_D \gg r_0 \approx n_0^{-1/3}$, therefore the shift in the ionization potential satisfies (Henry *et al.* 1982, Henry 1983)

$$\Delta I_{j+1} \ll k_B T \tag{8.45}$$

z_j which equals the charge (in units of e) of the ion is actually given by

$$z_j = j \tag{8.46}$$

and assuming an average charge $\bar{j}$

$$\bar{j} = \frac{n_e}{n} = \frac{N_e/V}{N/V} \tag{8.47}$$

and an average square of the charge

$$\sum_j N_j z_j^2 \equiv N \bar{j^2} = N \bar{j}^2 \tag{8.48}$$

one can derive the average shift in the ionization potential given in (8.42):

$$\overline{\Delta I} = 2 \left(\frac{n_e}{n} + 1 \right) e^3 \left[\frac{\pi n_e((n_e/n) + 1)}{k_B T} \right]^{1/2} \tag{8.49}$$

As an example let us consider a plasma with $n = 10^{13} \, \text{cm}^{-3}$, $n_e/n \approx 1$ and $k_B T \approx 1 \, \text{keV}$. This type of plasma is used in magnetic confinement fusion devices. The shift in the ionization potential is found from (8.49) to be negligible, $\overline{\Delta I} \approx 5.10^{-5} \, \text{eV}$. However, for an ionized gas (e.g. air) at $k_B T \approx 100 \, \text{eV}$ with an atmospheric density $n \approx 5.10^{19} \, \text{cm}^{-3}$ and an average ionization $\bar{j} \approx 3$ (obtained from the Saha equation for an ideal gas with $I \approx 50 \, \text{eV}$) one gets from (8.49) $\overline{\Delta I} \approx 4.2 \, \text{eV}$ which is already a significant value, $\overline{\Delta I}/I \approx 8.5\%$! Therefore one can realize that for the ionization processes of an atmospheric gas the Coulomb interactions can shift the ionization potential by as much as 10%.

We conclude this chapter by pointing out that the Debye–Hückel theory is used mainly for solutions of strong electrolytes (see, e.g., McQuarrie 1976), in appropriate astrophysical problems, as well as in ionized gases, where the electrostatic interactions cannot be neglected while at the same time they are not larger than the thermal energy (see, e.g., Brush S.G. 1967).

9

The Thomas–Fermi and related models

9.1 Overview

The statistical model for atomic electrons was first put forward by Thomas (1927) and Fermi (1928) and was originally introduced to study a multi-electron atom. Since then it has found important applications in molecular theory, solid state theory and also in determining the contribution from the electrons to the equation of state of matter at high pressures ($P \gtrsim 10^7$ atmospheres) – the last application being of considerable interest in inertial confinement fusion and in astrophysics problems. Indeed, among the various models which describe the electronic thermodynamic functions in highly compressed matter, the Thomas–Fermi model is the one most often used in hydrodynamic codes (e.g., simulation of inertial confinement fusion, astrophysics etc.). The advantages of the Thomas–Fermi model over other models comprise its simplicity, clarity and validity over a wide range of densities and temperatures. Although other models may give a more accurate description of the local phenomena of matter in certain domains, they are difficult to extend to other domains of pressure and density. If we refer to regions 4 and 5 in Fig. 1.3, the gas can be assumed to be composed of electrons and nuclei (rather than electrons and ions) – the electron gas being described by the Thomas–Fermi model and the nuclei can be assumed to obey classical statistics so that the pressure and energy associated with the nuclei are given by (3.14) and (3.15).

The main assumption of the Thomas–Fermi model is that the electrons in the atom are considered as a (degenerate) gas placed in a self-consistent electrostatic field, described by the electrostatic potential $\phi(r)$, such that $\phi(r)$ varies slowly in an electron wavelength so that a large number of electrons can be localized within a volume over which there is very little variation of the $\phi(r)$. In Section 9.2 we will discuss the basic physics of the Thomas–Fermi model and derive the Thomas–Fermi equation corresponding to

$T = 0$; we will then calculate the total energy and pressure associated with the electrons. In Section 9.3 we will take into account the effect of exchange interactions which will lead to the Thomas–Fermi–Dirac model. In Section 9.4 we will have a brief excursion with the virial theorem and in Section 9.5 we will discuss the Thomas–Fermi model at finite temperatures. In Section 9.6 we will discuss the effects of exchange and quantum corrections. Finally, in Appendix 5, we have given a large number of tables from which one can calculate the pressure and energy for a wide range of atomic densities and temperatures. The tables (adapted from McCarthy (1965)) also allow us to see the extent of the corrections due to exchange and quantum effects. Although the tables occupy considerable space, we feel it will be of significant interest to workers trying to calculate the total energy and pressure of a system.

9.2   The Thomas–Fermi model at $T = 0$

In Chapter 6 we have shown that for an ideal Fermi–Dirac gas, the number of particles per unit volume is given by

$$n = \frac{N}{V} = G \frac{1}{2\pi^3 \hbar^3} \int_0^\infty \frac{p^2 dp}{e^{(\varepsilon - \mu)/kT} + 1} \tag{9.1}$$

where μ is the chemical potential, ε the total energy of the electron, G is the degeneracy parameter ($G = 2$ for electrons) and use has been made of (A1.17) for the density of states. If the electron gas is in a potential field, Fermi–Dirac statistics can still be applied provided that (a) the field varies very little over a de-Broglie wavelength and (b) the field varies slowly enough so that we can consider a volume element $d\tau$ which contains a large number of particles and at the same time the field can be assumed to be approximately constant in this volume. Further, statistical mechanics tells us that (at thermal equilibrium) the chemical potential μ must be the same at all points. Thus replacing ε by

$$\frac{p^2}{2m} + V(r)$$

in (9.1) (where $V(r)$ is the potential energy function) we get

$$n(r) = \frac{1}{\pi^2 \hbar^3} \int_0^\infty \frac{p^2 dp}{\exp\{[(p^2/2m) + V(r) - \mu]/kT\} + 1} \tag{9.2}$$

where we have used $G = 2$. Now, at $T = 0$ the Fermi function is a step function (see Fig. 6.1) so that

$$n(r) = \frac{1}{\pi^2 \hbar^3} \int_0^{p_{max}} p^2 dp = \frac{1}{3\pi^2 \hbar^3} p_{max}^{\ 3} \tag{9.3}$$

where

$$\frac{p_{max}^2}{2m} = \mu - V(r) \tag{9.4}$$

Thus

$$n(r) = \frac{[2m(\mu + e\phi)]^{3/2}}{3\pi^2\hbar^3} \tag{9.5}$$

where $\phi(r)$ represents the electrostatic potential which is related to the potential energy function through the following relation

$$V(r) = -e\phi(r) \tag{9.6}$$

e being the magnitude of the electronic charge. Equation (9.5) represents the Thomas–Fermi relation between the electron density and the electrostatic potential. We rewrite (9.5) in the form[†]

$$n(r) = \frac{[2me(\phi - \phi_0)]^{3/2}}{3\pi^2\hbar^3} \tag{9.7}$$

where

$$\phi_0 = -\mu/e \tag{9.8}$$

Obviously, (9.7) can be used when $\phi > \phi_0$; for $\phi < \phi_0$, $n = 0$.

Now, for the theory to be self-consistent, the potential and the charge density should be related through the Poisson equation

$$\nabla^2\phi = -4\pi\rho(r) = 4\pi en(r) \tag{9.9}$$

where $\rho(r) (= -en(r))$ represents the electron charge density.[‡] Substituting for $n(r)$ from (9.7) we get

$$\nabla^2(\phi - \phi_0) = \frac{4e}{3\pi\hbar^3}(2me)^{3/2}(\phi - \phi_0)^{3/2} \tag{9.10}$$

We assume the nucleus (of charge Ze) to be at the origin of the coordinate

---

[†] It may be noted that $(p^2/2m) - e\phi$ (which represents the total energy of an electron) is always negative since otherwise the electron will not remain bound. Further, since $(p_{max}^2/2m) = \mu - V = e(\phi - \phi_0)$, $\phi_0 = 0$ for an isolated atom (see also (9.75)).

[‡] Actually ρ is the sum of the electron charge density $\rho_e (= -en(r))$ and the charge density associated with the nucleus ρ_n. We assume that the charge of the nucleus $(= Ze)$ is concentrated at the origin so that $\rho_n = Ze\delta(\mathbf{r})$ where $\delta(\mathbf{r})$ $[= \delta(x)\delta(y)\delta(z)]$ is the three-dimensional Dirac-delta function. Since $\delta(\mathbf{r}) = 0$ for all values of $\mathbf{r}$ (except at the origin), (9.9) is valid for all values of r except $r = 0$. Therefore, we solve (9.9) subject to the basic condition that $\phi(r) \to Ze/r$ as $r \to 0$.

system so that the potential is spherically symmetric and we may write

$$\nabla^2(\phi - \phi_0) = \frac{1}{r^2}\frac{d}{dr}\left[r^2\frac{d}{dr}(\phi - \phi_0)\right] \tag{9.11}$$

Further, as $r \to 0$, the potential will be essentially that due to the nucleus so that (as $r \to 0$) it will behave as Ze/r. This suggests that we introduce the variable $\chi(r)$ defined through the following equation:

$$\phi - \phi_0 = \frac{Ze}{r}\chi(r) \tag{9.12}$$

so that we have the boundary condition

$$\chi(0) = 1 \tag{9.13}$$

Using (9.11) and (9.12), (9.10) becomes

$$\frac{d^2\chi}{dr^2} = \frac{4}{3\pi\hbar^3}(2me^2)^{3/2}\frac{Z^{1/2}}{r^{1/2}}\chi^{3/2} \tag{9.14}$$

In order to put the above equation in a more convenient form we introduce the dimensionless variable x defined by the following equation:

$$r = \alpha x \tag{9.15}$$

where the parameter α is to be defined later. Substituting in (9.14) we get

$$\frac{d^2\chi(x)}{dx^2} = \left[\alpha^{3/2}\frac{4}{3\pi\hbar^3}(2me^2)^{3/2}Z^{1/2}\right]\frac{1}{x^{1/2}}\chi^{3/2} \tag{9.16}$$

We choose α so that the quantity inside the square brackets is unity:

$$\alpha = \frac{1}{2}\left(\frac{3\pi}{4}\right)^{2/3}\frac{\hbar^2}{me^2}\cdot\frac{1}{Z^{1/3}} \approx \frac{0.885\,34\,a_0}{Z^{1/3}} \tag{9.17}$$

where

$$a_0 = \frac{\hbar^2}{me^2} \approx 0.529 \times 10^{-8}\,\text{cm} \tag{9.18}$$

represents the Bohr radius. Thus (9.16) takes the form

$$\frac{d^2\chi}{dx^2} = \frac{1}{x^{1/2}}\chi^{3/2} \tag{9.19}$$

which is known as the dimensionless Thomas–Fermi equation; from now on, it will be abbreviated as the TF equation. For an isolated atom, the boundary conditions will be

$$\chi = 1 \quad \text{at} \quad x = 0 \quad \text{and} \quad \chi = 0 \quad \text{at} \quad x = \infty \tag{9.20}$$

In Section 9.2.2 we will briefly discuss the solution of the TF equation; however, before we do so we should mention that once $\chi(x)$ is known, $n(r)$ is known by using (9.7) and (9.12):

$$n(r) = \frac{1}{3\pi^2 \hbar^3} \left[2me \frac{Ze}{r} \chi(r) \right]^{3/2} \tag{9.21}$$

or, using (9.15) and (9.17) we get

$$n(x) = \frac{32}{9\pi^3} \frac{Z^2}{a_0^3} \left[\frac{\chi(x)}{x} \right]^{3/2} \tag{9.22}$$

Further, $\chi(x)$ should be such that

$$\int_0^\infty n(r) 4\pi r^2 \, dr = Z \tag{9.23}$$

which using (9.17) and (9.21) gives

$$\int_0^\infty x^{1/2} \chi^{3/2}(x) \, dx = 1 \tag{9.24}$$

Equation (9.24) can serve as a check on any numerical calculation. We should point out that since $\chi(0) = 1$, (9.22) predicts an infinite electron density at the origin; this is one of the basic errors of the TF model – see, e.g., Schwinger (1980). Indeed, for an atom the TF model is correct only in the domain

$$\frac{a_0}{Z} \lesssim r \lesssim a_0$$

where a_0 is the Bohr radius. The condition $a_0/Z \lesssim r$ comes from the fact that at distances smaller than a_0/Z the quasi-classical approximation (which is the basis of the TF model) breaks down. On the other hand, for $r \gtrsim a_0$, the de-Broglie wavelength of the electron becomes of the order of r^2/a_0 so that the quasi-classical approach is again violated (see Landau and Lifshitz 1965 §71, for a nice discussion on this point). We should mention that in a complex atom the majority of the electrons are in the domain $(a_0/Z) \lesssim r \lesssim a_0$, so that the TF model can be used to describe the complex atom system.

9.2.1   *Consideration of a gas of atoms*

The boundary conditions given by (9.20) correspond to an isolated atom. If we consider a gas of atoms then the gas is divided into cells, each cell associated with one atom. For simplicity, the cells are assumed to be spherical with radius r_0 given by

$$\frac{4\pi}{3} r_0{}^3 = \frac{1}{n_a}$$

where n_a ($= N/V$) represents the number of atoms per unit volume. Thus

$$r_0 = \left[\frac{3}{4\pi n_a} \right]^{1/3} \tag{9.25}$$

Now, the electric field at the boundary of the cell should vanish (because each cell is electrically neutral) and therefore at $r = r_0$

$$\frac{d\phi}{dr} = 0 \tag{9.26}$$

Using (9.12) and (9.15) we get

$$\left[\frac{x}{\chi(x)} \frac{d\chi}{dx} \right]_{x = x_0} = 1$$

Thus for a gas of atoms the boundary conditions for the TF equation are

$$\chi(0) = 1 \quad \text{and} \quad \left[\frac{x}{\chi(x)} \frac{d\chi}{dx} \right]_{x = x_0} = 1 \tag{9.27}$$

where

$$x_0 = \frac{r_0}{\alpha} \approx \frac{Z^{1/3}}{0.885\,34\, a_0} \left(\frac{3}{4\pi n_a} \right)^{1/3} \tag{9.28}$$

9.2.2   Solution of the Thomas–Fermi equation

The function $\chi(x)$ may be expanded in a power series of the form

$$\chi(x) = \sum_{k = 0,1,2,\ldots} a_k x^{k/2}$$

$$= 1 + a_1 x^{1/2} + a_2 x + a_3 x^{3/2} + \cdots \tag{9.29}$$

where we have used the fact the since $\chi(0) = 1$, a_0 must be equal to 1. Substituting in (9.19) we get

$$-\tfrac{1}{4} a_1 x^{-3/2} + \tfrac{3}{4} a_3 x^{-1/2} + 2 a_4 + \tfrac{15}{4} a_5 x^{1/2} \cdots$$

$$= \frac{1}{x^{1/2}} [1 + a_1 x^{1/2} + a_2 x + a_3 x^{3/2} + \cdots]^{3/2}$$

Since there is no term proportional to $x^{-3/2}$ on the right hand side, a_1 must vanish and therefore we have

$$\tfrac{3}{4}a_3 x^{-1/2} + 2a_4 + \tfrac{15}{4}a_5 x^{1/2} + 6a_6 x + \tfrac{35}{4}a_7 x^{3/2} + \cdots$$

$$= \frac{1}{x^{1/2}}\left[\, 1 + \tfrac{3}{2}(a_2 x + a_3 x^{3/2} + a_4 x^2 + \cdots) \right.$$

$$+ \tfrac{3}{8}a_2{}^2 x^2\left(1 + \frac{a_3}{a_2}x^{1/2} + \frac{a_4}{a_2}x + \cdots \right)^2$$

$$\left. - \tfrac{1}{16}a_2{}^3 x^3\left(1 + \frac{a_3}{a_2}x^{1/2} + \cdots \right)^3 + \cdots \right]$$

Equating coefficients of equal powers of x we readily get

$$\left.\begin{aligned}
&a_1 = 0 \quad a_3 = \tfrac{4}{3} \quad a_4 = 0 \quad a_5 = \tfrac{2}{5}a_2 \quad a_6 = \tfrac{1}{3}\\
&a_7 = \tfrac{3}{70}a_2 \quad a_8 = \tfrac{2}{15}a_2 \quad a_9 = \tfrac{2}{27} - \tfrac{1}{252}a_2{}^3\\
&a_{10} = \tfrac{1}{175}a_2{}^2 \quad a_{11} = \tfrac{31}{1485}a_2 + \tfrac{1}{1056}a_2{}^4, \text{ etc.}
\end{aligned}\right\} \qquad (9.30)$$

Fig. 9.1. The Thomas–Fermi function $\chi(x)$, as a function of x, for different value of the initial slope. The numbers on the curve represent the value of $-a_2$. The point $(x = x_0)$ at which the tangent to the curve passes through the origin represents the boundary of the atom. When $a_2 = -1.588\,08$, $\chi(x)$ approaches zero asymptotically and the solution corresponds to an isolated atom. The dotted curve represents the qualitative dependence of $\chi(x)$ on x when $|a_2|$ is greater than 1.588 08 and it corresponds to a positive ion. (Adapted from Slater and Krutter (1935).)

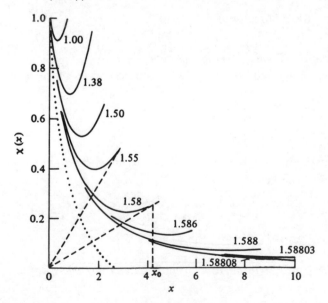

Thus we have

$$\chi(x) = 1 + a_2 x + \tfrac{4}{3}x^{3/2} + \tfrac{2}{5}a_2 x^{5/2} + \tfrac{1}{3}x^3 + \cdots \tag{9.31}$$

It may be noted that if a_2 is known, all the coefficients are determined and since

$$a_2 = \chi'(0) \tag{9.32}$$

the entire solution is determined if $\chi'(0)$ is known; of course, the power series expansion (9.31) is convergent for $x < 1$ and is useful (i.e., rapidly convergent) when $x \ll 1$. For $x \gtrsim 1$ one normally uses a numerical method to solve the TF equation.

Fig. 9.1 gives the numerical[†] solution of (9.19) for various values of the initial slope. When

$$a_2 = \chi'(0) = -1.588\,08$$

$\chi(x)$ approaches zero asymptotically and the solution corresponds to an isolated atom (9.20). When

$$|\chi'(0)| < 1.588\,08 \tag{9.33}$$

$\chi(x)$ decreases to a minimum and then starts increasing. These solutions correspond to atoms of finite size (see Section 2.1) and the point (on the curve) at which the tangent passes through the origin corresponds to

Table 9.1. *Dependence of x_0 on the initially chosen value of a_2*

a_2	x_0
− 1.00	1.19
− 1.38	1.69
− 1.50	2.20
− 1.55	2.80
− 1.58	4.23
− 1.586	5.85
− 1.588	8.59
− 1.588 03	11.3
− 1.588 06	16.0
− 1.588 08	∞

Source: adapted from Slater & Krutter (1935)

[†] Numerical methods for solving the TF equation have been discussed in many papers; see, e.g., Feynman, Metropolis and Teller (1949).

Table 9.2. *The Thomas–Fermi function* [χ(x)] *and its derivative corresponding to an isolated atom*

x	χ(x)	−χ'(x)	x	χ(x)	−χ'(x)
0.0000	0.100 000 00E + 01	0.158 807 10E + 01	1.0500	0.410 645 48E − 00	0.260 674 56E − 00
0.001	0.999 842 53E − 00	0.156 807 26E + 01	1.1000	0.397 925 30E − 00	0.248 278 12E − 00
0.0004	0.999 375 44E − 00	0.154 808 36E + 01	1.1500	0.385 803 79E − 00	0.236 714 47E − 00
0.0009	0.998 606 72E − 00	0.152 811 31E + 01	1.2000	0.374 241 23E − 00	0.225 908 59E − 00
0.0016	0.997 544 36E − 00	0.150 817 01E + 01	1.2500	0.363 201 41E − 00	0.215 794 13E − 00
0.0025	0.996 196 30E − 00	0.148 826 32E + 01	1.3000	0.352 651 28E − 00	0.206 312 18E − 00
0.0036	0.994 570 47E − 00	0.146 840 08E − 01	1.3500	0.342 560 53E − 00	0.197 410 263 − 00
0.0049	0.992 674 76E − 00	0.144 859 12E + 01	1.4000	0.332 901 37E − 00	0.189 041 43E − 00
0.0064	0.990 517 02E − 00	0.142 884 21E + 01	1.5000	0.314 777 46E − 00	0.173 738 80E − 00
0.0081	0.988 105 06E − 00	0.140 916 13E + 01	1.6000	0.298 097 71E − 00	0.160 115 01E − 00
0.0100	0.985 446 61E − 00	0.138 955 61E + 01	1.7000	0.282 706 44E − 00	0.147 933 39E − 00
0.0121	0.982 549 38E − 00	0.137 003 37E + 01	1.8000	0.268 469 51E − 00	0.136 998 44E − 00
0.0144	0.979 421 00E − 00	0.135 060 08E + 01	1.9000	0.255 270 65E − 00	0.127 147 29E − 00
0.0169	0.976 069 01E − 00	0.133 126 42E + 01	2.0000	0.243 008 51E − 00	0.118 243 19E − 00
0.0196	0.972 500 90E − 00	0.131 203 01E + 01	2.1000	0.231 594 32E − 00	0.110 170 54E − 00
0.0225	0.968 724 10E − 00	0.129 290 47E + 01	2.2000	0.220 949 98E − 00	0.102 830 98E − 00
0.0256	0.964 745 90E − 00	0.127 389 38E + 01	2.3000	0.211 006 50E − 00	0.961 403 50E − 01
0.0289	0.960 573 57E − 00	0.125 500 30E + 01	2.4000	0.201 702 70E − 00	0.900 262 76E − 01
0.0324	0.956 214 25E − 00	0.123 623 77E + 01	2.5000	0.192 984 12E − 00	0.844 261 87E − 01
0.0361	0.951 675 00E − 00	0.121 760 29E + 01	2.6000	0.184 802 15E − 00	0.792 857 63E − 01
0.0400	0.946 962 77E − 00	0.119 910 35E + 01	2.7000	0.177 113 23E − 00	0.745 576 47E − 01
0.0441	0.942 084 44E − 00	0.118 074 40E + 01	2.8000	0.169 878 26E − 00	0.702 003 88E − 01
0.0484	0.937 046 76E − 00	0.116 252 90E + 01	2.9000	0.163 062 01E − 00	0.661 775 80E − 01
0.0529	0.931 856 39E − 00	0.114 446 25E + 01	3.0000	0.156 632 67E − 00	0.624 571 31E − 01
0.0576	0.926 519 87E − 00	0.112 654 84E + 01	3.2000	0.144 822 25E − 00	0.558 130 27E − 01
0.0625	0.921 043 65E − 00	0.110 879 05E + 01	3.4000	0.134 247 00E − 00	0.500 771 16E − 01

0.0676	0.915434 06E − 00	0.109119 21E + 01	3.6000	0.124741 04E − 00	0.450976 30E − 01
0.0729	0.909697 31E − 00	0.107375 66E + 01	3.8000	0.116165 70E − 00	0.407527 38E − 01
0.0784	0.903839 50E + 00	0.105648 70E + 01	4.0000	0.108404 26E − 00	0.369437 58E − 01
0.0841	0.897866 63E − 00	0.103938 62E + 01	4.2000	0.101357 87E − 00	0.335900 97E − 01
0.0900	0.891784 56E − 00	0.102245 67E + 01	4.4000	0.949423 09E − 01	0.306254 44E − 01
0.1024	0.879315 73E − 00	0.989121 20E − 00	4.6000	0.890854 40E − 01	0.279948 61E − 01
0.1156	0.866477 61E − 00	0.956497 83E − 00	4.8000	0.837251 63E − 01	0.256525 43E − 01
0.1296	0.853312 90E − 00	0.924600 76E − 00	5.0000	0.788077 79E − 01	0.235600 75E − 01
0.1444	0.839862 38E − 00	0.893441 41E − 00	5.2000	0.742866 47E − 01	0.216850 62E − 01
0.1600	0.826164 92E − 00	0.863028 59E − 00	5.4000	0.701210 97E − 01	0.200000 50E − 01
0.1764	0.812257 47E − 00	0.833368 64E − 00	5.6000	0.662755 27E − 01	0.184816 57E − 01
0.1936	0.798175 11E − 00	0.804465 59E − 00	5.8000	0.627186 65E − 01	0.171098 84E − 01
0.2116	0.783951 04E − 00	0.776321 34E − 00	6.0000	0.594229 49E − 01	0.158675 50E − 01
0.2304	0.769616 64E − 00	0.748935 79E − 00	6.5000	0.521729 37E − 01	0.132356 07E − 01
0.2500	0.755201 47E − 00	0.722306 98E − 00	7.0000	0.460978 19E − 01	0.111425 32E − 01
0.2704	0.740733 34E − 00	0.696431 29E − 00	7.5000	0.409624 66E − 01	0.945826 46E − 02
0.2916	0.726238 34E − 00	0.671303 52E − 00	8.0000	0.365872 55E − 01	0.808860 30E − 02
0.3136	0.711740 89E − 00	0.646917 07E − 00	8.5000	0.328330 89E − 01	0.696416 38E − 02
0.3364	0.697263 79E − 00	0.623264 02E − 00	9.0000	0.295909 35E − 01	0.603307 47E − 02
0.3600	0.682828 26E − 00	0.600335 35E − 00	9.5000	0.267743 90E − 01	0.525603 00E − 02
0.4096	0.654159 31E − 00	0.556609 81E − 00	10.0000	0.243142 93E − 01	0.460288 19E − 02
0.4624	0.625874 54E − 00	0.515649 26E − 00	11.0000	0.202503 65E − 01	0.357981 52E − 02
0.5184	0.598092 91E − 00	0.477351 06E − 00	12.0000	0.170639 22E − 01	0.283053 64E − 02
0.5776	0.570913 67E − 00	0.441604 05E − 00	13.0000	0.145265 18E − 01	0.227052 46E − 02
0.6400	0.544418 17E − 00	0.408291 20E − 00	14.0000	0.124784 06E − 01	0.184450 14E − 02
0.7056	0.518671 63E − 00	0.377291 86E − 00	15.0000	0.108053 59E − 01	0.151532 31E − 02
0.7744	0.493724 84E − 00	0.348483 68E − 00	16.0000	0.942407 89E − 02	0.125743 53E − 02
0.8464	0.469615 81E − 00	0.321744 21E − 00	17.0000	0.827276 39E − 02	0.105288 68E − 02
0.9216	0.446371 26E − 00	0.296952 28E − 00	18.0000	0.730484 59E − 02	0.888831 11E − 03
1.0000	0.424008 05E − 00	0.273989 05E − 00	19.0000	0.648474 64E − 02	0.755921 42E − 03

Table 9.2. (*Contd.*)

x	$\chi(x)$	$-\chi'(x)$	x	$\chi(x)$	$-\chi'(x)$
20.0000	0.578 494 12E − 02	0.647 254 33E − 03	140.0000	0.395 741 39E − 04	0.788 564 47E − 06
21.0000	0.518 389 34E − 02	0.557 661 58E − 03	150.0000	0.326 339 64E − 04	0.609 139 95E − 06
22.0000	0.466 457 58E − 02	0.483 225 74E − 03	160.0000	0.272 310 37E − 04	0.478 074 16E − 06
23.0000	0.421 339 81E − 02	0.420 943 70E − 03	170.0000	0.229 613 51E − 04	0.380 511 34E − 06
24.0000	0.381 941 81E − 02	0.368 489 22E − 03	180.0000	0.195 421 02E − 04	0.306 644 32E − 06
25.0000	0.347 375 44E − 02	0.324 043 00E − 03	190.0000	0.167 712 48E − 04	0.249 929 01E − 06
26.0000	0.316 914 44E − 02	0.286 169 52E − 03	200.0000	0.145 018 03E − 04	0.205 753 23E − 06
27.0000	0.289 960 77E − 02	0.253 726 72E − 03	210.0000	0.126 250 79E − 04	0.170 938 68E − 06
28.0000	0.266 018 79E − 02	0.225 799 01E − 03	220.0000	0.110 595 15E − 04	0.143 199 26E − 06
29.0000	0.244 675 26E − 02	0.201 647 09E − 03	230.0000	0.974 309 01E − 05	0.120 875 22E − 06
30.0000	0.225 583 66E − 02	0.180 670 01E − 03	240.0000	0.862 806 70E − 05	0.102 744 31E − 06
32.0000	0.193 032 55E − 02	0.146 361 06E − 03	250.0000	0.767 729 08E − 05	0.878 946 80E − 07
34.0000	0.166 519 08E − 02	0.119 884 59E − 03	260.0000	0.686 154 84E − 05	0.756 379 15E − 07
36.0000	0.144 695 44E − 02	0.991 771 75E − 04	270.0000	0.615 765 44E − 05	0.654 485 28E − 07
38.0000	0.126 561 39E − 02	0.827 855 36E − 04	280.0000	0.554 704 71E − 05	0.569 213 13E − 07
40.0000	0.111 363 56E − 02	0.696 680 29E − 04	290.0000	0.501 474 64E − 05	0.497 408 61E − 07
42.0000	0.985 269 00E − 03	0.590 661 39E − 04	300.0000	0.454 857 20E − 05	0.436 594 96E − 07
44.0000	0.876 070 64E − 03	0.504 195 38E − 04	320.0000	0.377 646 38E − 05	0.340 493 89E − 07
46.0000	0.782 569 14E − 03	0.433 088 48E − 04	340.0000	0.316 996 37E − 05	0.269 470 14E − 07
48.0000	0.702 024 72E − 03	0.374 164 00E − 04	360.0000	0.268 689 55E − 05	0.216 058 08E − 07
50.0000	0.632 254 78E − 03	0.324 989 02E − 04	380.0000	0.229 735 43E − 05	0.175 262 91E − 07
52.0000	0.571 505 54E − 03	0.283 681 26E − 04	400.0000	0.197 973 26E − 05	0.143 668 23E − 07
54.0000	0.518 356 49E − 03	0.248 770 65E − 04	420.0000	0.171 815 18E − 05	0.118 890 56E − 07
56.0000	0.471 648 54E − 03	0.219 099 04E − 04	440.0000	0.150 076 45E − 05	0.992 371 54E − 08
58.0000	0.430 429 33E − 03	0.193 746 55E − 04	460.0000	0.131 860 86E − 05	0.834 862 85E − 08

60.0000	0.393 911 37E − 03	0.171 977 00E − 04	480.0000	0.116 481 92E − 05	0.707 431 70E − 08
65.0000	0.319 143 29E − 03	0.129 604 11E − 04	500.0000	0.103 407 72E − 05	0.603 436 34E − 08
70.0000	0.262 265 30E − 03	0.995 653 34E − 05	520.0000	0.922 217 65E − 06	0.517 885 89E − 08
75.0000	0.218 210 43E − 03	0.777 797 47E − 05	540.0000	0.825 947 95E − 06	0.446 987 33E − 08
80.0000	0.183 545 76E − 03	0.616 619 55E − 05	560.0000	0.742 641 55E − 06	0.387 827 77E − 08
85.0000	0.155 887 83E − 03	0.495 261 18E − 05	580.0000	0.670 185 90E − 06	0.338 148 50E − 08
90.0000	0.133 545 83E − 03	0.402 447 37E − 05	600.0000	0.606 868 77E − 06	0.296 182 25E − 08
95.0000	0.115 297 15E − 03	0.330 466 01E − 05	650.0000	0.479 984 46E − 06	0.216 546 95E − 08
100.0000	0.100 242 57E − 03	0.273 935 11E − 05	700.0000	0.386 176 52E − 06	0.161 983 22E − 08
105.0000	0.877 104 51E − 04	0.229 030 47E − 05	750.0000	0.315 325 80E − 06	0.123 583 41E − 08
110.0000	0.771 921 84E − 04	0.192 990 23E − 05	800.0000	0.260 813 73E − 06	0.959 243 86E − 09
115.0000	0.682 977 13E − 04	0.163 790 00E − 05	850.0000	0.218 186 65E − 06	0.755 927 24E − 09
120.0000	0.607 244 54E − 04	0.139 925 83E − 05	900.0000	0.184 372 42E − 06	0.603 766 18E − 09
125.0000	0.542 351 97E − 04	0.120 266 54E − 05	950.0000	0.157 205 04E − 06	0.488 057 17E − 09
130.0000	0.486 421 71E − 04	0.103 951 57E − 05	1000.0000	0.135 127 48E − 06	0.398 801 07E − 09

Note: Table adapted from Torrens (1972); the actual calculations are by P.J. Rijnierse—the original paper of Rijnierse was not available to the authors.

$x = x_0$. This follows immediately from (9.27):

$$\left(\frac{d\chi}{dx}\right)_{x=x_0} = \left(\frac{\chi(x)}{x}\right)_{x=x_0} \tag{9.34}$$

The values of x_0 corresponding to typical values of a_2 are given in Table 9.1. Finally, when

$$|\chi'(0)| > 1.588\,08 \tag{9.35}$$

$\chi(x)$ becomes zero at a particular value of x (see the dotted curve in Fig. 9.1). These solutions correspond to positive ions which have been discussed by Brillouin (1934); we will not be concerned here with these solutions.[†]

One of the important advantages of the TF model is the fact that the

Fig. 9.2. The electron distribution function $D(r)$ ($= 4\pi r^2 n(r)$) as a function of r for the mercury atom ($Z = 80$). The solid curve is obtained from the TF model (i.e., using Table 9.1) and the dashed curve represents the actual numerical results. (Adapted from Gombas (1949).)

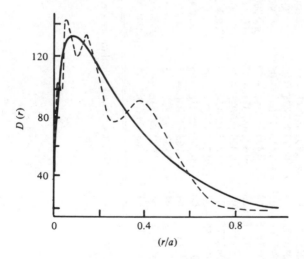

[†] This particular class of solutions (for which $\chi(x_0) = 0$) might be important or relevant also to calculate the ionization of compressed matter (e.g., in inertial confinement fusion problems, astrophysics etc.). Using Gauss' law, the charge inside a sphere of radius r_0 is given by

$$q \equiv eZ^* = \frac{1}{r\pi}\oint \mathbf{E}\cdot d\mathbf{a} = -\frac{1}{4\pi}\frac{\partial\phi}{\partial r}\bigg|_{r=r_0} (4\pi r_0^2) = Ze[\chi(r_0) - r\chi'(r_0)]$$

where we have used (9.12). If we now use the fact that $\chi(x_0) = 0$ we get the ionization of the compressed matter as

$$Z^* = Z[-x_0\chi'(x_0)]$$

(see also Moore (1981)).

curve corresponding to an isolated atom is universal, i.e. for a given value of Z one has to calculate α (see (9.17)) and simply scale the x-axis (using (9.15)) to get $\chi(r)$ as a function of r. Once $\chi(r)$ is known, $n(r)$ can be immediately determined by using (9.21). It is for this reason we have tabulated the numerical solution of the TF equation corresponding to an isolated atom (see Table 9.2). In Fig. 9.2 we have plotted the electron distribution function

$$D(r) = 4\pi r^2 n(r) \tag{9.36}$$

as a function of r for the mercury atom ($Z = 80$); the quantity $D(r)dr$ represents the number of electrons in the spherical shell whose inner and outer radii are r and $r + dr$ respectively. The solid curve is obtained from the TF model (i.e., using Table 9.2) and (9.15), the dashed curve represents the actual numerical results – the peaks approximately corresponding to the Bohr orbits.[†]

> [†] A very interesting application of the TF model is to (approximately) predict that the electrons in the p shell appear first in $Z = 5$ (boron), the d electrons for $Z = 21$ (scandium) and the f electrons for $Z = 58$ (cerium). We follow the analysis given by Landau and Lifshitz (1965) §73. We start with the fact that the electron (with orbital angular momentum l) in an atom moves with an 'effective potential energy' given by $V_l^{\mathrm{eff}} = -e\phi(r) + (l + \tfrac{1}{2})^2 \hbar^2/2mr^2$ where the second term is due to the fact that the centripetal force $m\omega^2 r \,(= L^2/mr^3, L = m\omega r^2$ being the angular momentum) is supplied by the potential energy to keep the particle in its orbit; this centripetal force adds to λ^2 the potential energy, $(l + \tfrac{1}{2})^2\hbar^2$, an additional term $L^2/2mr^3$. We have replaced L^2 by $(l + \tfrac{1}{2})\hbar^2$ to be consistent with the quasi-classical approximation. If $V_l^{\mathrm{eff}}(r) > 0$ for all r then there can be no electrons with that particular value of l because the total energy of the electron has to be negative. Now, for a given value of l, if we vary Z then we find that $V_l^{\mathrm{eff}}(r) > 0$ for all r when Z is sufficiently small. As we increase the Z value, we find that (for a given value of l) the r-axis is a tangent to the $V_l^{\mathrm{eff}}(r)$ curve and for larger Z values there is a region where $V_l^{\mathrm{eff}}(r) < 0$. The conditions for the r-axis to be a tangent to the curve are
>
> $$V_l^{\mathrm{eff}}(r) = -e\phi + (l + \tfrac{1}{2})^2\hbar^2/2mr^2 = 0$$
>
> and
>
> $$V_l^{\mathrm{eff}}(r) = -e\phi(r)' - (l + \tfrac{1}{2})^2\hbar^2/mr^3 = 0$$
>
> If we now use (9.12) and the fact that $\phi_0 = 0$ we would get
>
> $$Z^{3/2}\chi(x)/x = (\tfrac{4}{3}\pi)^{2/3}(l + \tfrac{1}{2})^2/x^2$$
>
> and
>
> $$Z^{2/3}[x\chi' - \chi]/x = -2(\tfrac{4}{3}\pi)^{2/3}(l + \tfrac{1}{2})^2 x^2$$
>
> Dividing the first equation by the second we get $x\chi'(x)/\chi(x) = -1$. We may now use Table 9.2 to plot $x\chi'(x)/\chi$ as a function of x and we find that it becomes -1 when $x \approx 2.1$. On substitution and using the value of $\chi(x)$ for $x \approx 2.1$ from Table 9.2 we get
>
> $$Z \approx 0.155 \,(2l + 1)^3$$
>
> which gives $Z \approx 4.2$, 19.4 and 53.2 corresponding to $l = 1$, 2 and 3 respectively. The correct values are 5, 21 and 58.

For a gas of atoms the procedure is rather involved: knowing the values of n_a and Z we first calculate x_0; however, for a numerical solution of the TF equation it is necessary to know the initial slope (vis. a_2). One can make calculations for a set of values of a_2 and each time determine x_0 and then by using an interpolation procedure determine the value of a_2 which will correspond to the given value of x_0. Obviously, it will be very convenient to have a relation giving the dependence of a_2 on x_0. Further, as will be seen later, the calculations of total energy, pressure etc. become very simple if χ_0 [$= \chi(x_0)$] is also known (see Sections 2.4 and 2.5). March (1957) has derived such relations which give accurately the values of a_2 and χ_0 (within about 1%) for given value of x_0:

$$B(\infty) - B(x_0) = a_2(x_0) - a_2(\infty) = \left[\sum_{n=2}^{7} A_n x_0{}^n + A_\lambda x_0{}^\lambda \right]^{-1} \quad (9.37)$$

and

$$\chi_0 = \left[\sum_{n=2}^{6} C_n x_0{}^{n/2} \right]^{-1} \quad (9.38)$$

where the coefficients A_n and C_n are tabulated in Table 9.3 and

$$B = -a_2 \quad (9.39)$$

representing the magnitude of the slope at $x = 0$; the 'infinity' in parenthesis implies that it corresponds to an isolated atom ($x_0 = \infty$). Thus[†]

$$B(\infty) = -a_2(\infty) = 1.588\,07 \quad (9.40)$$

Table 9.3. *Values of A_n and C_n appearing in the expressions for $a_2(x_0)$ and χ_0 (see (9.37) and (9.38) (adapted from March (1957))*

$A_2 = 4.8075 \times 10^{-1}$	$C_2 = 4.8075 \times 10^{-1}$
$A_3 = 4.3462 \times 10^{-1}$	$C_3 = 0$
$A_4 = 6.9203 \times 10^{-2}$	$C_4 = 6.934 \times 10^{-2}$
$A_5 = 5.9472 \times 10^{-2}$	$C_5 = 9.700 \times 10^{-3}$
$A_6 = -4.9688 \times 10^{-3}$	$C_6 = 3.3704 \times 10^{-3}$
$A_7 = 4.3386 \times 10^{-4}$	
$A_\lambda = 1.5311 \times 10^{-6}$	
$\lambda = \frac{1}{2}[\sqrt{73} + 7] = 7.7720$	

[†] Unfortunately, there is a discrepancy in the value of $B(\infty)$ given by various authors: Slater and Krutter (1935) give 1.588 08; according to Feynman *et al.* (1949) the value should lie between 1.588 74 and 1.588 84; Gombas (1949) gives 1.588 046 4 and the more recent value given by Rijnierse is 1.588 071 02 (quoted in Torrens 1972).

For a given value of x_0, a_2 can be determined by using (9.37) and (9.40) and Table 9.3, and knowing the initial slope one can numerically solve the TF equation.

We end this section by mentioning that there have been numerous attempts to obtain accurate analytical expressions describing the solution of the TF equation corresponding to an isolated atom (i.e., with boundary condition given by (9.20)). A useful expansion in inverse powers of x is given by

$$\chi^\infty(x) = \frac{144}{x^3} \sum_{k=0,1,2,\dots} c_k (Fx^\lambda)^k$$

$$= \frac{144}{x^3}\left[1 - \frac{13.27}{x^{0.772}} + \cdots \right] \tag{9.41}$$

where

$$\left. \begin{array}{l} F = 13.270\,973\,848 \\ \lambda = -0.772\,001\,872\,6 \end{array} \right\} \tag{9.42}$$

and the coefficients c_k are tabulated in Table 9.4; the superscript ∞ once again signifies that we are considering an isolated atom. The series in (9.41) converges rapidly for $x \gtrsim 10$. Thus when $x \ll 1$ we may use (9.31) (with $a_2 = -1.588\,08$) and for $x \gtrsim 10$ we may use (9.41) (with c_ks given by Table 9.4). For

Table 9.4. *Coefficients* c_k *in the asymptotic expansion of* $\chi^\infty(x)$ *(see (9.41)).*

k	c_k	k	c_k
0	0.100 000 000 000 000 E + 01	16	0.243 263 216 206 744 E − 06
1	− 0.100 000 000 000 000 E + 01	17	− 0.691 031 290 629 697 E − 07
2	0.625 697 497 782 349 E − 00	18	0.202 144 402 461 023 E − 07
3	− 0.313 386 115 073 309 E − 00	19	− 0.586 883 726 659 920 E − 08
4	0.137 391 276 719 371 E − 00	20	0.169 228 967 471 091 E − 08
5	− 0.550 834 346 641 491 E − 01	21	− 0.484 947 296 184 278 E − 09
6	0.207 072 584 991 917 E − 01	22	0.138 179 140 746 420 E − 09
7	− 0.741 452 947 849 571 E − 02	23	− 0.391 670 654 111 897 E − 10
8	0.255 553 116 794 870 E − 02	24	0.110 486 378 825 038 E − 10
9	− 0.854 165 377 806 924 E − 03	25	− 0.310 287 206 325 100 E − 11
10	0.278 373 839 349 473 E − 03	26	0.867 815 397 770 538 E − 12
11	− 0.888 226 094 136 591 E − 04	27	− 0.241 782 703 567 386 E − 12
12	0.278 359 915 878 395 E − 04	28	0.671 228 958 979 403 E − 13
13	− 0.858 949 083 219 027 E − 05	29	− 0.185 723 018 657 859 E − 13
14	0.261 505 871 330 370 E − 05	30	0.512 274 393 418 862 E − 14
15	− 0.786 798 261 946 265 E − 06		

Notes: Table adapted from Torrens (1972); the actual calculations are of P.J. Rijnierse – the original paper of Rijnierse was not available to authors.

intermediate values of x, it may be convenient to use the following formulae:

$$\chi^\infty(x) \approx 0.35\,e^{-0.3x} + 0.55\,e^{-1.2x} + 0.10\,e^{-6.0x} \tag{9.43}$$

and

$$\chi^\infty(x) \approx \left[1 + \left(\frac{x^3}{144}\right)^{\lambda/3}\right] \quad \text{with } \lambda = 0.772,\ x \gtrsim 7 \tag{9.44}$$

Equation (9.43) is due to Moliere (1947) and it gives an accurate description in the region $0 < x \lesssim 7$; (9.44) is due to Sommerfeld (1932) which gives an accurate description in the region $x \gtrsim 7$. There are many other formulae which are discussed in the book by Torrens (1972).

9.2.3 Derivation of the Thomas–Fermi equation using variational principle

Equation (9.7) can also be derived using variational method. Such a derivation is of considerable interest since it can be easily extended to incorporate the effects due to exchange interactions (see Section 9.3). We first calculate the total energy associated with the electrons.

The kinetic energy associated with the electrons in the volume element $d\tau$ is given by (see (6.28))

$$\frac{3}{5}\left[(3\pi^2)^{2/3}\frac{\hbar^2}{2m}n^{2/3}\right]n\,d\tau$$

Thus the total kinetic energy will be

$$E_{\text{kin}} = \frac{3\hbar^2}{10m}(3\pi^2)^{2/3}\int n^{5/3}\,d\tau \tag{9.45}$$

The potential energy will be given by

$$E_{\text{pot}} = -\int \frac{Ze^2 n(\mathbf{r})}{r}\,d\tau + \frac{1}{2}\int\int \frac{n(\mathbf{r}_1)n(\mathbf{r}_2)e^2}{|\mathbf{r}_1 - \mathbf{r}_2|}\,d\tau_1 d\tau_2 \tag{9.46}$$

where the first term (on the right hand side) is due to the interaction of electrons with the nucleus (which is assumed to be at the origin) and the second term is due to electron–electron interactions. Thus the total energy will be given by

$$E = c_k\int n^{5/3}\,d\tau - Ze^2\int \frac{n(\mathbf{r})}{r}\,d\tau + \frac{1}{2}\int\int \frac{n(\mathbf{r}_1)n(\mathbf{r}_2)e^2}{|\mathbf{r}_1 - \mathbf{r}_2|}\,d\tau_1 d\tau_2 \tag{9.47}$$

where

$$c_k = \frac{3\hbar^2}{10m}(3\pi^2)^{2/3} \tag{9.48}$$

Now, we require E to be minimum with respect to variations of the electron density n with the constraint that the total number of electrons given by

$$Z = \int n \, d\tau \tag{9.49}$$

remains constant. Thus we have

$$\delta(E - \mu Z) = 0 \tag{9.50}$$

where μ is the Lagrange multiplier. Now,

$$\delta E = \int \left[\tfrac{5}{3} c_k n^{2/3} - \frac{Ze^2}{r} \right] \delta n \, d\tau$$

$$+ \frac{e^2}{2} \int \int \frac{n(\mathbf{r}_1)\delta n(\mathbf{r}_2)}{|\mathbf{r}_1 - \mathbf{r}_2|} d\tau_1 d\tau_2 + \frac{e^2}{2} \int \int \frac{\delta n(\mathbf{r}_1) n(\mathbf{r}_2)}{|\mathbf{r}_1 - \mathbf{r}_2|} d\tau_1 d\tau_2$$

$$= \int \left[\tfrac{5}{3} c_k n^{2/3} + V(r) \right] \delta n \, d\tau \tag{9.51}$$

where

$$V(r) = -e\phi(r) = -\frac{Ze^2}{r} + \int \frac{n(\mathbf{r}_1)e^2}{|\mathbf{r} - \mathbf{r}_1|} d\tau_1 \tag{9.52}$$

represents the potential energy function. Now, (9.47), (9.49) and (9.50) give

$$0 = \delta(E - \mu Z) = \delta E - \mu \int \delta n \, d\tau$$

$$= \int \left[\tfrac{5}{3} c_k n^{2/3} - e\phi(r) - \mu \right] \delta n \, d\tau \tag{9.53}$$

Since δn is arbitrary, we have

$$\tfrac{5}{3} c_k n^{2/3} - e\phi - \mu = 0 \tag{9.54}$$

Substituting for c_k from (9.48) we get

$$n(r) = \frac{[2m(e\phi + \mu)]^{3/2}}{3\pi^2 \hbar^3} \tag{9.55}$$

or

$$n(r) = \frac{(2me)^{3/2}}{3\pi^2 \hbar^3} (\phi - \phi_0)^{3/2} \tag{9.56}$$

where $\phi_0 = -\mu/e$. The above equation is identical to (9.7).

9.2.4   *The kinetic and potential energies of an atom*

In this section we will use the TF model to calculate the kinetic and potential energies of an atom and will later compare the results for the total binding energy of an isolated atom with experimental data. This will enable us to see the errors involved in using the simple TF model.

We first calculate the kinetic energy associated with the electrons. This is given by (9.45)

$$E_{kin} = \frac{3\hbar^2}{10m}(3\pi^2)^{2/3} \int n^{5/3}(r)4\pi r^2 \, dr$$

$$= \frac{3}{5}\frac{e^2}{a_0}\frac{(2Z)^{7/3}}{(3\pi)^{2/3}} I \tag{9.57}$$

where

$$I = \int \frac{\chi^{5/2}(x)}{x^{1/2}} dx \tag{9.58}$$

and we have used (9.15), (9.17) and (9.22). We next evaluate the integral denoted by I.

Evaluation of the integral

In order to evaluate the integral, we introduce the variable

$$\psi(x) = \frac{\chi(x)}{x} \tag{9.59}$$

Thus $\psi(x)$ (which is proportional to $(\phi - \phi_0)$ – see (9.12)) will satisfy the equation (cf. (9.10))

$$\frac{d}{dx}\left[x^2\frac{d\psi}{dx}\right] = x^2\psi^{3/2} \tag{9.60}$$

Thus

$$I = \int \frac{\chi^{5/2}dx}{x^{1/2}} = \int x^2\psi^{5/2}dx$$

$$= \int \left[\frac{d}{dx}\left(x^2\frac{d\psi}{dx}\right)\right]\psi dx$$

$$= x^2\psi\frac{d\psi}{dx} - \int \frac{1}{x^2}\left[x^2\frac{d\psi}{dx}\right]^2 dx$$

where we have used (9.60) and in the last step we have carried out integration by parts. Integrating again by parts and then substituting for

$$\frac{d}{dx}\left(x^2\frac{d\psi}{dx}\right)$$

from (9.60), we find

$$I = x^2\psi\frac{d\psi}{dx} + \frac{1}{x}\left(x^2\frac{d\psi}{dx}\right)^2 - \int\frac{1}{x}2\left(x^2\frac{d\psi}{dx}\right)x^2\psi^{3/2}dx$$

$$= x^2\psi\frac{d\psi}{dx} + x^3\left(\frac{d\psi}{dx}\right)^2 - 2\int x^3\psi^{3/2}\frac{d\psi}{dx}dx \qquad (9.61)$$

Now

$$\int x^3\psi^{3/2}\frac{d\psi}{dx}dx = x^3(\tfrac{2}{5}\psi^{5/2}) - \frac{6}{5}\int x^2\psi^{5/2}dx$$

$$= \frac{2x^3}{5}\psi^{5/2} - \tfrac{6}{5}I$$

Substituting in (9.61) we get

$$I = \int_a^b\frac{\chi^{5/2}}{x^{1/2}}dx = -\frac{5}{7}\left[x^2\psi\frac{d\psi}{dx} + x^3\left(\frac{d\psi}{dx}\right)^2 - \tfrac{4}{5}x^3\psi^{5/2}\right]_a^b \qquad (9.62)$$

We set $a = 0$ and $b = x_0$ (the boundary of the atom). At the upper limit $d\psi/dx$ vanishes because of (9.26) and the fact that ψ is proportional to $(\phi - \phi_0)$. Thus

$$\left[x^2\psi\frac{d\psi}{dx} + x^3\left(\frac{d\psi}{dx}\right)^2 - \tfrac{4}{5}x^3\psi^{5/2}\right]_{x=x_0} = -\tfrac{4}{5}x_0^{1/2}\chi_0^{5/2}$$

where

$$\chi_0 \equiv \chi(x_0)$$

Now as $x \to 0$, χ varies at $1 + a_2x + a_3x^{3/2} + \cdots$ (see (9.31)) and therefore

$$\psi \approx \frac{1}{x} + a_2 + a_3x^{1/2} + \cdots \qquad (9.63)$$

and

$$\psi' \approx -\frac{1}{x^2}[1 - \tfrac{1}{2}a_3x^{3/2} + \cdots] \qquad (9.64)$$

Carrying out simple algebra we get

$$\int_0^{x_0}\frac{\chi^{5/2}(x)}{x^{1/2}}dx = -\tfrac{5}{7}[a_2 - \tfrac{4}{5}x_0^{1/2}\chi_0^{5/2}] \qquad (9.65)$$

where we must remember that a_2 is a negative quantity (see Table 9.1). Returning to (9.57), we then have

$$E_{kin} = \frac{3}{7}\frac{e^2}{a_0}\frac{(2Z)^{7/3}}{(3\pi)^{2/3}}[\beta + \tfrac{4}{5}x_0^{1/2}\chi_0^{5/2}]$$

$$= 0.4841\,Z^{7/3}\left(\frac{e^2}{a_0}\right)[\beta + \tfrac{4}{5}x_0^{1/2}\chi_0^{5/2}] \qquad (9.66)$$

where

$$B = -a_2 = -\frac{d\chi}{dx}\bigg|_{x=0} \tag{9.67}$$

Now, the potential energy is given by (see, e.g., Latter 1955):

$$E_{pot} = E_{pot}{}^{ee} + E_{pot}{}^{en} \tag{9.68}$$

where

$$E_{pot}{}^{ee} = \frac{1}{2}\int_0^{r_0}[-en(r)]\phi_e(r)4\pi r^2\,dr \tag{9.69}$$

represents the potential energy contribution due to electron–electron interactions ($\phi_e(r)$ being the potential arising from the electrons) and

$$E_{pot}{}^{en} = \int_0^{r_0}[-en(r)]\phi_n(r)4\pi r^2\,dr \tag{9.70}$$

represents the potential energy contribution due to electron–nucleus interactions ($\phi_n(r)(=Ze/r)$ is the potential arising from the nucleus). The quantity $[-en(r)]$ represents the charge density due to electrons and the factor $\frac{1}{2}$ outside the integral in (9.69) is due to the fact that the electron–electron interaction has been counted twice. We first calculate $E_{pot}{}^{en}$. Since $\phi_n(r) = Ze/r$, we have

$$E_{pot}{}^{en} = -4\pi ea^3\int_0^{x_0}dx\,x^2 n(x)\frac{Ze}{\alpha x}$$

$$= -\frac{4\pi Ze^2}{\alpha}\left[\frac{1}{8}\left(\frac{3\pi}{4}\right)^2\frac{a_0^3}{Z}\right]\left[\frac{32}{9\pi^3}\frac{Z^2}{a_0^3}\right]\int_0^{x_0}\frac{\chi^{3/2}(x)}{x^{1/2}}\,dx$$

where we have used (9.15), (9.17) and (9.22).

On simplification we get

$$E_{pot}{}^{en} = -\frac{Z^2e^2}{\alpha}\int_0^{x_0}\frac{\chi^{3/2}}{x^{1/2}}\,dx$$

Now, using the TF equation (9.19) we can write

$$\int_0^{x_0}\frac{\chi^{3/2}}{x^{1/2}}\,dx = \int\chi''(x)dx = [\chi'(x)]_0^{x_0} = \chi'(x_0) - \chi'(0)$$

$$= \frac{\chi(x_0)}{x_0} + B \tag{9.71}$$

where we have used (9.27) and (9.67). Thus

$$E_{pot}{}^{en} = -\frac{Z^2e^2}{\alpha}\left[B + \frac{\chi_0}{x_0}\right] \tag{9.72}$$

We next evaluate E_{pot}^{ee}. Since

$$\phi(r) = \phi_e(r) + \phi_n(r) = \phi_e(r) + \frac{Ze}{r}$$

we have

$$E_{pot}^{ee} = -\frac{e}{2}\int_0^{r_0} n(r)\left[\phi(r) - \frac{Ze}{r}\right]4\pi r^2 dr$$

$$= -2\pi e a^3 \int_0^{x_0} n(x)\left[\frac{Ze}{\alpha}\frac{\chi(x)}{x} + \phi_0 - \frac{Ze}{\alpha x}\right]x^2 dx$$

where we have used (9.12) and (9.15). Substituting the value of $n(x)$ from (9.22) we get

$$E_{pot}^{ee} = -\tfrac{1}{2}Ze\left[\frac{Ze}{\alpha}\int_0^{x_0}\frac{\chi^{5/2}}{x^{1/2}}dx + \phi_0\int_0^{x_0}x^{1/2}\chi^{3/2}dx\right.$$

$$\left. -\frac{Ze}{\alpha}\int_0^{x_0}\frac{\chi^{3/2}}{x^{1/2}}dx\right]$$

Since $\chi^{3/2} = x^{1/2}\chi''$ (see (9.19)) we have

$$\int_0^{x_0}x^{1/2}\chi^{3/2}(x)dx = \int_0^{x_0}x\chi'' dx = \int_0^{x_0}\frac{d}{dx}[x\chi'(x) - \chi(x)]dx$$

$$= [x\chi'(x) - \chi(x)]_0^{x_0}$$

$$= \chi(0) = 1 \tag{9.73}$$

where we have used the fact that the quantity $[x\chi'(x) - \chi(x)]$ vanishes at $x = x_0$ because of the second boundary condition in (9.27). Now, the potential at the boundary should vanish, i.e.,

$$\phi(r_0) = 0 \tag{9.74}$$

and therefore (9.12) gives us

$$-\phi_0 = \frac{Ze}{r_0}\chi(x_0) = \frac{Ze}{\alpha x_0}\chi_0 \tag{9.75}$$

(see also the first footnote on p. 106). We substitute the above relations in the expression for E_{pot}^{ee} and also use (9.65) and (9.71) to get

$$E_{pot}^{ee} = -\tfrac{1}{2}Ze\left[\frac{Ze}{\alpha}\tfrac{5}{7}(B + \tfrac{4}{5}x_0^{1/2}\chi_0^{5/2}) - \frac{Ze}{\alpha}\frac{\chi_0}{x_0} - \frac{Ze}{\alpha}\left(\frac{\chi_0}{x_0} + B\right)\right]$$

$$= -\frac{Z^2 e^2}{\alpha}\left[-\tfrac{1}{7}B + \tfrac{2}{7}x_0^{1/2}\chi_0^{5/2} - \frac{\chi_0}{x_0}\right] \tag{9.76}$$

Substituting for E_{pot}^{en} and E_{pot}^{ee} from (9.72) and (9.76) in (9.68) we get

$$E_{pot} = -\frac{6}{7}\frac{Z^2 e^2}{\alpha}[B + \tfrac{1}{3}x_0^{1/2}\chi_0^{5/2}]$$

$$= -\frac{6}{7}\frac{e^2}{a_0}\frac{(2Z)^{7/3}}{(3\pi)^{2/3}}[B + \tfrac{1}{3}x_0^{1/2}\chi_0^{5/2}] \tag{9.77}$$

For an isolated atom, we also have the relation

$$(E_{pot}^{en})^\infty = -7(E_{pot}^{ee})^\infty \tag{9.78}$$

where the superscript ∞ implies that we are referring to an isolated atom.

Fig. 9.3. The variation of the total energy (as given by the TF model) with scaled volume (ZV) at various values of $kT/Z^{4/3}$. The limiting curve (shown as the dashed curve) corresponds to $T = 0$. (Adapted from Latter (1955).)

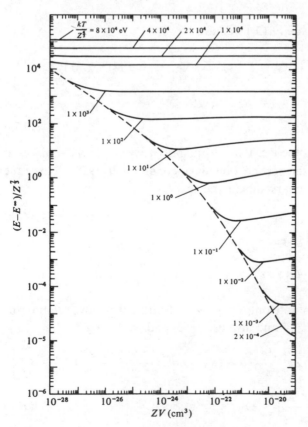

Now, using (9.66) and (9.77), the total energy will be given by

$$E = E_{kin} + E_{pot} = -\frac{e^2}{a_0}\frac{(2Z)^{7/3}}{(3\pi)^{2/3}}\left[\tfrac{3}{7}B - \tfrac{2}{35}x_0^{1/2}\chi_0^{5/2}\right] \tag{9.79}$$

For an isolated atom $x_0 = \infty$ and $\chi_0 = 0$ and we have

$$E^\infty = -\frac{e^2}{a_0}\frac{(2Z)^{7/3}}{(3\pi)^{2/3}}\tfrac{3}{7}B^\infty \tag{9.80}$$

We may therefore rewrite (9.79) as

$$E(x_0) - E(\infty) = \frac{e^2}{a_0}\frac{(2Z)^{7/3}}{(3\pi)^{2/3}}\left[\tfrac{3}{7}(B^\infty - B) + \tfrac{2}{35}x_0^{1/2}\chi_0^{1/2}\right] \tag{9.81}$$

In Fig. 9.3 we have plotted $[E(x_0) - E(\infty)]/Z^{7/3}$ as a function of ZV at different temperatures. The parameter ZV depends only on x_0 as can be seen from the following:

$$ZV = Z\frac{4\pi}{3}r_0^3 = Z\frac{4\pi}{3}\alpha^3 x_0^3$$

$$= \frac{3\pi^3}{32}a_0^3 x_0^3 \tag{9.82}$$

We must remember that (9.81) corresponds to $T = 0$ and it therefore corresponds to the dashed curve in Fig. 9.3. (The formula for $T > 0$ will be discussed in Section 9.5.) We may also use (9.37)–(9.40) and the numbers given in Table 9.3 to calculate $E(x_0)$ and if we are content with the approximations involved in (9.37) and (9.38) then it is not necessary to solve the TF equation to evaluate $E(x_0)$.

We next evaluate the total energy for an isolated atom and compare with experimental data; this will enable us to appreciate the errors involved in the TF model. We first note that for an isolated atom (for which $x_0 = \infty$ and $\chi_0 = 0$) we have[†]

$$E_{kin}(\infty) = -\tfrac{1}{2}E_{pot}(\infty) \approx 0.7687\, Z^{7/3}\left(\frac{e^2}{a_0}\right) \tag{9.83}$$

where

$$\frac{e^2}{a_0} \approx 4.36 \times 10^{-11}\,\text{ergs} \approx 27.2\,\text{eV} \tag{9.84}$$

[†] The relation $E_{kin} = -\tfrac{1}{2}E_{pot}$ (for an isolated atom) also follows from the virial theorem (see Section 9.4).

The total energy will therefore be given by

$$E(\infty) = E_{kin}(\infty) + E_{pot}(\infty) = -E_{kin}(\infty) \approx -0.7687\,Z^{7/3}\left(\frac{e^2}{a_0}\right)$$

$$(9.85)$$

The quantity $-E$ represents the total binding energy of the atom (i.e., the energy required to remove all the electrons to infinity). As an example, we consider the Hg atom ($Z = 80$). The experimental value for the total binding energy is

$$-E_{expt} = 1.813 \times 10^4\left(\frac{e^2}{a_0}\right) \qquad (9.86)$$

whereas the Thomas–Fermi value (9.85) is given by

$$-E_{TF} = 2.120 \times 10^4\left(\frac{e^2}{a_0}\right) \qquad (9.87)$$

which differs by about 17%. For lower Z values, the error gets bigger; for $Z = 12$, the error is 27%. As stated earlier, the TF model gets better with increase in the Z value.

Schwinger (1980) has given a correction factor to the TF expression:

$$-E = 0.7687\,Z^{7/3}F\left[\frac{e^2}{a_0}\right] \qquad (9.88)$$

where the correction factor F is given by

$$F = 1 - 0.6504\,Z^{-1/3} + 0.346\,Z^{-2/3} \qquad (9.89)$$

Using (9.88) and (9.89), the experimental results for the total binding energy, for $Z = 26$ down to $Z = 6$, are reproduced within 1%; for higher Z values, the error is less. For example, for $Z = 80$ one obtains

$$-E \approx 1.839 \times 10^4\left(\frac{e^2}{a_0}\right) \qquad (9.90)$$

which may be compared with (9.86). We may mention here that the term proportional to $Z^{-1/3}$ in (9.89) arises from compensating for the error of the TF model in giving the electrons an infinite density at the nucleus. The term proportional to $Z^{-2/3}$ takes into account exchange effects.

9.2.5   Calculation of pressure

According to the laws of thermodynamics the pressure is given by (see, e.g., Pathria 1972, p. 16):

$$P = -\left(\frac{\partial E}{\partial V}\right)_{N,S} \tag{9.91}$$

where E is the energy, V the volume, N represents the number of particles and S the entropy. In our case V will represent the atomic volume ($= (4\pi/3)r_0^3$). In Section 9.3.1 we will use (9.91) to derive an expression for the pressure when the effects of exchange are also taken into account and the formula derived in this section will be a special case of that. Here we will make use of kinetic theory considerations according to which the pressure is given by

$$P = \tfrac{2}{3}n(x_0)\bar{E}_{kin} \tag{9.92}$$

where $n(x_0)$ is the number of electrons per unit volume at the surface of the atomic boundary and $\bar{E}_{kin}$ is their mean kinetic energy. Now, at $x = x_0$, there are no Coulomb forces and therefore one can use the equation of state for the Fermi–Dirac gas (without interactions); thus, we have (see (6.28)):

$$\bar{E}_{kin} = \tfrac{3}{5}(3\pi^2)^{2/3}\frac{\hbar^2}{2m}[n(x_0)]^{2/3} \tag{9.93}$$

Thus

$$P = (3\pi^2)^{2/3}\frac{\hbar^2}{5m}[n(x_0)]^{5/3} \tag{9.94}$$

which is consistent with (6.29). Substituting for $n(x_0)$ from (9.22) we obtain

$$\frac{P}{Z^{10/3}} = \frac{128}{45\pi^3}\left(\frac{4}{3\pi}\right)^{2/3}\frac{e^2}{a_0^4}\left[\frac{\chi_0}{x_0}\right]^{5/2}$$

$$\approx 1.524 \times 10^{13}\left[\frac{\chi_0}{x_0}\right]^{5/2} \text{ dyne cm}^{-2} \tag{9.95}$$

For an isolated atom $x_0 = \infty$ and $\chi_0 = 0$ so that $P = 0$ (obviously). In general, for a given gas of atoms, we have to first calculate x_0 (9.28) and then use numerical methods to determine χ_0. In Fig. 9.4 we have plotted $P/Z^{10/3}$ as a function of ZV at different temperatures; the parameter ZV depends only on x_0 (9.82). We must remember that (9.95) corresponds to $T = 0$ and it therefore corresponds to the dashed curve in Fig. 9.4. (The formula for $T > 0$ will be discussed in Section 9.5.) Once again, we may also use (9.37)–(9.40) (and the values given in Table 9.3) to calculate P and, if we are content with the approximations involved, then it is not necessary to solve the TF equation. As an example, for $x_0 = 10$, $\chi_0 \approx 0.0550$ (using (9.38) and Table 9.3) and $P/Z^{10/3} \approx 3.42 \times 10^7$ dynes cm^{-2} and $ZV \approx 4.31 \times 10^{-22}$ cm^3 (see the curve corresponding to $T = 0$ in Fig. 9.4).

Regarding the validity of the equation of state derived from the TF model (viz. (9.95)) March (1957) writes '... no claim is made regarding the validity of such equations of state over a wide range of pressures;... it is only when dealing with pressures of the order of 10^7 atmospheres that the results will be at all generally applicable, and even then the effect of exchange should be considered' – we will consider the effect of exchange in Sections 9.3 and 9.6.

We may note here that at very high pressures, the results should be the same as for a free electron gas with electron density equal to Z/V where V $[=(4\pi/3)\alpha^3 x_0^3]$ represents the volume associated with an atom. In this approximation, the pressure will be given by (see (6.29))

$$P = (3\pi^2)^{2/3} \frac{\hbar^2}{5m} \left[\frac{Z}{\frac{4}{3}\pi\alpha^3 x_0^3} \right]^{5/3}$$

$$= \frac{128}{15\pi^3} \left(\frac{4}{3\pi} \right)^{2/3} \left(\frac{e^2}{a_0^4} \right) Z^{10/3} \frac{1}{x_0^5} \tag{9.96}$$

Fig. 9.4. The variation of pressure (as given by the TF model) with scaled volume (ZV) at various temperatures. The limiting curve (shown as the dashed curve) corresponds to $T = 0$. (Adapted from Latter (1955).)

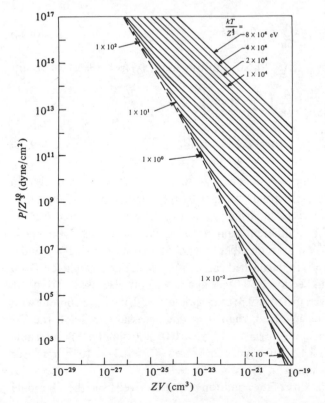

Comparing with (9.95) we get

$$\chi_0 = \frac{3^{2/3}}{x_0} \approx \frac{1}{0.480\,75\,x_0} \tag{9.97}$$

Indeed, starting from the TF equation March (1957) has shown that

$$\chi_0 \approx \frac{3^{2/3}}{x_0}\left[1 - \frac{3^{1/3}}{10}x_0 + \cdots\right] \tag{9.98}$$

which leads to the equation of state

$$PV = \frac{\hbar^2}{5m}(3\pi^2)^{2/3}\frac{Z^{5/3}}{V^{2/3}}\left[1 - \frac{1}{2\pi a_0}(4ZV)^{1/3} + \cdots\right] \tag{9.99}$$

When x_0 is small, (9.98) can be written in the form

$$\chi_0 \approx \frac{1}{0.480\,75\,x_0[1 + 0.144\,22\,x_0 + \cdots]} \tag{9.100}$$

which is consistent with (9.38) with C_ns given by Table 9.3. Indeed C_2, C_3 and C_4 were chosen in such a way that one obtains the correct behaviour for small values of x_0. For large values of x_0 (i.e., low pressures) Gilvarry (1954) has derived the following expression

$$\chi_0 \approx \frac{16(3 + 2\lambda)}{x_0{}^3} \tag{9.101}$$

with

$$\lambda = \tfrac{1}{2}(\sqrt{73} + 7) = 7.7720 \tag{9.102}$$

Indeed, for large values of x_0, (9.38) becomes (9.101).

It is of interest to mention that when all forces are derived from Coulomb's law, the kinetic and potential energies are related through the following equation

$$2E_{\text{kin}} + E_{\text{pot}} = 3PV \tag{9.103}$$

Equation (9.103) is known as the virial theorem and will be derived in Section 9.4. If we substitute for E_{kin} and E_{pot} from (9.66) and (9.77) we would get

$$PV = \frac{2}{15}\frac{e^2}{a_0}\frac{(2Z)^{7/3}}{(3\pi)^{2/3}}x_0{}^{1/2}\chi_0{}^{5/2} \tag{9.104}$$

Further, if we now use (9.82) we would get (9.95). Thus the virial theorem (which is exactly valid for a quantum mechanical system as well as for a classical system when the exact equations of motion of the electrons and nuclei are taken into account) remains valid in spite of the simplifying assumptions introduced in the TF model.

9.3   Inclusion of exchange interaction: the Thomas–Fermi–Dirac equation

The 'exchange energy' is the effect of the antisymmetrization of the electron wavefunctions on the electrostatic interaction energy. This exchange energy (per unit volume) for a system of free electrons was first evaluated by Bloch (1929) and is given by (see also Brush (1967)):

$$- c_e n^{4/3} \tag{9.105}$$

where

$$c_e = \tfrac{3}{4} e^2 \left(\frac{3}{\pi} \right)^{1/3} \tag{9.106}$$

Thus the total energy will be given by

$$E = c_k \int n^{5/3} d\tau - c_e \int n^{4/3} d\tau - Ze^2 \int \frac{n(\mathbf{r})}{r} d\tau$$

$$+ \frac{1}{2} \int \int \frac{n(\mathbf{r}_1) n(\mathbf{r}_2) e^2}{|\mathbf{r}_1 - \mathbf{r}_2|} d\tau_1 d\tau_2 \tag{9.107}$$

We now use the same procedure as used in Section 9.2.3 to obtain

$$\tfrac{5}{3} c_k n^{2/3} - \tfrac{4}{3} c_e n^{1/3} - (e\phi + \mu) = 0 \tag{9.108}$$

which is an equation quadratic in $n^{1/3}$. Solving the quadratic equation we get the following expression for the electron density

$$n^{1/3} = \frac{2 c_e}{5 c_k} + \frac{3}{10 c_k} [\tfrac{16}{9} c_e^2 + \tfrac{20}{3} c_k (e\phi + \mu)]^{1/2} \tag{9.109}$$

where we have disregarded the negative sign in front of the square root so that if we neglect the exchange term (i.e., substitute $c_e = 0$), (9.109) becomes the TF relation (viz., (9.5)); furthermore, it has been shown that the solution corresponding to the negative sign does not correspond to any physical situation (Plasskett 1953). Substituting for c_e and c_k and simplifying we get

$$n = \frac{(2m)^{3/2}}{3\pi^2 \hbar^3} \left[\left(\frac{me^4}{2\pi^2 \hbar^2} \right)^{1/2} + \left(\frac{me^4}{2\pi^2 \hbar^2} + (e\phi + \mu) \right)^{1/2} \right]^3 \tag{9.110}$$

The above equation is known as the Thomas–Fermi–Dirac (TFD) relation which when combined with (9.9) gives the TFD equation.

We rewrite (9.110) as

$$n(r) = \frac{(2me)^{3/2}}{3\pi^2 \hbar^3} [a + [\phi(r) - \phi_0 + a^2]^{1/2}]^3 \tag{9.111}$$

where

$$a \equiv \frac{(2me^3)^{1/2}}{2\pi\hbar} \tag{9.112}$$

and $\phi_0 = -\mu/e$. We substitute for $n(r)$ from (9.111) in (9.9) to obtain (cf. (9.10)):

$$\nabla^2(\phi - \phi_0) = \frac{4e}{3\pi\hbar^3}(2me)^{3/2}[a + (\phi - \phi_0 + a^2)^{1/2}]^3 \tag{9.113}$$

We use a procedure similar to that used in Section 9.2 and define $\Psi(r)$ through the following equation (cf. (9.12)):

$$\phi(r) - \phi_0 + a^2 = \frac{Ze}{r}\Psi(r) \tag{9.114}$$

Substituting in (9.113) and carrying out simple manipulations we get

$$\frac{d^2\Psi}{dx^2} = x\left[\varepsilon + \left(\frac{\Psi(x)}{x}\right)^{1/2}\right]^3 \tag{9.115}$$

where $x = r/\alpha$ (see 9.15) and

$$\varepsilon = \left[\frac{3}{32\,\pi^2 Z^2}\right]^{1/3} \approx \frac{0.211\,873}{Z^{2/3}} \tag{9.116}$$

Equation (9.115) is known as the Thomas–Fermi–Dirac (TFD) equation.[†] If we substitute $\varepsilon = 0$ then (9.115) becomes the TF equation (9.19). Substituting (9.114) in (9.111) and carrying out simple manipulations we get

$$n(x) = \frac{32}{9\pi^3}\frac{Z^2}{a_0^3}\left[\varepsilon + \left(\frac{\Psi(x)}{x}\right)^{1/2}\right]^3 \tag{9.117}$$

which may be compared with (9.22). It can be easily seen that the boundary conditions prove to be as in the TF model, viz.

$$\Psi(0) = 1 \quad \text{and} \quad \left[\frac{x}{\Psi(x)}\frac{d\Psi}{dx}\right]_{x=x_0} = 1 \tag{9.118}$$

which may be compared with (9.27). Furthermore, $\Psi(x)$ can be expanded in a series of the form (cf. (9.29))

$$\Psi(x) = \sum_{k=0,1,\ldots} b_k x^{k/2}$$

$$= 1 + b_1 x^{1/2} + b_2 x + \cdots \tag{9.119}$$

where we have used the fact that since $\Psi(0) = 1$, b_0 must be equal to 1. Using

[†] An alternative derivation of the TFD equation has been given by Slater and Krutter (1935).

a procedure similar to that used in Section 9.2.2 we find

$$
\left.
\begin{aligned}
& b_1 = 0 \quad b_3 = \tfrac{4}{3} \quad b_4 = \tfrac{3}{2}\varepsilon \quad b_5 = \tfrac{2}{5}a_2 + \tfrac{4}{5}\varepsilon^2 \\
& b_6 = \tfrac{1}{3} + \tfrac{1}{2}b_2\varepsilon + \tfrac{1}{6}\varepsilon^3 \quad b_7 = \tfrac{6}{35}b_2\varepsilon^2 + \tfrac{3}{70}b_2{}^2 + \tfrac{5}{7}\varepsilon \\
& b_8 = \tfrac{2}{18}b_2 + \tfrac{77}{120}\varepsilon^2 \\
& b_9 = \tfrac{2}{27} - \tfrac{1}{252}b_2{}^3 + \tfrac{11}{35}b_2\varepsilon - \tfrac{1}{42}b_2{}^2\varepsilon^2 + \tfrac{10}{63}\varepsilon^3 + \tfrac{16}{105}\varepsilon^4
\end{aligned}
\right\}
\qquad (9.120)
$$

etc. It can be seen immediately that for a given value of ε, the complete solution is known if we know the value of the initial slope $(= b_2)$. The numerical solution of the TFD equation has been discussed by several workers – see, e.g., Slater and Krutter (1935), Jensen *et al.* (1938), Feynman

Table 9.5. *Summary of the numerical results for Carbon* $(Z = 6)$ *(for* $\rho \equiv 2.25\,\mathrm{g\,cm^{-3}}$, $x_0 \approx 6.28$*).*

$-b_2$	x_0	$\psi(x_0)$
1.674 0	3.261 7	0.231 94
1.680 0	3.715 3	0.169 46
1.680 0	4.378 4	0.108 16
1.685 8	5.167 8	0.061 24
1.686 3	5.729 1	0.039 42
1.686 7	7.008 4	0.010 79

Notes: Table adapted from Feynman, Metropolis & Teller (1949).

Table 9.6. *Summary of the numerical results for Uranium* $(Z = 92)$ *(for* $\rho = 19.0\,\mathrm{g\,cm^{-3}}$, $x_0 \approx 16.47$*).*

$-b_2$	x_0	$\psi(x_0)$
1.603 94	6.513 9	0.101 64
1.604 44	7.487 7	0.074 95
1.604 84	10.214 3	0.035 07
1.604 88	11.332 6	0.026 23
1.604 90	12.403 7	0.019 91
1.604 91	13.386 2	0.015 51
1.604 91	14.294 1	0.012 28
1.604 92	15.608 4	0.008 71
1.604 92	19.430 1	0.002 86
1.605 00[†]		0

Notes: Table adapted from Feynman, Metropolis & Teller (1949).

et al. (1949), etc. We should mention that whereas for the TF equation a set of universal solutions was found which could be applied to any value of Z by a simple change of scale, in the case of TFD equation, it is necessary however to obtain separate solutions for each Z. For example, Slater and Krutter (1935) have reported numerical solutions for $Z = 3, 11, 29$, Jensen *et al.* (1938) for $Z = 18, 36, 54$ and Feynman *et al.* (1949) for $Z = 6$ and $Z = 92$. In Tables 9.5 and 9.6 we give the values of x_0 and $\Psi(x_0)$ for different initial slopes for $Z = 6$ and $Z = 92$.

We should hasten to point out some of the difficulties of the TFD model. Firstly, there is a solution of (9.115) for which the x-axis is tangent to the Ψ-curve; however, the tangent occurs at a *finite* value of x. (We may recall that for the TF equation this tangency occurred at $x = \infty$ and therefore the solution corresponded to an isolated atom.) Brillouin (1934) suggested that the solution which is tangential to the x-axis should be taken as the 'best' representation for isolated atoms. However, there has been lot of controversy on this point – see, e.g., March (1957). Fortunately, the value of x

Table 9.7. *Initial value of the slope* $-b_2$, *in the Thomas–Fermi–Dirac case for which the x-axis is tangent to the* $\Psi(x)$ *curve.*

Atom	Z	ε	$-b_2$
Thomas–Fermi		0	1.588 08
Cu	29	0.022 437	1.622 5
Na	11	0.042 818	1.653 6
Li	3	0.101 815	1.742 7

Note: (Table adapted from Slater & Krutter (1935).)

Fig. 9.5. The variation of the initial slope $(-b_2)$ with ε corresponding to the solutions of the TFD equation for which the x-axis is tangent to the Ψ-curve. (Adapted from Slater and Krutter (1935).)

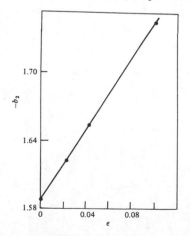

at which the x-axis becomes a tangent to the Ψ-curve is usually much greater than the interatomic distance. In Table 9.7 we have tabulated the values of the initial slope for which the x-axis becomes a tangent to the Ψ-curve. A plot between $-b_2$ and ε is very close to a straight line (see Fig. 9.5).

Another difficulty with the TFD equation (which has not been satisfactorily resolved) is the fact that $\Psi = 0$ does not imply the electron density to be zero (9.117) and therefore the solution leads to a discontinuity in the charge density at the boundary – once again, we refer to March (1957) for a discussion on this point.

9.3.1   Calculation of pressure
We will use the relation

$$P = -\left(\frac{\partial E}{\partial V}\right)_N = -\frac{1}{4\pi r_0{}^2}\left(\frac{\partial E}{\partial r_0}\right)_N \tag{9.121}$$

for the calculation of pressure. In (9.121), $V\ (= 4\pi/3(r_0{}^3))$ represents the atomic volume and

$$N = \int_0^{r_0} n4\pi r^2 dr \tag{9.122}$$

represents the total number of particles. It should be noted that in carrying out the differentiation with respect to r_0 (in (9.121)), N should remain constant, i.e., n must vary so that

$$\frac{dN}{dr_0} = 0 \tag{9.123}$$

Thus, according to (9.122)[†]

$$\frac{dN}{dr_0} = n(r_0)4\pi r_0{}^2 + \int_0^{r_0} \frac{dn}{dr_0}4\pi r^2 dr = 0 \tag{9.124}$$

Now, the expression for the total energy is (9.107)

$$E = c_k \int_{r=0}^{r_0} n^{5/3}d\tau - c_e \int_{r=0}^{r_0} n^{4/3}d\tau - Ze^2 \int_{r=0}^{r_0} \frac{n}{r}d\tau$$

$$+ \frac{e^2}{2}\int_{r=0}^{r_0}\int_{r=0}^{r_0} \frac{n(r)n(r')}{|\mathbf{r}-\mathbf{r}'|}d\tau d\tau' \tag{9.125}$$

[†] If

$$F(x) = \int_0^x f(y)dy$$

then

$$dF/dx = f(x)$$

Further, the second term in (9.124) comes from the dependence of $n(r)$ on r_0.

where $d\tau = 4\pi r^2 dr$. Thus carrying out elementary differentiations we get

$$\frac{dE}{dr_0} = \Gamma_1 + \Gamma_2 \tag{9.126}$$

where

$$\Gamma_1 = \int_0^{r_0} d\tau \frac{dn}{dr_0} \left[\tfrac{5}{3}c_k n^{2/3} - \tfrac{4}{3}c_e n^{1/3} - \frac{Ze^2}{r} + e^2 \int_0^{r_0} \frac{n(r')}{|\mathbf{r} - \mathbf{r}'|} d\tau' \right] \tag{9.127}$$

and

$$\Gamma_2 = 4\pi r_0^2 \left[c_k n^{5/3} - c_e n^{4/3} - \frac{Ze^2}{r} n + e^2 n(r) \int_0^{r_0} \frac{n(r')}{|\mathbf{r} - \mathbf{r}'|} d\tau' \right]_{r=r_0} \tag{9.128}$$

If we now use (9.52) and (9.108) we would readily get

$$\Gamma_1 = \mu \int_0^{r_0} d\tau \frac{dn}{dr_0}$$

$$= -\mu n(r_0) 4\pi r_0^2 \tag{9.129}$$

where in the last step we have used (9.124). We again use (9.108) to obtain

$$\Gamma_1 = -\left[\tfrac{5}{3}c_k n^{2/3} - \tfrac{4}{3}c_e n^{1/3} - e\phi \right] n(r_0) 4\pi r_0^2$$

$$= -\left[\tfrac{5}{3}c_k n^{2/3} - \tfrac{4}{3}c_e n^{1/3} - \frac{Ze^2}{r} + \int \frac{n(r')e^2}{|\mathbf{r} - \mathbf{r}'|} d\tau' \right]_{r=r_0} n(r_0) 4\pi r_0^2 \tag{9.130}$$

where in the last step we have used (9.52). We should point out that the quantity inside the square brackets is simply μ and therefore independent of r — so we may evaluate it at any value of r. We finally substitute for Γ_1 and Γ_2 from (9.130) and (9.128) in (9.126) to obtain

$$\frac{dE}{dr_0} = -4\pi r_0^2 \left[\tfrac{2}{3}c_k n^{5/3} - \tfrac{1}{3}c_e n^{4/3} \right]_{r=r_0} \tag{9.131}$$

Substituting in (9.121) we get

$$P = \left\{ \tfrac{2}{3}c_k [n(r_0)]^{5/3} - \tfrac{1}{3}c_e [n(r_0)]^{4/3} \right\} \tag{9.132}$$

If we neglect exchange effects (i.e., put $c_e = 0$) and substitute for c_k from (9.48), the above expression for P will become identical to (9.94) (obviously!). Using (9.48), (9.106) and (9.117) we get after simplifications

$$P = \frac{128}{45\pi^3} \left(\frac{4}{3\pi} \right)^{2/3} \left(\frac{e^2}{a_0^4} \right) Z^{10/3} \left[\left(\frac{\Psi(x_0)}{x_0} \right)^{1/2} + \varepsilon \right]^5$$

$$\cdot \left[1 - \frac{5\varepsilon/4}{(\Psi(x_0)/x_0)^{1/2} + \varepsilon} \right] \tag{9.133}$$

which may be compared with (9.95). Following Jensen (1938) we write (9.133) in the form

$$P = \tfrac{2}{3}c_k(f\bar{n})^{5/3} = \bar{P}f^{5/3} \tag{9.134}$$

where

$$\bar{n} = \frac{Z}{V} = \frac{Z}{(4\pi/3)r_0{}^3} \tag{9.135}$$

$$\bar{P} = \tfrac{2}{3}c_k\bar{n}^{5/3} = (3\pi^2)^{2/3}\frac{\hbar^2}{5m}\bar{n}^{5/3} \tag{9.136}$$

and

$$f = \frac{x_0{}^3}{3}\left[\left(\frac{\Psi(x_0)}{x_0}\right)^{1/2} + \varepsilon\right]^3\left[1 - \frac{5\varepsilon/4}{(\Psi(x_0)/x_0)^{1/2} + \varepsilon}\right]^{3/5} \tag{9.137}$$

Obviously, $\bar{P}$ represents the pressure resulting from a uniform distribution of all the electrons throughout the volume (see (6.29)). Now, for a given gas of atoms $\bar{n}$ (and hence $\bar{P}$) can easily be calculated and therefore if we know the dependence of f on x_0 and Z then the value of the pressure can readily be determined. Jensen (1938) and Feynman *et al.* (1949) plot f as a function

Fig. 9.6. The variation of f with ξ for different values of Z. The parameter f is related to the pressure corresponding to the TFD model (at $T = 0$) and the parameter ξ is related to the atomic density (see (9.138)). The curves allow easy interpolation for intermediate values of Z. (Adapted from Feyman, Metropolis and Teller (1949).)

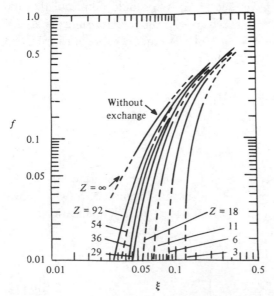

of a parameter (ξ) which is proportional to $(Z/V)^{1/3}$; they choose

$$\xi = Z^{-2/3} a_0 \left[\frac{Z}{V} \right]^{1/3} = Z^{-2/3} a_0 \left[\frac{Z}{(4\pi/3)\alpha^3 x_0{}^3} \right]^{1/3}$$

$$= \frac{2}{\pi} \left(\frac{4}{3} \right)^{1/3} \frac{1}{x_0} \approx \frac{0.7007}{x_0} \tag{9.138}$$

The factor $Z^{-2/3} a_0$ was chosen such that ξ depends only on x_0. Figure 9.6 gives the dependence of $f(\xi)$ on ξ for $Z = 3, 6, 11, 18, 29, 36, 54$ and 92. From these curves it is very straightforward to carry out an interpolation procedure to determine the value of f for given values of x_0 and Z. We have also plotted the curve corresponding to the TF model to which the TFD curves approach in the limit of $Z \to \infty$. Obviously, the effect of the exchange is to lower the pressure and energy of the TF model. Once again the expression for the pressure given by (9.133) (or, (9.134)) is expected to be valid at rather high pressure ($\gtrsim 10\,\mathrm{Mbar} = 10^{13}\,\mathrm{dyne\,cm^{-2}}$) when there is very little effect of the outer shell structure of the atoms.

9.4   Derivation of equation (9.103) using the virial theorem

In this section we will use the virial theorem to derive (9.103). The derivation will closely follow the treatment given by Latter (1955). We start with the virial theorem according to which the total kinetic energy of a system of particles is given by

$$2E_{\mathrm{kin}} = I \tag{9.139}$$

where

$$I = -\left\langle \sum_i \mathbf{r}_i \cdot \mathbf{F}_i \right\rangle \tag{9.140}$$

is known as the virial; $\mathbf{F}_i$ denotes the force acting on the ith electron whose position is given by $\mathbf{r}_i$ and angular brackets denote a space (or time average). The virial theorem is proved in Section 15.6. Now, the force acting on an electron may be due to electron–electron interaction, electron–nucleus interaction or collisions with the atomic boundary, the interactions will be designated by superscripts ee, en and eb respectively. We therefore write

$$I = I^{\mathrm{ee}} + I^{\mathrm{en}} + I^{\mathrm{eb}} \tag{9.141}$$

Since the force applied by the boundary is directed towards the centre of the sphere and occurs at the atomic boundary where $r = r_0$, we have

$$I^{\mathrm{eb}} = -\left\langle \sum_i \mathbf{r}_i \cdot \mathbf{F}_i{}^{\mathrm{eb}} \right\rangle$$

$$= r_0 \left\langle \sum_i F_i{}^{\mathrm{eb}} \right\rangle = r_0 [4\pi r_0{}^2 P] \tag{9.142}$$

where in the last step we have used the fact that the time average of the boundary force is just the pressure P times the area of the boundary ($= 4\pi r_0{}^2$). Since $V = (4\pi/3)r_0{}^3$ we have

$$I^{eb} = 3PV \tag{9.143}$$

We next consider the electron nuclear interaction for which

$$\mathbf{F}_i{}^{en} = -\frac{Ze^2}{r_i{}^3}\mathbf{r}_i \tag{9.144}$$

the negative sign showing that force is directed towards the origin. Thus

$$I^{en} = \left\langle \sum_i \frac{Ze^2}{r_i} \right\rangle = -E_{pot}{}^{en} \tag{9.145}$$

where $E_{pot}{}^{en}$ is the potential energy arising out of electron–nuclear interactions. Finally, we consider the force due to electron–electron interactions which will be given by

$$\mathbf{F}_i{}^{ee} = e^2 \sum_j \frac{\mathbf{r}_i - \mathbf{r}_j}{|\mathbf{r}_i - \mathbf{r}_j|^3} \tag{9.146}$$

where the summation is over all possible values of j excepting $j = i$. Thus

$$\begin{aligned}
I^{ee} &= -e^2 \left\langle \sum_i \sum_j \frac{\mathbf{r}_i \cdot (\mathbf{r}_i - \mathbf{r}_j)}{|\mathbf{r}_i - \mathbf{r}_j|^3} \right\rangle \\
&= -\frac{e^2}{2} \left\langle \sum_i \sum_j \frac{(\mathbf{r}_i - \mathbf{r}_j) \cdot (\mathbf{r}_i - \mathbf{r}_j)}{|\mathbf{r}_i - \mathbf{r}_j|^3} \right\rangle \\
&= -\frac{e^2}{2} \sum_i \sum_j \frac{1}{|\mathbf{r}_i - \mathbf{r}_j|} = -E_{pot}{}^{ee}
\end{aligned} \tag{9.147}$$

where $E_{pot}{}^{ee}$ represents the potential energy due to electron–electron interactions. Substituting the expressions for I^{eb}, I^{en} and I^{ee} in (9.141) and then substituting for I in (9.139) we get

$$2E_{kin} + E_{pot}{}^{en} + E_{pot}{}^{ee} = 3PV \tag{9.148}$$

which is identical to (9.103) since the total potential energy is given by (9.68). We may mention here that based on similarity considerations, Feynman *et al.* (1949) have given a very general proof of the virial theorem. Furthermore, the virial theorem can be applied equally well to calculations with exchange effects; this follows from the fact that the exchange energy, as all other potential energies, is proportional to e^2/r. Thus (9.148) takes the form

$$2E_{kin} + E_{pot}^{en} + E_{pot}^{ee} + E_{ex} = 3PV \tag{9.149}$$

9.5 The Thomas–Fermi model at finite temperatures

Until now our considerations have been for $T = 0$. In this section we extend the analysis to finite temperatures[†] – our treatment closely follows the analysis of Feynman, Metropolis & Teller (1949) and Latter (1955). We start with (9.2) and rewrite it in the form

$$n(r) = \int_0^\infty \rho(r, p)\mathrm{d}p \tag{9.150}$$

where

$$\rho(r, p) = \frac{1}{4\pi^3\hbar^3} \frac{4\pi p^2}{\exp\left[\dfrac{p^2}{2mkT} - \dfrac{e(\phi - \phi_0)}{kT}\right] + 1} \tag{9.151}$$

represents the number of electrons (per unit volume) whose momenta lie between p and $p + \mathrm{d}p$. We introduce the dimensionless variable

$$\xi = \frac{r}{r_0} \tag{9.152}$$

where $r = r_0$ represents the atomic boundary (9.25). Furthermore, we introduce the variable $\Psi(\xi)$ defined through the following equation (cf. (9.12)):

$$\Psi(\xi) = \frac{e(\phi - \phi_0)}{kT} \quad \xi = \frac{e(\phi - \phi_0)}{kT}\frac{r}{r_0} \tag{9.153}$$

If we now use the Poisson equation (9.9) and introduce the variable

$$y = \frac{p^2}{2mkT} \tag{9.154}$$

we would get

$$\frac{kT}{e}r_0\frac{1}{r^2}\frac{\mathrm{d}}{\mathrm{d}r}\left[r^2\frac{\mathrm{d}}{\mathrm{d}r}\left(\frac{\Psi(r)}{r}\right)\right]$$
$$= \frac{4\pi e}{4\pi^3\hbar^3}\int_0^\infty\frac{4\pi[2mkTy][(2mkT)^{1/2}\frac{1}{2}y^{-1/2}\mathrm{d}y]}{\exp[y - (\Psi(\xi)/\xi)] + 1}$$

[†] The analysis for the Thomas–Fermi–Dirac model for finite temperatures is fairly involved; for a perturbation treatment we refer to Umeda and Tomishima (1953), a more rigorous analysis has been given by Cowan and Ashkin (1957).

On simplification we obtain

$$\frac{d^2\Psi(\xi)}{d\xi^2} = a\xi I_{1/2}\left[\frac{\Psi(\xi)}{\xi}\right] \tag{9.155}$$

where

$$a = \left(\frac{r_0}{c}\right)^2 \tag{9.156}$$

$$c = \left(\frac{\pi\hbar^3}{4me^2(2mkT)^{1/2}}\right)^{1/2} = \frac{1.602 \times 10^{-9}\,\text{cm}}{T_{kv}^{1/4}} \tag{9.157}$$

(where T_{kv} is the temperature measured in kilovolts) and

$$I_n(\eta) \equiv \int_0^\infty \frac{y^n dy}{e^{y-\eta}+1} \tag{9.158}$$

is known as the Fermi–Dirac function. The functions $I_{3/2}(\eta), I_{1/2}(\eta)$ and the derivatives of $I_{1/2}(\eta)$ appear not only in the calculation of the equation of state but also in other areas of statistical physics. McDougall and Stoner (1938) have given detailed tables of these functions; we reproduce these tables in Appendix 3.

Now, as $r \to 0$, $\phi(r)$ should behave as Ze/r and we immediately have

$$\Psi(0) = \frac{Ze^2}{kTr_0} \tag{9.159}$$

Further, the boundary condition at $r = r_0$ (i.e., at $\xi = 1$) gives us (see (9.26))

$$\Psi'(1) = \Psi(1) \tag{9.160}$$

where primes denote differentiation with respect to ξ. The electron pressure is given by (cf. (9.92))

$$P = \frac{2}{3}\int_0^\infty \left(\frac{p^2}{2m}\right)\rho(r_0, p)dp \tag{9.161}$$

where the electron density is evaluated at the atomic boundary. Substituting for ρ from (9.151) and using (9.153) and (9.154) we get

$$P = \frac{(2mkT)^{5/2}}{6\pi^2\hbar^3 m} I_{3/2}(\Psi(1)) \tag{9.162}$$

which Latter (1955) writes in the following form:

$$P = \frac{ZkT2}{V}\frac{a}{9}\frac{1}{\Psi(0)} I_{3/2}(\Psi(1)) \tag{9.163}$$

$V [= (4\pi/3)r_0^3]$ represents the atomic volume. Now, the kinetic energy per electron is $p^2/2m$, therefore the total kinetic energy will be given by

$$E_{kin} = \int_0^\infty 4\pi r^2 dr \int_0^\infty \rho(r,p)\frac{p^2}{2m} dp \qquad (9.164)$$

Using (9.151) and (9.158) we get

$$E_{kin} = \frac{(2mkT)^{5/2}r_0^3}{\pi m h^3} \int_0^1 d\xi \xi^2 I_{3/2}\left(\frac{\Psi(\xi)}{\xi}\right) \qquad (9.165)$$

which can be rewritten in the form (Latter 1955):

$$E_{kin} = ZkT\frac{a}{\Psi(0)}\int_0^1 d\xi \xi^2 I_{3/2}\left[\frac{\Psi(\xi)}{\xi}\right] \qquad (9.166)$$

In order to calculate the potential energy we must know $n(r)$ and if we use (9.150)–(9.154) we readily get

$$n(r) = \frac{(2mkT)^{3/2}}{2\pi^2\hbar^3} I_{1/2}\left[\frac{\Psi(\xi)}{\xi}\right] \qquad (9.167)$$

Thus, using the fact that $\phi_n(r) = Ze/r$, (9.70) gives

$$E_{pot}{}^{en} = -4\pi\frac{(2mkT)^{3/2}}{2\pi^2\hbar^3} r_0^3 \int_0^1 d\xi \xi^2 I_{1/2}\left[\frac{\Psi(\xi)}{\xi}\right]\left[\frac{Ze}{r_0\xi}\right] \qquad (9.168)$$

Further, since

$$\phi_e(r) = \phi(r) - \phi_n(r) = \phi(r) - \frac{Ze}{r}$$

$$= \frac{kT}{e\xi}\Psi(\xi) + \phi_0 - \frac{Ze}{r} \qquad (9.169)$$

(9.69) becomes

$$E_{pot}{}^{ee} = -2\pi e\frac{(2mkT)^{3/2}}{2\pi^2\hbar^3} r_0^3 \int_0^1 d\xi \xi^2 I_{1/2}\left[\frac{\Psi(\xi)}{\xi}\right]$$

$$\times \left[\frac{kT}{e\xi}\Psi(\xi) + \phi_0 - \frac{Ze}{r_0\xi}\right] \qquad (9.170)$$

Since $\phi(r_0) = 0$, we have

$$\Psi(1) = -\frac{e\phi_0}{kT} \qquad (9.171)$$

Thus

$$E_{pot} = E_{pot}{}^{ee} + E_{pot}{}^{en} = -\frac{(2mkT)^{5/2}r_0{}^3}{2\pi mh^3}\int_0^1 d\xi I_{1/2}\left[\frac{\Psi(\xi)}{\xi}\right]$$

$$\times \xi[\Psi(\xi) - \xi\Psi(1) + \Psi(0)]$$

$$= -ZkT\frac{a}{2\Psi(0)}\int_0^1 d\xi I_{1/2}\left[\frac{\Psi(\xi)}{\xi}\right]\xi[\Psi(\xi) - \xi\Psi(1) + \Psi(0)]$$

$$(9.172)$$

Using (9.163), (9.166) and (9.170) it is possible to derive (9.148) – the analysis is a bit involved and we have given it in Appendix 4. This again shows that the virial theorem remains valid inspite of the simplifying assumptions introduced in the TF model.

Latter (1955) has derived an expression for the entropy which is given by

$$\frac{S}{Zk} = \frac{1}{ZkT}[\tfrac{5}{3}E_{kin} + 2E_{pot}{}^{ee} + E_{pot}{}^{en}] - \Psi(1) \qquad (9.173)$$

Finally, Feynman *et al.* (1949), making use of similarity considerations, have derived the following relation

$$\tfrac{7}{3}(E_{pot} + E_{kin}) = E_{pot}{}^{en} - ZE_w + PV + \tfrac{4}{3}Tc_v \qquad (9.174)$$

where c_v is the specific heat per atom at constant volume and E_w represents the ionization energy of the outermost electron. If we now use the virial theorem (viz. $2E_{kin} = 3PV - E_{pot}$) and the relation $E_{pot} = E_{pot}{}^{ee} + E_{pot}{}^{en}$ we would get

$$\tfrac{5}{2}PV + \tfrac{7}{6}E_{pot}{}^{ee} + \tfrac{1}{6}E_{pot}{}^{en} = -ZE_w + \tfrac{4}{3}Tc_v \qquad (9.175)$$

We may note that for an isolated atom at $T = 0$, we get

$$(E_{pot}{}^{en})_{T=0}{}^{\infty} = -7(E_{pot}{}^{ee})_{T=0}{}^{\infty}$$

which is identical to (9.78).

9.5.1   Calculation of thermodynamic functions

In principle, the procedure for the numerical evaluation of thermodynamic functions is simple: for a given value of a, the TF equation (9.155) is first solved subject to the boundary conditions given by (9.159) and (9.160). Once $\Psi(\xi)$ is known, all quantities like P, E_{kin}, E_{pot}, S etc. can be evaluated. We should mention that if we substitute the values of r_0 and c (from (9.28) and (9.157)) in (9.156) then the parameter a can be written in the form

$$a = \frac{3}{4}\left(\frac{9\pi^2}{2}\right)^{1/6}\frac{1}{(e^2/\hbar c)}\frac{1}{(mc^2)^{1/2}}\left[\frac{kT}{Z^{4/3}}\right]^{1/2}x_0{}^2 \qquad (9.176)$$

Since $e^2/\hbar c \approx 1/137$ and $mc^2 \approx 0.511 \text{MeV} = 0.511 \times 10^6 \text{eV}$ we get

$$a \approx 0.2705 \, \Gamma^{1/2} x_0^2 \tag{9.177}$$

where

$$\Gamma \equiv \frac{kT}{Z^{4/3}} \tag{9.178}$$

kT being measured in electron volts. If we use (9.82) then we get

$$a \approx 4.7426 \times 10^{15} \, \Gamma^{1/2} (VZ)^{2/3} \tag{9.179}$$

where $V \, (= (4\pi/3) r_0^3)$ is measured in cm^3. Now, in terms of the parameter Γ, (9.162) can be written in the form

$$\frac{P}{Z^{10/3}} = \frac{(2m)^{3/2}}{3\pi^2 \hbar^3} \Gamma^{5/2} I_{3/2}(\Psi(1)) \tag{9.180}$$

Thus, in Fig. 9.4 we have plotted $P/Z^{10/3}$ as a function of (VZ) for different values of the parameter Γ. Latter (1955) has also plotted the variation of $P/Z^{10/3}$ with (VZ) for different values of total energy and also for different values of entropy. In Fig. 9.3 we have given the dependence of the total energy on ZV.

Finally, we outline the numerical method for solving (9.155). We first note that it is more convenient to obtain a numerical solution if we put the equation in the following form[†]

$$\Psi(\xi) = \Psi(1)\xi + a \int_\xi^1 d\eta (\eta - \xi) \eta I_{1/2}\left[\frac{\xi(\eta)}{\eta}\right] \tag{9.181}$$

If we write the above equation as

$$\Psi(\xi) = \Psi(1)\xi - a\left[\int_1^\xi d\eta \eta^2 I_{1/2}\left[\frac{\Psi(\eta)}{\eta}\right] \right.$$
$$\left. - \xi \int_1^\xi d\eta \eta I_{1/2}\left[\frac{\Psi(\eta)}{\eta}\right] \right]$$

[†] Equation (9.181) immediately follows from the fact that if we integrate the differential equation twice

$$\frac{d^2\Psi}{d\xi^2} = f(\xi)$$

then we would readily obtain

$$\Psi(\xi) = \int^\xi (\xi - \eta) f(\eta) d\eta + c\xi$$

where c is a constant.

then

$$\Psi'(\xi) = \Psi(1) - a\left[\xi^2 I_{1/2}\left[\frac{\Psi(\xi)}{\xi}\right]\right.$$

$$\left. - \int_1^\xi \mathrm{d}\eta\,\eta I_{1/2}\left[\frac{\Psi(\eta)}{\eta}\right] - \xi^2 I_{1/2}\left[\frac{\Psi(\xi)}{\xi}\right]\right]$$

$$= \Psi(1) + a\int_1^\xi \mathrm{d}\eta\,\eta I_{1/2}\left[\frac{\Psi(\eta)}{\eta}\right] \tag{9.182}$$

Fig. 9.7. The dependence of $\Psi(0)$ on $\Psi(1)$ for various values of a. (Adapted from Latter (1955). For a more detailed graph see Latter (1955).)

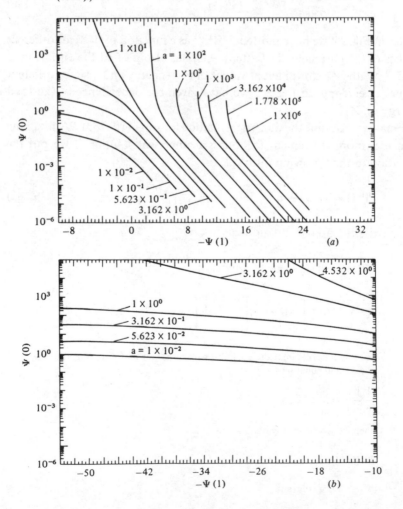

from which it immediately follows that $\Psi'(1) = \Psi(1)$; thus, (9.181) satisfies the boundary condition given by (9.160). If we differentiate (9.182) once more we would get (9.155). Furthermore, from (9.181) it readily follows that

$$
\begin{aligned}
\Psi(0) &= a\int_0^1 d\eta\, \eta^2 I_{1/2}\left[\frac{\Psi(\eta)}{\eta}\right] \\
&= \int_0^1 d\eta\, \eta \Psi''(\eta) \\
&= \left[\eta\Psi'(\eta)\Big|_0^1 - \int_0^1 \Psi'(\eta)d\eta\right] \\
&= \Psi'(1) - \Psi(1) + \Psi(0) = \Psi(0)
\end{aligned}
$$

where we have used (9.155) and (9.160). The numerical integration of (9.181) becomes much easier if we introduce the following variables:

$$
\xi = u^2 \quad \text{and} \quad \eta = v^2 \tag{9.183}
$$

In these variables, (9.181) becomes

$$
\Psi(u) = \Psi(1)u^2 + 2a\int_v^1 dv(v^2 - u^2)v^3 I_{1/2}\left(\frac{\Psi(v)}{v^2}\right) \tag{9.184}
$$

In principle, the procedure is to start with particular values of $\Psi(1)$ and a and then iterate (9.184) to obtain $\Psi(u)$. The result of such a calculation is shown in Fig. 9.7. Latter (1955) has shown that the difference equations corresponding to (9.184) can be written in the following form:

$$
\begin{aligned}
\Psi_n = \Psi(1)u_n^2 &+ 2a\{\tfrac{1}{3}\Delta u[(4g_{n+1} + g_{n+2}) - u_n^2(4f_{n+1} + f_{n+2})] \\
&+ Q_{n+2} - u_n^2 P_{n+2}\}, \quad n = N-2, N-3,\dots
\end{aligned} \tag{9.185}
$$

where

$$
Q_n = \frac{\Delta u}{3}[g_n + 4g_{n+1} + g_{n+2}] + Q_{n+2} \tag{9.186}
$$

$$
P_n = \frac{\Delta u}{3}[f_n + 4f_{n+1} + f_{n+2}] + P_{n+2} \tag{9.187}
$$

$$
g_n = u_n^5 I_{1/2}\left[\frac{\psi_n}{u_n^2}\right] \tag{9.188}
$$

$$
f_n = u_n^3 I_{1/2}\left[\frac{\Psi_n}{u_n^2}\right] \tag{9.189}
$$

and

$$
\Delta u = 1/N \tag{9.190}
$$

N being the total number of integration points. In order to use the above equations it is necessary to know Ψ_N, Q_N, P_N and $\Psi_{N-1}, Q_{N-1}, P_{N-1}$ which are given by the following equations:

$$\Psi_N = 1 \tag{9.191}$$

$$Q_N = 0 \tag{9.192}$$

$$P_N = 0 \tag{9.193}$$

and

$$\Psi_{N-1} = \Psi(1)u_{N-1}{}^2 + a \sum_{k=0}^{3} \frac{A_k}{k!}$$
$$\times \left[\frac{(1-u_{N-1}{}^2)^{k+2}}{(k+1)(k+2)} - \frac{(1-u_{N-1}{}^2)^{k+3}}{(k+2)(k+3)} \right] \tag{9.194}$$

$$P_{N-1} = \frac{1}{2} \sum_{k=0}^{3} \frac{A_k}{k!} \left[\frac{(1-u_{N-1}{}^2)^{k+1}}{k+1} - \frac{(1-u_{N-1}{}^2)^{k+2}}{k+2} \right] \tag{9.195}$$

$$Q_{N-1} = -\frac{1}{2} \sum_{k=0}^{3} \frac{A_k}{k!} \left[\frac{(1-u_{N-1}{}^2)^{k+2}}{k+2} - \frac{(1-u_{N-1}{}^2)^{k+3}}{k+3} \right]$$
$$+\frac{1}{2} \sum_{k=0}^{3} \frac{A_k}{k!} \left[\frac{(1-u_{N-1}{}^2)^{k+1}}{k+1} - \frac{(1-u_{N-1}{}^2)^{k+2}}{k+2} \right] \tag{9.196}$$

where

$$\left.\begin{aligned}
A_0 &= I_{1/2}[\Psi(1)] \\
A_1 &= 0 \\
A_2 &= aI_{1/2}'[\Psi(1)]I_{1/2}[\Psi(1)] \\
\end{aligned}\right\} \tag{9.197}$$

and

$$A_3 = 2aI_{1/2}'[\Psi(1)]I_{1/2}[\Psi(1)]$$

We may use the Table given in Appendix 3 to obtain the numerical solution; alternatively, it would be very convenient to have simple analytical forms of $I_{1/2}(\eta)$ which have been given by Latter (1955) – we give them below:

(a) $-\infty < \eta \leqslant -2.0$:

$$I_{1/2}(\eta) = \frac{\sqrt{\pi}}{2} e^{\eta} \left[1 - \frac{e^{\eta}}{2^{3/2}} + \frac{e^{2\eta}}{3^{3/2}} - \frac{e^{3\eta}}{4^{3/2}} + \frac{e^{4\eta}}{5^{3/2}} - \frac{e^{5\eta}}{6^{3/2}} + \frac{e^{6\eta}}{7^{3/2}} \right]$$

(b) $-2.0 \leqslant \eta \leqslant 0$:

$$I_{1/2}(\eta) = 0.678\,091 + 0.536\,196\,67\eta + 0.169\,097\,48\eta^2$$
$$+ 0.018\,780\,823\eta^3 - 0.002\,357\,544\,6\eta^4$$
$$- 0.000\,639\,610\,797\eta^5$$

(c) $0 \leqslant \eta \leqslant 3.0$:

$$I_{1/2}(\eta) = 0.678\,091 + 0.536\,38\eta + 0.166\,823\,5\eta^2$$
$$+ 0.020\,606\,7\eta^3 - 0.006\,014\,91\eta^4 + 0.000\,490\,398\eta^5$$

(d) $3.0 \leqslant \eta \leqslant 10.0$:

$$I_{1/2}(\eta) = 0.757\,064\,709 + 0.392\,288\,8\eta + 0.270\,552\,5\eta^2$$
$$- 0.016\,829\,33\eta^3 + 0.000\,825\,836\,4\eta^4$$
$$- 0.000\,018\,197\,71\eta^5$$

(e) $10.0 \leqslant \eta \leqslant 10^5$:

$$I_{1/2}(\eta) = \tfrac{2}{3}\eta^{3/2}\left[1 + \frac{1.233\,700\,5}{\eta^2} + \frac{1.065\,411\,9}{\eta^4} + \frac{9.701\,518\,5}{\eta^6}\right.$$
$$\left. + \frac{242.715\,02}{\eta^8} + \frac{12\,313.691}{\eta^{10}}\right]$$

(f) $10^5 \leqslant \eta < \infty$:

$$I_{1/2}(\eta) = \tfrac{2}{3}\eta^{3/2} \qquad\qquad (9.198)$$

The fractional error in the fitting of the above functions with the exact numerical values of $I_{1/2}(\eta)$ was found to be never greater than 6×10^{-5} and except for a few isolated regions, less than 1×10^{-5} (Latter 1955). Once $\Psi(\xi)$ is known, all thermodynamic functions can easily be evaluated.

9.6   Exchange and quantum corrections to the Thomas–Fermi model
     In Section 9.3 we discussed the effect of the electron exchange interaction which led to the Thomas–Fermi–Dirac (TFD) model. The effect of the exchange was found to lower the pressure and energy of the TF model. Kompaneets and Pavlovskii (1957) showed that there is an additional correction term (usually referred to as the quantum correction) which is of the same order as that due to exchange effects. Kirzhnits (1957, 1959) has evaluated finite temperature, quantum and exchange corrections. The model is usually abbreviated as the TFC (quantum and exchange corrected) model and the basic physics of the model has also been discussed by McCarthy (1965). The theoretical details of the model are rather involved; we give here the final results. The corrected expression for the electron density is given by

$$n(\zeta) = n_{\mathrm{TF}}(\zeta_{\mathrm{TF}}) + \left[\frac{dn}{d\zeta}\right]_{\zeta=\zeta_{\mathrm{TF}}}(\zeta - \zeta_{\mathrm{TF}})$$
$$+ \delta_1 n(\zeta_{\mathrm{TF}}) + \delta_2 n(\zeta_{\mathrm{TF}}) \qquad\qquad (9.199)$$

where

$$\zeta = \frac{e(\phi - \phi_0)}{kT} = \frac{\Psi(\xi)}{\xi} \tag{9.200}$$

$$n_{\text{TF}} = \frac{(2mkT)^{3/2}}{2\pi^2\hbar^3} \, I_{1/2}(\zeta_{\text{TF}}) = \frac{\sqrt{2}}{\pi^2} \frac{\theta^{3/2}}{N} I_{1/2}(\zeta_{\text{TF}}) \tag{9.201}$$

$$\delta_1 n(\zeta) = \frac{\sqrt{2}}{24\pi^2} \left(\frac{4\pi}{3}\right)^{2/3} Z^{-2/3} \frac{\theta^{1/2}}{N^{1/3}}$$
$$\times \left[I_{1/2}'''(\zeta)(\nabla\zeta)^2 + 2I_{1/2}''(\zeta)\nabla^2\zeta \right] \tag{9.202}$$

$$\delta_2 n(\zeta) = \frac{2}{\pi^2 N} Z^{-2/3} [I_{1/2}'(\zeta)]^2 \tag{9.203}$$

$$\theta = Z^{-4/3} \frac{kT}{(e^2/a_0)} \tag{9.204}$$

$$N = \frac{a_0^3}{(4\pi/3)r_0^3 Z} \tag{9.205}$$

Notice that (9.201) is the same as (9.167). Thus one has first to solve the TF equation (viz. (9.155)) and then use that solution to evaluate the corrections. The subscripts 1 and 2 (i.e., $\delta_1 n$ and $\delta_2 n$) refer to quantum and exchange corrections respectively. Now,

$$\zeta(\xi) = \zeta_{\text{TF}}(\xi) + \delta_1\zeta(\xi) + \delta_2\zeta(\xi) \tag{9.206}$$

where the corrections to the potential ($\delta_1\zeta$ and $\delta_2\zeta$) are obtained by solving the Poisson equation with the approximate expression for the electron density. Indeed, if we write

$$\delta_1\zeta(\xi) = \frac{\sqrt{2}}{6\pi} \frac{Z^{-2/3}}{\theta^{1/2}} [I_{1/2}' + u_1(\xi)] \tag{9.207}$$

$$\delta_2\zeta(\xi) = \frac{\sqrt{2}}{6\pi} \frac{Z^{-2/3}}{\theta^{1/2}} u_2(\xi) \tag{9.208}$$

then u_1 and u_2 satisfy the following equations

$$\nabla^2 u_1(\xi) - aI_{1/2}' u_1(\xi) = a[(I_{1/2}')^2 + I_{1/2}'' I_{1/2}] \tag{9.209}$$

$$\nabla^2 u_2(\xi) - aI_{1/2}' u_2(\xi) = 6a[(I_{1/2}')^2] \tag{9.210}$$

with the boundary conditions

$$\left.\begin{array}{l} u_i(0) = 0 \\ u_i'(1) = 0 \end{array}\right\} \quad \begin{array}{l} i = 1\text{: quantum correction} \\ i = 2\text{: exchange correction} \end{array} \tag{9.211}$$

The argument of *all* Fermi–Dirac integrals (like $I_{1/2}'$, $I_{1/2}$ etc.) is ζ_{TF}. The numerical technique of solving (9.209) and (9.210) has been discussed by McCarthy (1965) and as before, it is convenient to introduce the variable $\eta = \xi^2$.

The pressure is given by

$$P = P_{TF}(\zeta_{TF}) + \delta_1 P + \delta_2 P \tag{9.212}$$

where

$$P_{TF}(\zeta_{TF}) = \frac{(2mkT)^{5/2}}{6\pi^2 \hbar^3 m} I_{3/2}[\zeta_{TF}^{(1)}]$$

$$= \frac{2\sqrt{2}}{3\pi^2} \theta^{5/2} I_{3/2}[\zeta_{TF}(1)] \tag{9.213}$$

Fig. 9.8. A comparison of the equation of state calculations using the TF, TFD and TFC models corresponding to lead at $T = 0\,\mathrm{K}$. The experimental zero temperature isotherm is from the Soviet Hugoniot data (Altshuler *et al.* (1962)). (Figure adapted from McCarthy (1962).)

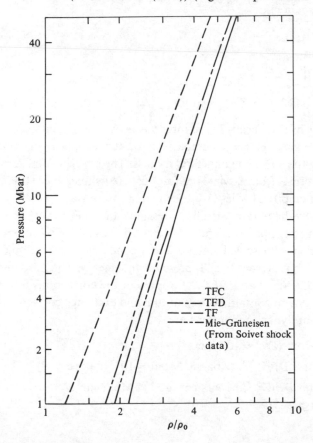

$$\delta_1 P = \tfrac{3}{2} P_{TF} \frac{I_{1/2}(\zeta_{TF}(1))}{I_{3/2}(\zeta_{TF}(1))} \delta_1 \zeta(1) \tag{9.214}$$

$$\delta_2 P = \frac{\sqrt{2}}{4\pi} P_{TF} \frac{Z^{-2/3}}{\theta^{1/2} I_{3/2}[\zeta_{TF}(1)]}$$

$$\times \left\{ 6 \int_{-\infty}^{\zeta(1)} (I_{1/2}')^2 dt + I_{1/2}[\zeta_{TF}(1)] u_2(1) \right\} \tag{9.215}$$

$\xi = 1$ represents the atomic boundary. Finally, the correction to the energy of the atom is

$$\delta E = \delta_1 E + \delta_2 E \tag{9.216}$$

where

$$\delta_i E = \frac{Z}{3\pi} \left[\frac{e^2}{2a_0} kT \right]^{1/2} u_i(0)$$

$$+ \frac{4R^2}{3\pi} \int_0^1 d\xi \xi^2 [\tfrac{1}{2} I_{1/2} u_1(\xi) + \Gamma_i(\xi)] + c \tag{9.217}$$

$$\Gamma_1(\xi) = I_{1/2} I_{1/2}' \tag{9.218}$$

$$\Gamma_2(\xi) = 6 \int_{\infty}^{\xi_{TF}(\xi)} (I_{1/2}')^2 dt \tag{9.219}$$

$$c(z) = 7.342\,396\,61\, Z^{5/3}\, eV/atom \tag{9.220}$$

and the argument of the Fermi–Dirac integrals is the TF potential ζ_{TF}.

In Fig. 9.8 we have given a comparison of the equation of state calculations using the TF, TFD and TFC models. The experimental zero-temperature isotherm is from Soviet Hugoniot data (Altshuler *et al.* (1962)) employing a semi-empirical Mie–Grüneisen equation of state.

In Appendix 5 we have tabulated TF pressures and energy with their corrections for scaled temperatures from 10^{-2} to 10^{-5} eV. The first column represents ZV which is the scaled volume. The second and fourth columns $(P/Z^{10/3}$ and $E/Z^{7/3})$ represent the TF pressure and energy. The third and the fifth columns $(DP/Z^{8/3}$ and $DE/Z^{5/3})$ represent the corrections to the pressure and energy and contain both quantum and exchange corrections. The units of thermodynamic functions are

ZV: Å^3/atom $(1\text{Å}^3 = 10^{-24}\, cm^3)$

$P/Z^{10/3}$ and $DP/Z^{8/3}$: Mbars $(1\text{Mbar} = 10^{12}\, dynes\, cm^{-2})$

$E/Z^{7/3}$ and $DE/Z^{5/3}$: Mbars cm^3 g^{-1} $(1\, Mbar\, cm^3\, g^{-1}$
$= 10^{12}\, ergs\, g^{-1})$

10

Grüneisen equation of state

10.1 Introduction

The equation of state is described by a functional relationship between the thermodynamic variables defined for a system in equilibrium. If one neglects the electron–phonon interactions then the thermodynamic functions can be expressed as a superposition of terms. In particular one can write the energy E and the pressure P in the following form

$$E = E_c + E_{Ta} + E_{Te} \tag{10.1}$$

$$P = P_c + P_{Ta} + P_{Te} \tag{10.2}$$

where E_c and P_c are the energy and the pressure at zero temperature ($T = 0\,\text{K}$), E_{Ta} and P_{Ta} are the contributions of the atom vibrations (about their mean positions) to the energy and the pressure, while E_{Te} and P_{Te} are the appropriate electron thermal contributions. In this chapter we discuss the relation between E_{Ta} and P_{Ta}.

The Grüneisen equation of state is given by (Grüneisen 1926)[†]

$$P_{Ta} = \frac{\gamma}{V} E_{Ta} \tag{10.3}$$

where γ is called the Grüneisen ratio (or parameter) and it is assumed to be a function only of the specific volume V. The derivation of (10.3) is the main topic of this chapter.

We assume a solid of N atoms to be described by a system of harmonic oscillators with $3N$ independent modes of vibrations. The vibrations are quantized so that the energy of the j mode of vibration is given by

$$E_j = (n_j + \tfrac{1}{2})h\nu_j \quad j = 1, 2, \ldots, 3N \tag{10.4}$$

[†] This equation is occasionally called the Mie–Grüneisen equation of state.

where h is Planck's constant, n_j is an integer and v_j is the frequency of vibration of the j mode. Denoting by E_0 the potential energy of the solid at zero degrees Kelvin ($T = 0$) where all the amplitudes of the oscillations vanish, the energy of the solid crystal is described by

$$E_i \equiv E_{n_1 n_2 \ldots n_{3N}} = \sum_{j=1}^{3N} (n_j + \tfrac{1}{2}) h v_j + E_0 \tag{10.5}$$

or equivalently one can write

$$E_i = \sum_{j=1}^{3N} n_j h v_j + E_c \tag{10.6}$$

where

$$E_c = \frac{1}{2} \sum_{j=1}^{3N} h v_j + E_0 \tag{10.7}$$

The main assumption of this model is that E_c, γ_j for $j = 1, \ldots, 3N$ and $E_{n_1 n_2 \ldots n_{3N}}$ may be a function of the volume V but not of the temperature T.

The partition function Q of the oscillator system, with an energy given by (10.6), is

$$Q = \sum_i e^{-\beta E_i} = \sum_{n_1=0}^{\infty} \cdots \sum_{n_{3N}=0}^{\infty} \exp\left(-\sum_{j=1}^{3N} \beta n_j h v_j - \beta E_c \right) \tag{10.8}$$

where as usual

$$\beta = \frac{1}{kT} \tag{10.9}$$

and k is the Boltzmann constant. The exponential sum of (10.8) can be rewritten as a product of terms where each product has a single value of an index of summation, namely (see also Chapter 3 for an explicit example)

$$Q = e^{-\beta E_c} \sum_{n_1=0}^{3N} e^{-\beta n_1 h v_1} \sum_{n_2=0}^{3N} e^{-\beta n_2 h v_2} \cdots \sum_{n_{3N}=0}^{3N} e^{-\beta n_{3N} h v_{3N}} \tag{10.10}$$

Since each sum in the product is a geometric series, one can write

$$\sum_{n_j=0}^{3N} e^{-\beta n_j h v_j} = (1 - e^{-\beta h v_j})^{-1} - \sum_{n_j=3N+1}^{\infty} e^{-\beta n_j h v_j} \tag{10.11}$$

This series converges since $\beta h v_j > 0$ for every j. Moreover, since N is very large (of the order of 10^{23}!), it is possible to neglect the second term on the right hand side of (10.11), so that the partition function can be written in the following way:

$$Q = e^{-\beta E_c} \prod_{j=1}^{3N} [1 - \exp(-\beta h v_j)]^{-1} \tag{10.12}$$

The Helmoltz free energy F is given by

$$F = -\beta^{-1} \ln Q = E_c + \beta^{-1} \sum_{j=1}^{3N} \ln[1 - \exp(-\beta h v_j)] \qquad (10.13)$$

The average energy E of the solid and its pressure P are obtained using the appropriate thermodynamic relations,

$$E = F - T\left(\frac{\partial F}{\partial T}\right)_V = E_c + \sum_{j=1}^{3N} \frac{h v_j}{e^{\beta h v_j} - 1} \qquad (10.14)$$

$$P = -\left(\frac{\partial F}{\partial V}\right)_T = -\frac{dE_c}{dV} + \frac{1}{V} \sum_{j=1}^{3N} \gamma_i h v_i / [e^{\beta h v_j} - 1] \qquad (10.15)$$

where in the last equation γ_j is defined by

$$\gamma_j = -\frac{V}{v_j}\left(\frac{\partial v_j}{\partial V}\right)_T = -\left(\frac{\partial \ln v_j}{\partial \ln V}\right)_T \qquad (10.16)$$

The first term in (10.15) is the pressure at zero temperature, $T = 0$. The second term describes the thermal pressure. This thermal term differs from zero since γ_j does not vanish, implying that the vibrational frequencies depend on volume. The dependence of the frequency on the specific volume is explained by the fact that the compression of a solid makes it harder and thus the restoring forces become greater which in turn implies an increase in the vibrational frequencies. Therefore one would expect v_j to increase with decreasing volume and γ_j to have positive values.

In order to derive the Grüneisen equation of state, (10.3), from (10.14)–(10.16), one has to introduce at this stage a further assumption. Two models which lead to the Grüneisen equation are discussed (Brush, S.G. 1967).

10.2 The Einstein model of solids

Einstein (1907, 1911) developed a simple model in order to explain the experimental observations that the solid heat capacity decreases at low temperatures below the Dulong–Petit value of $3R$ (per mole). Einstein assumes that the thermal properties of the vibrations of a solid can be described by a lattice of N atoms vibrating as a set of $3N$ independent harmonic oscillators in one dimension where each oscillator has a frequency v. The energy of the oscillators is quantized in a similar way to Planck's quantization of the theory of black body radiation.

The main assumption of this model is that all frequencies are equal

$$v_j = v \quad j = 1, 2, \ldots, 3N \qquad (10.17)$$

Substituting this equation in (10.14)–(10.16) gives

$$E = E_c + 3Nh\nu/(e^{\beta h \nu} - 1) \tag{10.18}$$

$$P = P_c + 3Nh\nu\gamma/[V(e^{\beta h \nu} - 1)] \tag{10.19}$$

where the cold pressure P_c is defined by

$$P_c = -\frac{dE_c}{dV} \tag{10.20}$$

and the Grüneisen parameter in this model is

$$\gamma = -\frac{V}{\nu}\left(\frac{d\nu}{dV}\right) \tag{10.21}$$

Defining the atomic thermal contribution to the energy and pressure by

$$E_{Ta} = E - E_c \tag{10.22}$$

$$P_{Ta} = P - P_c \tag{10.23}$$

one obtains from (10.18) and (10.19) the Grüneisen equation of state, (10.3).

In this model the specific heat at constant volume c_v is obtained from (10.18)

$$c_v = \left(\frac{\partial E}{\partial T}\right)_V = 3Nk\left(\frac{h\nu}{kT}\right)^2 \frac{\exp(h\nu/kT)}{[\exp(h\nu/kT) - 1]^2} \tag{10.24}$$

or equivalently for a mole of atoms one has

$$c_v = 3R\left(\frac{T_E}{T}\right)^2 e^{T_E/T}/(e^{T_E/T} - 1)^2 \tag{10.25}$$

where $R = Nk$ is the gas constant and N in this case is Avogadro's number. The characteristic temperature T_E (the Einstein temperature) is defined by

$$h\nu = kT_E \tag{10.26}$$

and for most of the solids it can be fitted in the range of $T_E \approx 100$–$300\,\mathrm{K}$. The temperature $T_E = 300\,\mathrm{K}$ is equivalent (from (10.26)) to $\nu \approx 5 \times 10^{12}\,\mathrm{s}^{-1}$, which is a typical magnitude of atomic oscillation frequency. Einstein's model predicts $c_v \rightarrow 0$ as $T \rightarrow 0$, however, the experimental data are found to be only in qualitative agreement with (10.24). The decrease of c_v with temperature was experimentally slower at low temperature than predicted by Einstein's (10.24). It has been suggested by Einstein himself (1911) that his formula can be improved by assuming a large number of different frequencies which are associated with elastic sound waves. This model was developed by Debye (1912).

10.3 The Debye model of solids

In Debye's theory the single frequency assumed by Einstein is replaced by a spectrum of frequencies. The atoms are coupled and vibrate collectively. Debye assumed the solid to be a continuous medium and he calculated the number of standing waves in this medium. Since the solid contains N atoms, only the $3N$ lowest frequencies occur. Therefore the number of standing waves $g(v)dv$ which can be fitted in a solid of specific volume V between the frequencies v and $v + dv$ is given by (see Appendix 1 for a typical calculation of this problem)

$$g(v)dv = 4\pi v^2 dv \, V\left(\frac{1}{u_l^3} + \frac{2}{u_t^3}\right) \tag{10.27}$$

where u_l is the longitudinal wave velocity and u_t is the transverse wave velocity. The maximum frequency v_D is determined by the requirement that there are only $3N$ oscillating modes possible, i.e.

$$3N = 4\pi V\left(\frac{1}{u_l^3} + \frac{2}{u_t^3}\right)\int_0^{v_D} v^2 dv \tag{10.28}$$

implying

$$3N = \frac{4\pi}{3} V v_D^3 \left(\frac{1}{u_l^3} + \frac{2}{u_t^3}\right) \tag{10.29}$$

Comparing (10.27) and (10.29) one can write the number density

$$g(v)dv = \frac{9Nv^2 dv}{v_D^3} \tag{10.30}$$

It is customary to express v_D in terms of the Debye characteristic temperature T_D defined in a similar way to Einstein's temperature (10.26)

$$hv_D = kT_D \tag{10.31}$$

The Debye temperature range between $100\,\mathrm{K}$ and $400\,\mathrm{K}$ for most of the metals, e.g. $T_D = 390\,\mathrm{K}$ for Al, $T_D = 150\,\mathrm{K}$ for Na, etc. Using the notation

$$\xi = \frac{hv}{kT} \quad d\xi = \frac{hdv}{kT} \tag{10.32}$$

the atomic thermal energy E_{Ta} is obtained from (10.14) by using (10.30) as

$$E_{Ta} = \int_0^{v_D} hvg(v)dv = \frac{9Nh}{v_D^3}\int_0^{v_D} \frac{v^3 dv}{e^{hv/kT} - 1} \tag{10.33}$$

or by using the notation (10.32) for a mole of atoms

$$E_{Ta} \equiv E_D = 9RT \left(\frac{T}{T_D}\right)^3 \int_0^{T_D/T} \frac{\xi^3 d\xi}{e^\xi - 1} \tag{10.34}$$

where E_D is defined as the thermal energy of one mole in Debye's theory of a solid. The specific heat per mole in this model is

$$c_v = \left(\frac{\partial E_D}{\partial T}\right)_V = 9R \left(\frac{T}{T_D}\right)^3 \int_0^{T_D/T} \frac{e^\xi \xi^4 d\xi}{(e^\xi - 1)^2} \tag{10.35}$$

At high temperatures $T_D/T \ll 1$ and one can approximate $T \to \infty$ in (10.34) and (10.35) so that $\xi \to 0$, $e^\xi - 1 \to \xi$ implying

$$E_D = 3RT \tag{10.36}$$

$$c_v = 3R \tag{10.37}$$

which is the Dulong–Petit value for c_v. At very low temperatures $T_D/T \gg 1$, the upper limit of the integral in (10.34) is taken to be infinite and the integral can be exactly calculated,

$$E_D = 9RT \left(\frac{T}{T_D}\right)^3 \int_0^\infty \frac{\xi^3 d\xi}{e^\xi - 1} \tag{10.38}$$

Since $e^{-\xi} < 1$ one expands $(e^\xi - 1)^{-1}$ to the power of $e^{-\xi}$

$$\int_0^\infty \frac{\xi^3 d\xi}{e^\xi - 1} = \sum_{n=1}^\infty \int_0^\infty \xi^3 e^{-n\xi} d\xi = \sum_{n=1}^\infty \frac{6}{n^4} = \frac{\pi^4}{15} \tag{10.39}$$

so that the Debye energy in this case is

$$E_D = \frac{3\pi^4 RT^4}{5T_D^3} \tag{10.40}$$

and the specific heat per mole is

$$c_v = \frac{12\pi^4 R}{5} \left(\frac{T}{T_D}\right)^3 = 233.78 R \left(\frac{T}{T_D}\right)^3 \tag{10.41}$$

This is the famous T^3 law derived by Debye in order to explain satisfactorily the low temperature specific heat.

The Grüneisen equation of state (10.3) in the Debye model (Grüneisen 1926, Slater 1939, Kittel 1956)

$$P_{Ta} = \frac{\gamma}{V} E_D \tag{10.42}$$

is obtained from equations (10.14) and (10.15) by making the assumption

$$\gamma_j = \gamma \quad j = 1, 2, \ldots, 3N \tag{10.43}$$

In order to understand this statement and the meaning of γ it is possible to assume that v_j depends on the volume by the simple relation

$$v_j = \frac{c_j}{V^\gamma} \tag{10.44}$$

since v_j increase with decreasing volume. c_j is a constant, therefore from (10.44) one gets

$$\frac{-\text{d} \ln v_j}{\text{d} \ln V} = \gamma$$

so that the γ defined in (10.16) is independent of j and the Grüneisen assumption, (10.43) is justified.

In a more general way one can derive (10.42) using the following thermodynamic identities, (Kittel 1956)

$$E = F - T\left(\frac{\partial F}{\partial T}\right)_V = \left[\frac{\partial(F/T)}{\partial(1/T)}\right] \tag{10.45}$$

$$P = -\left(\frac{\partial F}{\partial V}\right)_T \tag{10.46}$$

and the approximation

$$F = F_c(V) + F_D(T, V) \tag{10.47}$$

where F_D is the contribution of the lattice vibrations to the free energy in the Debye model. It is assumed that F_D can be specified by

$$F_D = F_D(T, T_D(V))$$

so that (10.45)–(10.47) yield

$$P = -\frac{\text{d}E_c}{\text{d}V} - \frac{\partial F_D}{\partial T_D}\frac{\partial T_D}{\partial V} \tag{10.48}$$

In the Debye model (10.34) one has

$$E_D = T h(T_D/T) \tag{10.49}$$

therefore using (10.45) F_D must be of the same form

$$F_D = T f(T_D/T) \tag{10.50}$$

implying

$$\frac{\partial F_D}{\partial T_D} = f' = T_D^{-1} \left[\frac{\partial (F_D/T)}{\partial (1/T)} \right] = \frac{E_D}{T_D} \qquad (10.51)$$

Substituting this result in (10.48) one gets the Grüneisen equation of state

$$P = P_c + \frac{\gamma E_D}{V} \qquad (10.52)$$

where the Grüneisen parameter is defined by

$$\gamma = -\frac{V}{T_D} \frac{dT_D}{dV} = -\frac{d\ln T_D}{d\ln V} \qquad (10.53)$$

10.4   The Grüneisen relation

Taking the derivative with respect to temperature of (10.52) for a constant specific volume one gets

$$\left(\frac{\partial P}{\partial T} \right)_V = \frac{\gamma c_v}{V} \qquad (10.54)$$

The linear expansion coefficient α is (one third of the volume expansion coefficient)

$$\alpha = \frac{1}{3V} \left(\frac{\partial V}{\partial T} \right)_P \qquad (10.55)$$

Using the thermodynamic identity

$$\left(\frac{\partial V}{\partial T} \right)_P = - \left(\frac{\partial P}{\partial T} \right)_V \left(\frac{\partial V}{\partial P} \right)_T \qquad (10.56)$$

and the definition of the isothermal compressibility κ,

$$\kappa = -\frac{1}{V} \left(\frac{\partial V}{\partial P} \right)_T \qquad (10.57)$$

one gets from (10.55)–(10.57)

$$\alpha = \frac{\kappa}{3} \left(\frac{\partial P}{\partial T} \right)_V \qquad (10.58)$$

Using the Grüneisen equation of state as given by (10.54) and (10.58) one obtains the Grüneisen relation

$$\gamma = \frac{3\alpha V}{\kappa c_v} \qquad (10.59)$$

The quantities on the right hand side of this relation can be measured experimentally and are given in Table 10.1. In this table a comparison is also made with the γ values calculated by Slater and described in the next section. The agreement between Slater's calculations and the Grüneisen relation given by (10.59) seems to be fairly good.

10.5 Slater–Landau calculation of γ

According to Debye's theory the frequency spectrum is determined by the maximum allowed frequency, denoted by v_D (see (10.30)). If v_D changes then all other oscillations will change their frequencies in the same ratio, so that the Grüneisen's assumption that γ is the same for all frequencies is justified. Using this justification Slater (1939) assumes

$$\gamma = -\frac{d \ln v_D}{d \ln V} \tag{10.60}$$

From (10.29) one has

$$v_D = \text{const } V^{-1/3} \left(\frac{1}{u_l^3} + \frac{2}{u_t^3} \right)^{-1/3} \tag{10.61}$$

From the theory of elasticity one obtains the velocities of sound propa-

Table 10.1. *Values of Grüneisen parameter γ as obtained from (10.59) and (10.72) with the appropriate values of P_1 and P_2*

	P_1 (cgs)	P_2 (cgs)	$-\dfrac{2}{3}+\dfrac{P_2}{P_1}$ (10.72)	γ Grüneisen (10.59)
Li	0.115×10^{12}	0.149×10^{12}	0.63	1.17
Na	0.064	0.160	1.83	1.25
Al	0.757	1.46	1.27	2.17
K	0.028	0.090	2.55	1.34
Mn	1.24	8.1	5.8	2.42
Fe	1.73	4.0	1.7	1.60
Co	1.85	5.1	2.1	1.87
Ni	1.90	5.4	2.2	1.88
Cu	1.39	3.6	1.9	1.96
Pd	1.93	5.9	2.4	2.23
Ag	1.01	3.2	2.5	2.40
W	3.40	8.	1.7	1.62
Pt	2.78	11.	3.3	2.54
Pb	0.42	1.30	2.42	2.73

Source: Taken from J.C. Slater *Introduction to Chemical Physics*, McGraw Hill, New York, 1939, p. 451.

gation as

$$u_1 = \left[\frac{3(1-\sigma)}{\kappa\rho(1+\sigma)} \right]^{1/2} \tag{10.62}$$

$$u_t = \left[\frac{3(1-2\sigma)}{2\kappa\rho(1+\sigma)} \right]^{1/2} \tag{10.63}$$

where κ is the volume compressibility (10.57), σ is the Poisson ratio and ρ is the density. Assuming that the Poisson ratio is independent of the specific volume V, one can write from (10.61)–(10.63)

$$v_D = \text{const } V^{1/6}\kappa^{-1/2} \tag{10.64}$$

where the constants are the volume independent terms and we have used the fact that the density is inversely proportional to the specific volume V. The relation (10.64) yields

$$\ln v_D = \tfrac{1}{6}\ln V - \tfrac{1}{2}\ln \kappa + \text{const} \tag{10.65}$$

so that the Grüneisen coefficient which is given by (10.60) can be obtained from the relation

$$\gamma = -\frac{1}{6} + \frac{1}{2}\frac{d\ln\kappa}{d\ln V} \tag{10.66}$$

Using for the compressibility the definition of (10.57) one has

$$\ln\kappa = -\ln V - \ln\left[-\left(\frac{\partial P}{\partial V}\right)_T \right] \tag{10.67}$$

implying, by (10.66)

$$\gamma = -\frac{1}{6} + \frac{1}{2}\left\{ -1 - \frac{d\ln|-(\partial P/\partial V)_T|}{d\ln V} \right\}$$

$$\gamma = -\frac{2}{3} - \frac{1}{2}V\left(\frac{\partial^2 P}{\partial V^2}\right)_T \bigg/ \left(\frac{\partial P}{\partial V}\right)_T \tag{10.68}$$

The derivatives of (10.68) are estimated from the following expansion

$$P = P_0 + P_1\left[\frac{V_0 - V}{V_0} \right] + P_2\left[\frac{V_0 - V}{V_0} \right]^2 \tag{10.69}$$

where in general P_0, P_1 and P_2 have positive values and are functions of temperature. P_0 is the pressure one has to apply to a solid in order to reduce its volume to V_0, which is the volume at $T = 0$ under no pressure (note: $P_0 = 0$ at $T = 0$). From (10.69)

$$\left(\frac{\partial P}{\partial V}\right)_T = -\frac{P_1}{V_0} + \frac{2P_2}{V_0}\left(\frac{V_0 - V}{V_0}\right) \tag{10.70}$$

$$\left(\frac{\partial^2 P}{\partial V^2}\right)_T = \frac{2P_2}{V_0^2} \tag{10.71}$$

Calculating the Grüneisen value of γ at $T = 0$ (i.e. for $V = V_0$) one gets from (10.68), (10.70) and (10.71)

$$\gamma(V_0) = -\frac{2}{3} + \frac{P_2}{P_1} \tag{10.72}$$

The values of P_2 and P_1 can be calculated from the experiments by using the following approximation,

$$\frac{V_0 - V}{V_0} = -a_0 + a_1 P - a_2 P^2 + \cdots \tag{10.73}$$

where a_0, a_1 and a_2 are in general functions of temperature and they can be fitted experimentally. Substituting (10.73) into (10.69)

$$\begin{aligned}
P = P_0 &+ P_1(-a_0 + a_1 P - a_2 P^2 + \cdots) \\
&+ P_2(-2a_0 a_1 P + 2a_0 a_2 P^2 + a_1^2 P^2 + \cdots)
\end{aligned} \tag{10.74}$$

where a_0^2 was neglected as a small quantity ($a_0 = 0$ for $V = V_0$ and $P = 0$, while a_1 is approximately the compressibility κ which is in general much larger than a_0), and equating the coefficients of the different powers in P one has

$$a_0 = \frac{P_0}{P_1}$$

$$a_1 = \frac{1}{P_1 - 2a_0 P_2} \approx \frac{1}{P_1}\left(1 + \frac{2P_0 P_2}{P_1^2}\right)$$

$$a_2 = \frac{P_2 a_1^2}{P_1 - 2a_0 P_2} \approx \frac{P_2}{P_1^3}\left(1 + \frac{6P_0 P_2}{P_1^2}\right) \tag{10.75}$$

Solving for P_0, P_1 and P_2 one gets

$$P_0 = \frac{a_0}{a_1}$$

$$P_1 = \frac{1}{a_1}\left(1 + \frac{2a_0 a_2}{a_1^2}\right) \tag{10.76}$$

$$P_2 = \frac{a_2}{a_1^3}$$

Since the values of a are measured experimentally, P_0, P_1 and P_2 can be calculated. In Table 10.1 the values of P_1 and P_2 are given and the Grüneisen coefficient is calculated from (10.72).

10.6   Results and discussion

Equation (10.68) (or (10.72)) is not the only model that has been suggested in order to calculate the Grüneisen coefficient γ. The main assumption which seems to be unjustified in the Slater–Landau model is the statement that the Poisson ratio does not change with volume. Therefore several other models have been developed in the literature and their results are summarized by the following equations (Altshuler 1965)

$$\gamma(V) = -\left(\frac{2}{3} - \frac{t}{3}\right) - \frac{V}{2}\frac{d^2}{dV^2}(P_c V^{2t/3}) \bigg/ \frac{d}{dV}(P_c V^{2t/3}) \tag{10.77}$$

The most popular values for t are $t = 0$ or 1 or 2, where the value $t = 0$ corresponds to (10.68) (note γ does not depend on T so that one can take the derivatives at $T = 0$). The value of $t = 1$ was suggested by Dugdale and McDonald (1953). These three models for γ give almost equal values as the pressure is increased and they differ mainly at low pressures. More recently (Romain *et al.* 1979) a generalization of (10.77) has been proposed where t is a characteristic parameter for each material.

The Slater–Landau and Dugdale–MacDonald formulae have been extensively used in analyzing shock wave Hugoniot experiments (see Chapter 13). It has been observed from these shock wave data that to a good approximation one can write

$$\frac{\gamma_0}{V_0} = \frac{\gamma}{V} \tag{10.78}$$

This empirical result together with the fact that at very high compressions the limiting value for γ for all materials is $\frac{2}{3}$ suggested an interpolation formula for $\gamma(V)$,

$$\gamma = \frac{\gamma_0 V}{V_0} + \frac{2}{3}\left(1 - \frac{V}{V_0}\right) \tag{10.79}$$

which is used in the SESAME library of equations of state (1978). It is possible to construct other interpolation formulae in order to fit the experimental data, and other theories for the solid state equation of state have been developed (e.g., Born & Huang 1954).

11

An introduction to fluid mechanics in relation to shock waves[†]

11.1 Fluid equations of motion

The state of a moving fluid or gas can be defined in terms of its velocity ($\mathbf{u}$), density (ρ), and pressure (P) as functions of position ($\mathbf{r}$), and time (t). These functions are obtained by integral equations or differential equations which are derived from the conservation laws of *mass, momentum* and *energy*.

11.1.1 Mass conservation equation

We define $\rho = \rho(\mathbf{r}, t)$ to be the mass density at a point ($\mathbf{r}, t$) and $\mathbf{u}(\mathbf{r}, t)$ its velocity. The change of mass in a volume Γ is $(\partial/\partial t)\int_\Gamma \rho \, dV$. This change is balanced by the mass flow $\oint_S \rho \mathbf{u} \cdot d\mathbf{A}$ ($d\mathbf{A}$ is an area vector) across the surface S enclosing the volume Γ. Therefore, the mass conservation can be written in an integral form by

$$\frac{\partial}{\partial t} \int_\Gamma \rho \, dV + \oint_S \rho \mathbf{u} \cdot d\mathbf{A} = 0 \qquad (11.1)$$

Using Gauss' theorem, one can write (11.1) in the form

$$\frac{\partial}{\partial t} \int_\Gamma \rho \, dV + \int_\Gamma \boldsymbol{\nabla} \cdot (\rho \mathbf{u}) \, dV = 0 \qquad (11.2)$$

Since this last relation is true for any arbitrary volume Γ, we obtain the following equation describing the conservation of mass

$$\frac{\partial \rho}{\partial t} + \boldsymbol{\nabla} \cdot (\rho \mathbf{u}) = 0 \qquad (11.3)$$

The above equation is known as the equation of continuity.

[†] A detailed account of the physics of shock waves can be found in Zel'Dovich and Raizer (1966, 1967).

11.1.2   Momentum conservation equation

It is convenient to consider a moving volume whose surface encloses a constant amount of fluid. For this purpose one defines $\partial/\partial t$ as a partial derivative with respect to time at a given point in space, and D/Dt as a total derivative describing the change in time of a moving fluid volume. This fluid mass ρdV, sometimes called a 'fluid particle' has a vector velocity $\mathbf{u}$. The relation between the two derivatives is given by

$$\frac{D}{Dt} = \frac{\partial}{\partial t} + \mathbf{u} \cdot \nabla \tag{11.4}$$

Denoting the surface forces by a tensor f, with dimensions Newton/m² in m.k.s. units, and the body forces by a vector $\mathbf{F}$, dimensions Newton/m³, one can write the Newton's second law as

$$\int_\Gamma \rho \frac{D\mathbf{u}}{Dt} dV = \oint_S \mathbf{f} \cdot d\mathbf{A} + \int_\Gamma \mathbf{F} dV \tag{11.5}$$

The surface forces are usually described by the stress tensor f_{ij} where i, j ($= 1, 2, 3$) denote three orthogonal directions in space, i.e., $x-y-z$ directions,

$$f_{ij} = -(P + \tfrac{2}{3}\eta \nabla \cdot \mathbf{u})\delta_{ij} + \eta \left(\frac{\partial u_i}{\partial x_j} + \frac{\partial u_j}{\partial x_i} \right) \tag{11.6}$$

where η is the fluid viscosity and P the pressure. Since in this chapter the viscosity contributions are neglected, it is assumed that

$$\eta = 0 \tag{11.7}$$

so that the total force acting on the surface in the ith direction is

$$\oint_S \sum_{j=1}^{3} f_{ij} dA_j = -\oint_S P dA_i = -\int_\Gamma \frac{\partial P}{\partial x_i} dV \tag{11.8}$$

where Gauss' theorem has been used in the last step of (11.8). Thus, using (11.7) and (11.8), (11.5) can be written in the form

$$\int_\Gamma \rho \frac{D\mathbf{u}}{Dt} dV = -\int_\Gamma \nabla P dV + \int_\Gamma \mathbf{F} dV \tag{11.9}$$

Since the last equation is true for an arbitrary volume Γ, the differential equation describing the motion of the fluid can be obtained from (11.9)

$$\rho \frac{D\mathbf{u}}{Dt} = -\nabla P + \mathbf{F} \tag{11.10}$$

The body forces $\mathbf{F}$ are acting on the volume element of the fluid and differ from one problem to another. For example, the electromagnetic force due to the motion of free charges is equal to $\mathbf{F} = \rho_e \mathbf{E} + \mathbf{J} \times \mathbf{B}$ where $\mathbf{E}$ and $\mathbf{B}$ are the electric and magnetic fields respectively, $\mathbf{J}$ is the sum of the conduction current and the current flow due to the convective transport of charges and ρ_e is the electric charge density. In a similar way, for appropriate processes the polarization force, the magnetization force, the gravitational force or any other force acting on the volume elements should be taken into account. In the discussions of this chapter, the volume forces will be neglected so that

$$\mathbf{F} = 0 \tag{11.11}$$

Therefore, the equation of motion (11.10), with the assumption (11.11), takes the form as suggested by Euler:

$$\rho \frac{\partial \mathbf{u}}{\partial t} + \rho(\mathbf{u} \cdot \nabla)\mathbf{u} = - \nabla P \tag{11.12}$$

where we have used (11.4). This differential equation describes the conservation of momentum if the viscosity and the volume forces $\mathbf{F}$ are neglected.

11.1.3 Energy conservation equation

In order to discuss energy conservation, we consider the energy in the frame of reference of the moving fluid. The increase of the fluid particle energy is given by the rate of change of its kinetic energy $\int_\Gamma \frac{1}{2}\rho(Du^2/Dt)dV$(Joule s^{-1}) and its internal energy $\int_\Gamma \rho(D\varepsilon/Dt)dV$, where ε is the volume density energy of the fluid (Joule m^{-3}). This energy should be balanced by the external energy sources Q_{EX}(Joule s^{-1}) entering the system, the energy rate of heat conduction Q_H, and the work (per unit time) done by the volume forces (W_F) and by the surface forces (W_f). The energy conservation can be written in this case by

$$\int_\Gamma \rho \frac{D(u^2/2)}{Dt} dV + \int_\Gamma \rho \frac{D\varepsilon}{Dt} dV = Q_{EX} + Q_H + W_F + W_f \tag{11.13}$$

The rate of work done by the surface forces is obtained by integrating $\mathbf{u} \cdot \mathbf{f}$ over the surface of the fluid particle,

$$W_f = \oint_S \mathbf{u} \cdot \mathbf{f} \cdot d\mathbf{A} = \oint_S \sum_{i=1}^3 \sum_{j=1}^3 u_i f_{ij} dA_j = - \oint_S P\mathbf{u} \cdot d\mathbf{A},$$

where (11.6) has been used with $\eta = 0$. Applying Gauss' theorem, the rate of work is given by

$$W_f = - \int_\Gamma \nabla \cdot (P\mathbf{u}) dV \,[\text{Joule s}^{-1}] \tag{11.14}$$

The rate of work done by the volume forces is

$$W_F = \int_\Gamma \mathbf{F} \cdot \mathbf{u} dV \,[\text{Joule s}^{-1}] \tag{11.15}$$

and the energy entering per unit time by heat conduction is

$$Q_H = - \int_\Gamma \mathbf{\nabla} \cdot (\kappa \mathbf{\nabla} T) dV \,[\text{Joule s}^{-1}] \tag{11.16}$$

where κ is the coefficient of thermal conductivity and T is the temperature. The rate at which external energy enters the fluid particle depends on the system under consideration. For example, electric (ohm) deposition of energy is given by $Q_{EX} = \int_\Gamma \mathbf{E} \cdot \mathbf{J} dV$. However, in this chapter we assume

$$Q_{EX} = 0 \quad Q_H = 0 \text{ (i.e. } \kappa = 0) \quad W_F = 0 \text{ (i.e. } \mathbf{F} = 0) \tag{11.17}$$

therefore, under these assumptions the energy conservation equation (11.12) becomes

$$\rho \frac{D}{Dt} (\varepsilon + \tfrac{1}{2} u^2) = - \mathbf{\nabla} \cdot (P\mathbf{u}) \tag{11.18}$$

where (11.14) has been used. Applying the transformation of (11.4) on the left hand side (LHS) of (11.18) and using the mass conservation equation (viz. (11.3)), one gets

$$\begin{aligned}
\text{LHS of (11.18)} &= \rho \frac{\partial}{\partial t} (\varepsilon + \tfrac{1}{2} u^2) + \rho \mathbf{u} \cdot \mathbf{\nabla} (\varepsilon + \tfrac{1}{2} u^2) \\
&= \frac{\partial}{\partial t} (\rho \varepsilon + \tfrac{1}{2} \rho u^2) - (\varepsilon + \tfrac{1}{2} u^2) \frac{\partial \rho}{\partial t} \\
&\quad + \mathbf{\nabla} \cdot [(\rho \varepsilon + \tfrac{1}{2} \rho u^2) \mathbf{u}] - (\varepsilon + \tfrac{1}{2} u^2) \mathbf{\nabla} \cdot (\rho \mathbf{u}) \\
&= \frac{\partial}{\partial t} [\rho (\varepsilon + \tfrac{1}{2} u^2)] + \mathbf{\nabla} \cdot [\rho \mathbf{u} (\varepsilon + \tfrac{1}{2} u^2)] \tag{11.19}
\end{aligned}$$

Substituting this relation in (11.18), we obtain the following differential equation which describes the conservation of energy.

$$\frac{\partial}{\partial t} [\rho (\varepsilon + \tfrac{1}{2} u^2)] = - \mathbf{\nabla} \cdot [\rho \mathbf{u} (\varepsilon + \tfrac{1}{2} u^2) + P\mathbf{u}] \tag{11.20}$$

where we have neglected viscosity, thermal conductivity, volume forces and external energy sources.

In conclusion, in our introductory discussion of shock waves, our starting point will be the equations of mass conservation (11.3), momentum conservation (11.12) and energy conservation (11.20). In general the study of flow phenomena is based on these macroscopic equations of motion.

11.2   Sound waves and Rieman invariants

From now on, we shall consider only waves in one dimension with planar symmetry. This approach is justified if we are far from the source so that the wave form is nearly plane. For such a case, the variables describing the wave phenomena are functions only of the displacement x and the time t, i.e., the pressure P, the density ρ and the flow velocity u are in general functions of x and t. The physics of sound waves is described by the motion of the fluid and the changes of density of the medium caused by a pressure change. Assuming an equilibrium pressure P_0 and density ρ_0, then for the case of a sound wave, the changes in pressure ΔP and density $\Delta\rho$ due to the existence of sound waves are extremely small.

$$\left.\begin{aligned}\rho = \rho_0 + \Delta\rho \quad & \frac{\Delta\rho}{\rho_0} \ll 1 \\[2ex] P = P_0 + \Delta P \quad & \frac{\Delta P}{P_0} \ll 1\end{aligned}\right\} \tag{11.21}$$

We assume now that the undisturbed flow velocity u_0 is zero and the flow speed caused by the pressure disturbance is Δu, so that

$$u = u_0 + \Delta u \quad u_0 \equiv 0 \quad \frac{\Delta u}{c} \ll 1 \tag{11.22}$$

where c is the speed of sound defined later in (11.30). The continuity equation (viz., (11.3)) and the momentum conservation condition (viz., (11.12)) can be written in the form

$$\frac{\partial(\Delta\rho)}{\partial t} = -\rho_0 \frac{\partial u}{\partial x} \tag{11.23}$$

$$\rho_0 \frac{\partial u}{\partial t} = -\frac{\partial P}{\partial x} \tag{11.24}$$

where we have neglected second order quantities in $\Delta\rho$, ΔP and u. Taking into account that the particle motion in a sound wave is isentropic, i.e.,

$$S(x) = \text{const} \tag{11.25}$$

one can relate the pressure changes to the density changes by

$$\Delta P = \left(\frac{\partial P}{\partial\rho}\right)_S \Delta\rho \equiv c^2 \Delta\rho \tag{11.26}$$

Differentiating (11.23) with respect to time and (11.24) with respect to space (x) and adding the equations one gets the wave equation

$$\frac{\partial^2(\Delta\rho)}{\partial t^2} = c^2 \frac{\partial^2(\Delta\rho)}{\partial x^2} \tag{11.27}$$

where (11.26) has been used. From (11.27) it is evident that c is the velocity of the $\Delta\rho$ disturbance which is defined as the velocity of the sound wave. The pressure change ΔP and the fluid velocity u satisfy an equation similar to (11.27). Equation (11.27) has two families of solutions

$$\left.\begin{array}{r}\Delta\rho\\\Delta P\\\Delta u\end{array}\right\} \approx f(x-ct) \tag{11.28}$$

$$\left.\begin{array}{r}\Delta\rho\\\Delta P\\\Delta u\end{array}\right\} \approx g(x+ct) \tag{11.29}$$

where

$$c = +\left(\frac{\partial P}{\partial\rho}\right)_s^{1/2} \tag{11.30}$$

Disturbances of the first type are propagating in the positive x direction while the second type are propagating in the negative x direction. Using the solutions (11.28) and (11.29) in (11.23) and (11.26), one gets

$$\frac{\Delta u}{c} = \pm\frac{\Delta\rho}{\rho_0} = \pm\frac{\Delta P}{\rho_0 c^2} \tag{11.31}$$

In order to get an idea of how the temperature varies in a sound wave, we calculate the relation between the sound velocity and the temperature for an ideal gas. As mentioned above, the pressure variation with density in a sound wave does not allow heat flow so that the process is adiabatic. The equation of state for an adiabatic process in an ideal gas is

$$P = P_0\left(\frac{\rho}{\rho_0}\right)^\gamma \tag{11.32}$$

Thus, (11.32) gives us

$$c = \left(\frac{\gamma P}{\rho}\right)^{1/2} \tag{11.33}$$

where γ denotes the ratio of c_p/c_v. Introducing temperature through the ideal equation of state

$$P = \frac{NkT}{V} = \frac{kT\rho}{m} \tag{11.34}$$

where k is Boltzmann's constant, T is the temperature and m is the mass of a molecule of the gas, we get

$$c = \left(\frac{\gamma k T}{m}\right)^{1/2} \tag{11.35}$$

Since γ does not usually depend strongly on the pressure or density of the gas, the sound velocity in a gas is proportional to the square root of the temperature, therefore for a given T the value of c does not depend on the pressure. It is important to note that from (11.35) one can easily see that the velocity of sound in a gas is of the same order of magnitude as the mean thermal velocities of the molecules.

If the undisturbed gas or fluid is not stationary, but moves with a constant velocity u, then the waves are carried by the flow-stream. By making a transformation from the coordinate system moving with the flow to the laboratory (rest) frame of reference, one finds that the sound wave is travelling with a velocity $u + c$ for waves going in the $+ x$ direction and with a velocity $u - c$ for waves travelling in the $- x$ direction (u is in the $+ x$ direction). The curves in the x–t plane in the direction of propagation of small disturbances are called characteristic curves. For isentropic flow there are two families of characteristics described by[†]

$$C_+ : \frac{dx}{dt} = u + c$$

$$C_- : \frac{dx}{dt} = u - c \tag{11.36}$$

so that the variables P and ρ do not change along these stream-lines. In general, the $C_\pm$ characteristics are curved lines in the x–t plane because u and c are, in general, functions of x and t.

Defining a curve $x = \phi(t)$ in the x–t plane, the derivative of a function $f(x, t)$ along the curve ϕ is given by

$$\left(\frac{df}{dt}\right)_\phi = \frac{\partial f}{\partial t} + \frac{\partial f}{\partial x}\frac{dx}{dt} = \frac{\partial f}{\partial t} + \phi'\frac{\partial f}{\partial x} \tag{11.37}$$

For example, along $x = $ constant $\phi' = 0$; along a stream-line, $\phi' = u$, along a characteristic C_+, $\phi' = u + c$, etc. Using the definition (11.37) we now derive the fluid equation of motion along the characteristics C_+ and C_-.

[†] There is a third possible characteristic for an adiabatic flow which is not isentropic, i.e., $S(x) \neq$ constant but $DS/Dt = 0$. This case is possible if the entropy of a particle does not change with time but different particles do not have the same entropy. This type of disturbance in entropy does not change the other variables P, ρ and u, since the disturbances in S are localized within the particle along a stream-line described by the third characteristic C_0 which satisfy the equation $dx/dt = u$.

In general the density ρ is a function of two variables, e.g. P and S (entropy). However, for an isentropic process one has

$$\rho = \rho(P) \tag{11.38}$$

so that the ρ derivatives can be expressed in terms of the P derivatives

$$\frac{\partial \rho}{\partial t} = \left(\frac{\partial \rho}{\partial P}\right)_S \frac{\partial P}{\partial t} = \frac{1}{c^2} \frac{\partial P}{\partial t}$$

$$\frac{\partial \rho}{\partial x} = \frac{1}{c^2} \frac{\partial P}{\partial x} \tag{11.39}$$

If we now use the continuity equation (viz. (11.3)) and momentum conservation equation (viz. (11.12)), we get

$$\left. \begin{array}{l} \dfrac{1}{\rho c} \dfrac{\partial P}{\partial t} + \dfrac{u}{\rho c} \dfrac{\partial P}{\partial x} + c \dfrac{\partial u}{\partial x} = 0 \\[3mm] \dfrac{\partial u}{\partial t} + u \dfrac{\partial u}{\partial x} + \dfrac{1}{\rho} \dfrac{\partial P}{\partial x} = 0 \end{array} \right\} \tag{11.40}$$

Adding and subtracting these two relations we get

$$\left[\frac{\partial u}{\partial t} + (u + c)\frac{\partial u}{\partial x}\right] + \frac{1}{\rho c}\left[\frac{\partial P}{\partial t} + (u + c)\frac{\partial P}{\partial x}\right] = 0 \tag{11.41}$$

$$\left[\frac{\partial u}{\partial t} + (u - c)\frac{\partial u}{\partial x}\right] - \frac{1}{\rho c}\left[\frac{\partial P}{\partial t} + (u - c)\frac{\partial P}{\partial x}\right] = 0 \tag{11.42}$$

Equations (11.41) and (11.42) are derivatives along the characteristics C_+ and C_- respectively which can be rewritten as

$$\left. \begin{array}{l} dJ_+ \equiv du + \dfrac{1}{\rho c} dP = 0 \quad \text{along} \quad C_+ \quad \dfrac{dx}{dt} = u + c \\[3mm] dJ_- \equiv du - \dfrac{1}{\rho c} dP = 0 \quad \text{along} \quad C_- \quad \dfrac{dx}{dt} = u - c \end{array} \right\} \tag{11.43}$$

Equations (11.43) are the dynamic equations describing isentropic flow in one dimension with planar symmetry; $dJ_\pm$ are total differentials and can be integrated

$$\left. \begin{array}{l} J_+ = u + \displaystyle\int \dfrac{dP}{\rho c} = u + \displaystyle\int c \dfrac{d\rho}{\rho} \\[3mm] J_- = u - \displaystyle\int \dfrac{dP}{\rho c} = u - \displaystyle\int c \dfrac{d\rho}{\rho} \end{array} \right\} \tag{11.44}$$

where (11.26) has been used $(dP = c^2 d\rho)$. Since $dJ_\pm = 0$ along the curves $C_\pm$ one has

$$\left.\begin{array}{l} J_+ = \text{constant along } C_+ \\ J_- = \text{constant along } C_- \end{array}\right\} \tag{11.45}$$

J_+ and J_- are called *Riemann invariants* and are usually used to solve the equations of flow given of (11.43) following the invariants. Note, however, that for non-isentropic flow ρ and c are usually functions of two variables, i.e., $\rho = \rho(P, S)$ and $c = c(P, S)$, so that $dJ_\pm$ are no longer total differentials and $J_\pm$ have no physical meaning.

As an example we calculate the Riemann invariants J_+ and J_- for an ideal gas. The equation of state for an acoustic disturbance is described by (11.32) and the appropriate sound velocity by (11.33). From these two equations the following relation can be derived which relates the density ρ and the sound velocity c

$$\frac{c}{c_0} = \left(\frac{\rho}{\rho_0}\right)^{\gamma - 1/2} \tag{11.46}$$

Substituting the above in (11.44) and performing the integrations, we get

$$J_\pm = u \pm \frac{2}{\gamma - 1}(c - c_0) \tag{11.47}$$

where c_0 is the sound speed for $u = 0$. The Riemann invariants are determined within an arbitrary constant, which in the above example was chosen as c_0. Equation (11.47) enables us to express the flow velocity and the sound velocity in terms of the Riemann's invariants

$$\begin{array}{l} u = \tfrac{1}{2}(J_+ + J_-) \\ c - c_0 = \tfrac{1}{4}(\gamma - 1)(J_+ - J_-) \end{array} \tag{11.48}$$

The characteristics $C_\pm$ are given by

$$C_+: \frac{dx}{dt} = u + c = \frac{\gamma + 1}{4}J_+ + \frac{3 - \gamma}{4}J_- + c_0 \equiv F_+(J_+, J_-)$$

$$C_-: \frac{dx}{dt} = u - c = \frac{3 - \gamma}{4}J_+ + \frac{\gamma + 1}{4}J_- - c_0 \equiv F_-(J_+, J_-) \tag{11.49}$$

Since J_+ is a constant along C_+, the change in the slope (i.e., in dx/dt) of a characteristic is determined only by the change in J_-.

11.3 Rarefaction waves

We now consider the case where the pressure is suddenly dropped in an isentropic process. If one follows the variation in time for a given gas element (i.e., a fluid particle in the previous section) one gets

$$\frac{D\rho}{Dt} < 0 \quad \frac{DP}{Dt} < 0 \tag{11.50}$$

In general we have shown in the previous section that disturbances with small amplitudes satisfy the wave equation (11.27) and the solutions are of the form $f(x - ct)$ and $g(x + ct)$ (see (11.28) and (11.29)). The forward $(+x)$ propagating waves are described by

$$f(x - ct) \quad dP = \rho c \, du \tag{11.51}$$

while the backward $(-x)$ propagating waves are

$$g(x + ct) \quad dP = -\rho c \, du \tag{11.52}$$

where c is the velocity at which the small disturbances are moving into a material at rest.

For example, one can imagine the motion of a gas caused by a receding piston (moving in the $-x$ direction). In this case a forward $(+x$ direction) rarefaction wave is produced so that the gas is continually rarefied as it flows in the $-x$ direction. This type of motion is schematically described in

Fig. 11.1 A forward moving rarefaction wave. The propagation velocity at each point is $u + c$.

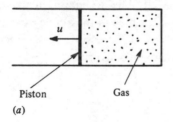

Piston　　　Gas

(a)

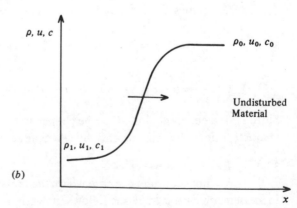

(b)

Fig. 11.1. One can consider the rarefaction wave to be represented by a sequence of jumps $d\rho$, du, etc. so that (11.51) can be integrated to obtain a relation between P and u for the points in the rarefaction front of Fig. 11.1

$$u - u_0 = \int_{P_0}^{P} \frac{dP}{\rho c} \tag{11.53}$$

The equivalent relation for a rarefaction wave moving in the $-x$ direction (backward) is obtained from (11.52)

$$u - u_0 = -\int_{P_0}^{P} \frac{dP}{\rho c} \tag{11.54}$$

From (11.53) and (11.54) one can see that the rarefaction waves satisfy

$$u \pm \int_{P_0}^{P} \frac{dP}{\rho c} = u_0 = \text{constant} \tag{11.55}$$

where the $-$ and $+$ signs correspond to forward and backward moving rarefaction waves respectively. Comparing (11.55) with (11.44) we get the relation between Riemann invariants and a rarefaction wave. Thus, the forward rarefaction wave moving into a material in the undisturbed state described by pressure (P_0), density (ρ_0) and fluid flow (u_0) satisfies the condition

$$J_- = u_0 = u - \int_{P_0}^{P} \frac{dP}{\rho c} \tag{11.56}$$

Over the entire physical part of the x–t plane.

Assuming now the case of an ideal gas with a constant γ, one can use (11.47) to obtain J_-,

$$J_- = u - \frac{2}{(\gamma - 1)}(c - c_0) = u_0 \tag{11.57}$$

From this relation one has

$$c = c_0 + \tfrac{1}{2}(\gamma - 1)(u - u_0) \tag{11.58}$$

In the above example the piston is moving in the $-x$ direction and at the gas–piston interface, the gas velocity is the same as the piston velocity which has a negative value.[†] Therefore from (11.58) one can see that the

---

[†] The negative gas flow velocity is limited, from the equation $c_0 + \tfrac{1}{2}(\gamma - 1)(u - u_0) = 0$, which is derived using the fact that the sound velocity is always positive (by definition, see (11.30)). For $u_0 = 0$ the maximum absolute value of the flow velocity is $|u| = 2c_0/(\gamma - 1)$.

speed of sound is decreased since u is negative (i.e., $c < c_0$). This implies, using (11.46), that the density and similarly the pressure (see (11.32)) are decreasing as expressed mathematically in (11.50).

11.4   Shock waves and the Hugoniot relation

The smooth disturbances previously discussed in this chapter are evolving into discontinuous quantities as a result of the nonlinear nature of the conservation equations (11.3), (11.12) and (11.20). The development of these nonlinear effects into shock waves had already been discussed by Riemann, Rankine and Hugoniot in the second half of the last century (1860–1890).

One can see from (11.46) that different disturbances of density travel with

Fig. 11.2. Wave steepening due to nonlinear effects.

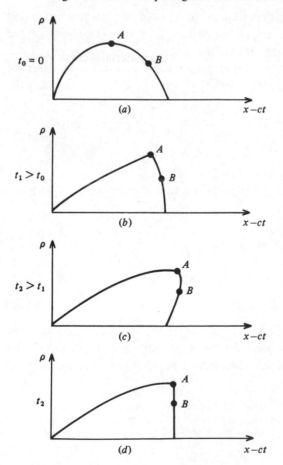

different velocities, so that for larger ρ the wave travels faster. Therefore, an initial profile $\rho(x, 0)$ becomes distorted, with time. This is true not only for the density but also for the pressure $P(x, 0)$ (see (11.32)), for the flow velocity $u(x, 0)$, etc. A schematic picture of wave distortion is given in Fig. 11.2. The density at point A of Fig. 11.2 is higher than the density at point B, therefore from (11.46) one gets a propagation velocity of point A higher than for point B, so that at a time $t_1 > t_0$ the shape of Fig. 11.2(a) has changed into that of Fig. 11.2(b). This type of nonlinearity develops in time until Fig. 11.2(c) might occur. However, the situation illustrated in Fig. 11.2(c) is not physically possible since in this case the density is not sufficiently defined in space. Mathematically, this situation can be visualized as intersecting characteristics (of C_+ in this case). Therefore, the final physical picture which can develop is given in Fig. 11.2(d). This type of singularity in the flow of a gas or a fluid is defined as a shock wave.

We assume a gas initially at rest, $u_0 = 0$, having a constant density ρ_0 and pressure P_0. The gas is compressed by a piston moving into it with a constant velocity u. A shock wave starts moving into the material with a velocity denoted by D. Behind the shock front the material is compressed to a density ρ_1 and a pressure P_1. The gas flow velocity in this region is equal to the piston velocity u. This case is schematically illustrated in Fig. 11.3. The initial volume that is being compressed in a time t is equal to ADt where A is the cross-sectional area of the tube. At the end of the compression this volume is $A(D - u)t$ as can be seen from Fig. 11.3. The mass of the gas that is set into motion is $\rho_0 ADt$, which by using the conservation of mass equals $\rho_1 A(D - u)t$. Therefore one gets

$$\rho_0 D = \rho_1 (D - u) \tag{11.59}$$

The momentum of the gas put into motion is $\rho_0 ADtu$, which is equal to the

Fig. 11.3. Shock wave created by the motion of a piston.

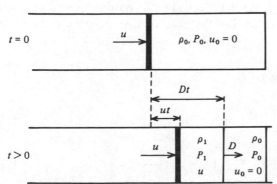

impulse due to the pressure forces $(P_1 - P_0)At$, so that by Newton's second law (i.e., momentum conservation) one has

$$\rho_0 Du = P_1 - P_0 \tag{11.60}$$

The increase of internal energy and of kinetic energy due to the above described motion is $(\rho_0 ADt)(E_1 - E_0 + \frac{1}{2}u^2)$, where E is the internal energy per unit mass. By conservation of energy, the work done by the piston, $P_1 A ut$, should equal the energy increase in the gas, so that

$$\rho_0 D\left(E_1 - E_0 + \frac{u^2}{2} \right) = P_1 u \tag{11.61}$$

Now making a transformation to the shock surface frame of reference, so that the undisturbed gas flows into the shock discontinuity with a velocity $-D$ and leaves this discontinuity with a velocity $-(D-u)$, we have (in this frame of reference):

$$u_0 = -D \quad u_1 = -(D-u) \tag{11.62}$$

Substituting the transformation (11.62) in the conservation equations (11.59), (11.60) and (11.61), a symmetric set of equations is obtained: mass conservation:

$$\rho_1 u_1 = \rho_0 u_0 \tag{11.63}$$

momentum conservation:

$$P_1 + \rho_1 u_1^{\,2} = P_0 + \rho_0 u^2 \tag{11.64}$$

energy conservation:

$$E_1 + \frac{P_1}{\rho_1} + \frac{u_1^{\,2}}{2} = E_0 + \frac{P_0}{\rho_0} + \frac{u_0^{\,2}}{2} \tag{11.65}$$

Equations (11.63), (11.64) and (11.65) can be obtained more rigorously from the fluid equations (11.3), (11.12) and (11.20). Writing these fluid conservation equations explicitly for the case of one dimension with planar symmetry one gets

$$\frac{\partial \rho}{\partial t} = -\frac{\partial}{\partial x}(\rho u) \tag{11.66}$$

$$\frac{\rho}{\partial t}(\rho u) = -\frac{\partial}{\partial x}(P + \rho u^2) \tag{11.67}$$

$$\frac{\partial}{\partial t}\left(\rho \varepsilon + \rho \frac{u^2}{2} \right) = -\frac{\partial}{\partial x}\left(\rho \varepsilon u + \rho \frac{u^3}{2} + Pu \right) \tag{11.68}$$

Performing the integration of Equations (11.66)–(11.68) over the shock wave layer and assuming that the thickness of the layer tends to zero (i.e., for two points x_0 and x_1 lying on both sides of the discontinuity one assumes that $x_1 - x_0$ goes to zero) one can write

$$\lim_{x_1 \to x_0} \int_{x_0}^{x_1} \frac{\partial}{\partial t} [\quad] dx \to 0$$

$$\lim_{x_1 \to x_0} \int_{x_0}^{x_1} \frac{\partial}{\partial x} [\quad] dx = [\quad]_{x_1} - [\quad]_{x_0} \tag{11.69}$$

where [ ] represent the different terms of (11.66)–(11.68). Using (11.69) in (11.66)–(11.68) the conservation equations (11.63)–(11.65) are obtained.

We now derive some algebraic relations using the conservation equations (11.63)–(11.65). It is convenient to define the specific volume $V[cm^3/g$ in c.g.s. units]

$$V = \frac{1}{\rho} \tag{11.70}$$

and to solve (11.63) and (11.64) for u_0 and u_1 in terms of V_0 and V_1

$$\frac{V_0}{V_1} = \frac{\rho_1}{\rho_0} = \frac{u_0}{u_1} \tag{11.71}$$

$$u_0 = V_0 \left(\frac{P_1 - P_0}{V_0 - V_1} \right)^{1/2} \tag{11.72}$$

$$u_1 = V_1 \left(\frac{P_1 - P_0}{V_0 - V_1} \right)^{1/2} \tag{11.73}$$

The particle velocity (i.e., the flow velocity) inside the compressed shock region in the laboratory (rest) frame of reference is given by u,

$$|u| = u_0 - u_1 = [(P_1 - P_0)(V_0 - V_1)]^{1/2} \tag{11.74}$$

If the shock wave is created by the motion of the piston, as described in the above example, then $|u|$ is equal to the piston velocity with respect to the undisturbed gas. Substituting (11.72) and (11.73) in the energy conservation equation (11.65), one gets the so-called 'Hugoniot relation'

$$E_1(P_1, V_1) - E_0(P_0, V_0) = \tfrac{1}{2}(P_1 + P_0)(V_0 - V_1) \tag{11.75}$$

The thermodynamic relation $E(P, V)$ is the equation of state of the material under consideration, and assuming this function to be known, one gets from (11.75) a curve

$$P = P_H(V, P_0, V_0) \tag{11.76}$$

known in the literature as the Hugoniot curve. P_H is shown schematically in Fig. 11.4. The Hugoniot curve is a two-parameter (P_0, V_0) family of curves, so that in order to plot all possible curves P_H it is necessary to start with an 'infinity squared' of points corresponding to all possible values of the parameters P_0 and V_0. In this respect, the Hugoniot curve is different from the isentropic curves of the pressure P_S,

$$P = P_S(V; S) \tag{11.77}$$

which are a one-parameter family of curves, where the single parameter is the conserved quantity of entropy S.

We now consider the shock wave relations for an ideal gas with constant specific heats. As mentioned above, in order to plot the Hugoniot curve a knowledge of the equation of state is necessary. For an ideal gas this equation can be written as

$$E = \frac{PV}{\gamma - 1} \tag{11.78}$$

Substituting (11.78) in the Hugoniot relation (11.75) the Hugoniot curve is obtained for an ideal gas

$$\frac{P_1}{P_0} = \frac{(\gamma + 1)V_0 - (\gamma - 1)V_1}{(\gamma + 1)V_1 - (\gamma - 1)V_0} \tag{11.79}$$

From (11.79) it is evident that the compression ratio $V_0/V_1 = \rho_1/\rho_0$ does not increase above a certain value determined by the value of γ. In particular, for a very strong shock wave where the pressure of the shock

Fig. 11.4. The Hugoniot curve.

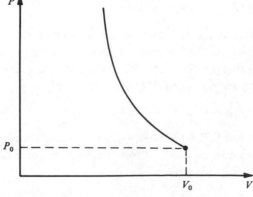

rises to infinity one gets

$$\left(\frac{\rho_1}{\rho_0}\right)_{max} = \left(\frac{V_0}{V_1}\right)_{min} = \frac{\gamma+1}{\gamma-1} \tag{11.80}$$

The maximum density compression caused by a planar shock wave in an ideal monatomic gas is

$$\gamma = \frac{5}{3} \quad \left(\frac{\rho_1}{\rho_0}\right)_{max} = 4, \tag{11.81}$$

in a diatomic gas with the vibration frozen, i.e. assuming that the vibrational modes are not excited, one has

$$\gamma = \frac{7}{5} \quad \left(\frac{\rho_1}{\rho_0}\right)_{max} = 6 \tag{11.82}$$

while in a diatomic gas with vibrational modes

$$\gamma = \frac{9}{7} \quad \left(\frac{\rho_1}{\rho_0}\right)_{max} = 8 \tag{11.83}$$

Solving (11.72)–(11.73) for an ideal gas described by (11.78), one gets for the fluid velocities

$$u_0{}^2 = \frac{V_0}{2}[(\gamma-1)P_0 + (\gamma+1)P_1] \tag{11.84}$$

$$u_1{}^2 = \frac{V_0}{2}\left(\frac{[(\gamma+1)P_0 + (\gamma-1)P_1]^2}{[(\gamma-1)P_0 + (\gamma+1)P_1]}\right) \tag{11.85}$$

while for the velocity of sound (see (11.33)) one has

$$c^2 = \gamma P V \tag{11.86}$$

The ratio of the flow velocity to the sound velocity is known as the Mach number M, and in the process of a shock wave flow in an ideal gas the Mach numbers are given by,

$$M_0{}^2 \equiv \left(\frac{u_0}{c_0}\right)^2 = \frac{(\gamma-1) + (\gamma+1)P_1/P_0}{2\gamma} \tag{11.87}$$

$$M_1{}^2 \equiv \left(\frac{u_1}{c_1}\right)^2 = \frac{(\gamma-1) + (\gamma+1)P_0/P_1}{2\gamma} \tag{11.88}$$

In the limit of a weak shock wave, defined by $P_1 \approx P_0$, one has

$$M_1 \approx M_0 \approx 1 \quad \text{(weak shock wave)} \tag{11.89}$$

while for a strong shock wave, defined by $P_1 \gg P_0$,

$$M_0 > 1 \quad M_1 < 1 \quad \text{(strong shock wave)} \tag{11.90}$$

The meaning of the last relation is that in the shock frame of reference, the gas flows into the discontinuity at a supersonic velocity ($M_0 > 1$) and flows out at a subsonic velocity ($M_1 < 1$). In the laboratory, rest frame of the undisturbed gas, one has $D = -u_0$ for the shock velocity, and therefore one gets the 'well known' result that the shock wave propagates at a supersonic speed with respect to the undisturbed gas and at a subsonic speed with respect to the compressed gas behind it.

The entropy of an ideal gas is given by

$$S = c_v \ln PV^\gamma + S_0 \tag{11.91}$$

where the specific heat at constant volume is

$$c_v = \frac{PV}{(\gamma - 1)T} \tag{11.92}$$

The increase in entropy during a shock wave process is given by

$$\begin{aligned}
S_1 - S_0 &= c_v \ln \frac{P_1 V_1{}^\gamma}{P_0 V_0{}^\gamma} \\
&= \frac{P_0 V_0}{(\gamma - 1)T_0} \ln \left\{ \frac{P_1}{P_0} \left[\frac{(\gamma - 1)P_1/P_0 + (\gamma + 1)}{(\gamma + 1)P_1/P_0 + (\gamma - 1)} \right]^\gamma \right\}
\end{aligned} \tag{11.93}$$

From this expression one can see that for a weak shock wave ($P_1 \approx P_0$) the entropy does not change significantly, $S_1 \approx S_0$, while for a strong shock wave there is an increase in the entropy. The increase in entropy indicates that a shock wave is not a reversible process, but rather a dissipative one. The entropy jump of a fluid compressed by a shock wave increases with the strength of the shock wave. As seen above, the change in entropy ΔS, in pressure ΔP, in density $\Delta \rho$, etc. are determined solely by the *conservation laws*, however, the mechanisms of these changes are described by viscosity and thermal conductivity. This subject and the related relaxation phenomena will not be discussed here.

We end this chapter with the formulation of achieving high compression by applying a sequence of shock waves to an ideal gas. As explained by (11.80), one single planar shock wave can reach a fixed maximum compression given by that equation. Assuming now, that a sequence of shock waves satisfying

$$\frac{\rho_1}{\rho_0} = \frac{\rho_2}{\rho_1} = \frac{\rho_3}{\rho_2} = \cdots = \frac{\rho_n}{\rho_{n-1}} \equiv r \tag{11.94}$$

are applied to a material in such a way that shocks do not overtake earlier ones and that all the shock waves arrive together at some time t at the place to be highly compressed. The shock created at time $t_j (j = 1, \ldots, n)$ overtakes the one formed at time t_{j-1} at time t if

$$D_j(t - t_j) = D_{j-1}(t - t_{j-1}) \quad j = 1, \ldots, n \tag{11.95}$$

where D_j is the velocity of the jth shock wave. The set of equations (11.95) has the solution

$$t = (D_n t_n - D_1 t_1)/(D_n - D_1). \tag{11.96}$$

In order to calculate the pressure and the compression obtained in this case, we use (11.79) and (11.94)

$$\frac{P_j - P_{j-1}}{P_{j-1}} \equiv \Delta = \frac{2\gamma(r-1)}{(1+\gamma) - r(\gamma - 1)} \quad j = 1, \ldots, n \tag{11.97}$$

$$\frac{P_n}{P_0} = (1 + \Delta)^n \tag{11.98}$$

$$\frac{\rho_n}{\rho_0} = r^n \tag{11.99}$$

As a numerical example we take $r = 2$ for a sequence of 10 shock waves ($n = 10$). In this case one has a compression of 2^{10} ($\approx 1000!$) obtained only with a pressure of $P_n/P_0 \approx 2.8 \times 10^5!$, i.e., for an initial pressure of 1 atmosphere, one can achieve a compression of a 1000 with pressures less than a megabar if the pressure is 'tailored' in space, time, and intensity (i.e., shaped) in a proper way.

12

Derivation of hydrodynamics from kinetic theory

12.1 Foundations of hydromechanics

In the previous chapter we discussed how the state of a moving fluid or gas can be described by macroscopic (hydrodynamic) quantities like velocity, density and temperatures under the action of pressure and other forces given electric or magnetic fields or by gravitation. The hydrodynamic equations were derived from Newton's mechanics and from the geometry and mathematics of macroscopic field quantities. In order to provide a closer link with the preceding chapters on statistical mechanics, the derivation of the macroscopic hydrodynamic equations (Chapter 11) from microscopic kinetic concepts is given here.

The description of a gas is possible by studying the mechanical motion of its atoms or molecules whose number is N. The equations of motion for each mass m_n (in most cases of the same value) with coordinates x_n, y_n, z_n and with forces are the following set of $3N$ differential equations

$$
\left.
\begin{aligned}
m\frac{d^2 x_n}{dt^2} &= F_{x_n}\left(x_1, \ldots z_N, \frac{\partial x_1}{\partial t}, \ldots \frac{\partial z_N}{\partial t} \right) \\[2mm]
m\frac{d^2 y_n}{dt^2} &= F_{y_n}\left(x_1, \ldots z_N, \frac{\partial x_1}{\partial t}, \ldots \frac{\partial z_N}{\partial t} \right) \\[2mm]
m\frac{d^2 z_n}{dt^2} &= F_{z_n}\left(x_1, \ldots z_N, \frac{\partial x_1}{\partial t}, \ldots \frac{\partial z_N}{\partial t} \right)
\end{aligned}
\right\}
\tag{12.1}
$$

for $n = 1 \cdots N$

The computation of the state of a gas or of a fully ionized plasma for one million particles with large computers using Eq. (12.1) is well established (Valeo and Kruer (1972)) where simplifications on the range of the forces beyond a certain number of neighbour particles were necessary. Such multi-particle simulation reproduced the main properties of the state (with some restriction due to the range of the forces), however, not all interesting properties such as Coulomb collisions have yet been included in these simulations. The generation of large electric fields and inhibition of thermal conduction could have been derived from these simulations immediately in the same way as from an experiment, but the neglect of collisions and other simplifications caused difficulties. The other problem was that the general models and their computer output had to be analyzed on the questions of interest. So it was that the electric fields in laser produced plasmas and the derivation of thermal conduction were first derived from macroscopic theory (Hora *et al.* 1983) in agreement with experiments (Eliezer and Ludmirsky 1983) before the same properties were 'detected' in the computer output of multi-particle computations.

The kinetic theory or the theory of kinetic equations simplifies the N-particle problem with its $3N$ coordinates and equations (12.1) by defining a continuous function $f(x, y, z, v_x, v_y, v_z)$ which describes how many particles are to be located in an element of the phase volume $dxdydzdv_xdv_ydv_z$. A mathematical discussion of this distribution function then arrives at a very detailed description of the microscopic state of the plasma, though some of the information of the most general N-particle problem was lost. Still, the distribution function covers the description of non-thermalized states, contrary to the more purified macroscopic (hydrodynamic) picture which is produced from the subsequent integration of expressions of the f-function.

The kinetic theory, therefore is a link between the single particle description of the microscopic theory and the macroscopic continuum theory of hydrodynamic quantities such as particle densities, net velocities and temperatures. Any temperature is an approximation even at equilibrium conditions for which the kinetic theory can give a more realistic description e.g. for cases where the microscopic state of the fluid or gas cannot be determined by one or several 'temperatures' of the plasma components.

12.2 Distribution functions and the Boltzmann equation

For a straight-forward understanding, an ad-hoc description of the elements of the kinetic theory is given in this section. We have first to underline that distribution functions do not necessarily relate in a philosophical sense – to probabilities, uncertainties, or inaccuracies. If one has to find the average value of a set of N integers a_i:

$$M = \frac{\sum_{i=1}^{N} a_i N}{N} \tag{12.2}$$

one can arrange the numbers in sub-sets $j = 1, \ldots, n$, where f_j determines the number of elements in each micro-set. The task of finding the average value in Eq. (12.2) is then solved by

$$M = \left(\sum_{i=1}^{n} f_i a_i \right) \Big/ \left(\sum_{i=1}^{n} f_i \right) \tag{12.3}$$

Therefore, a set of f_i or, continuously, a distribution function $f(x)$ can then determine the average value of a quantity $a(x)$

$$\langle a \rangle = \int f(x) a(x) \mathrm{d}x \Big/ \int f(x) \mathrm{d}x \tag{12.4}$$

The same procedure with distribution functions is used to reduce the microscopic N-particle problem with its $3N$ differential equations (12.1) as follows: The geometric space is subdivided into cells of volume $\mathrm{d}^3 r = \mathrm{d}x_1 \mathrm{d}x_2 \mathrm{d}x_3$ about $\mathbf{x}$ and the velocity space is subdivided into cells $\mathrm{d}^3 w = \mathrm{d}w_1 \mathrm{d}w_2 \mathrm{d}w_3$ about $\mathbf{w}$. Instead of describing each of the N particles by its differential equation (12.1) in time, we ask, how many particles are distributed into the cells, depending on time t, given by the distribution function

$$f(x_1, x_2, x_3, w_1, w_2, w_3, t) \mathrm{d}^3 x \mathrm{d}^3 w \tag{12.5}$$

The density of the particles (number of particles per cm^3) is then from this definition with the subscript e or i (electrons or ions)

$$n_{\mathrm{e,i}}(\mathbf{r}, t) = \int \int \int_{-\infty}^{+\infty} f(\mathbf{r}, \mathbf{w}, t) \mathrm{d}^3 w \tag{12.6}$$

Similarly to (12.4) we find the value of any physical quantity $Q(\mathbf{r}, \mathbf{w}, t)$ depending on the velocity $\mathbf{w}$ at a point $\mathbf{r}$ at a time t as

$$\overline{Q(\mathbf{r}, t)} = \int \int \int_{-\infty}^{+\infty} Q(\mathbf{r}, \mathbf{w}, t) f(\mathbf{r}, \mathbf{w}, t) \mathrm{d}^3 w \Big/ \int \int \int_{-\infty}^{+\infty} f(\mathbf{r}, \mathbf{w}, t) \mathrm{d}^3 w \tag{12.7}$$

The average velocity of the N particles at $\mathbf{r}$ and t is then

$$\mathbf{v}(\mathbf{r}, t) = \int \int \int_{-\infty}^{+\infty} \mathbf{w} f(\mathbf{r}, \mathbf{w}, t) \mathrm{d}^3 w \Big/ \int \int \int_{-\infty}^{+\infty} f(\mathbf{r}, \mathbf{w}, t) \mathrm{d}^3 w \tag{12.8}$$

If $Q(\mathbf{w})$ does not depend on $\mathbf{r}$ and t, we find the following relations, using (12.6)

$$\int Q(\mathbf{w}) \frac{\partial f}{\partial t} \mathrm{d}^3 w = \frac{\partial}{\partial t} \int Q(\mathbf{w}) f \mathrm{d}^3 w \tag{12.9}$$

$$= \frac{\partial}{\partial t} (n \bar{Q})$$

$$\int Q(\mathbf{w})\mathbf{w}\cdot\nabla f\,\mathrm{d}^3w = \nabla\cdot\int Q(\mathbf{w})\mathbf{w}f\,\mathrm{d}^3w = \nabla(n\,\overline{\mathbf{w}Q}) \qquad (12.10)$$

where the vector operator $\nabla = (\partial/\partial x_1; \partial/\partial x_2; \partial/\partial x_3)$ is called 'del' or nabla. Using a general vectorial function $\mathbf{F}$ and the operator $\nabla_w = (\partial/\partial w_1; \partial/\partial w_2; \partial/\partial w_3)$

$$\int Q(\mathbf{w})\mathbf{F}(\mathbf{r},\mathbf{w})\cdot\nabla_w f\,\mathrm{d}^3w = -\int f\nabla_w\cdot\{\mathbf{F}(\mathbf{r},\mathbf{w})Q(\mathbf{w})\}\,\mathrm{d}^3w$$

and using the fact at partial integration of $f(\pm\infty) = 0$ (for reasons of convergence of (12.6)),

$$\int Q(\mathbf{w})F(\mathbf{r},\mathbf{w},t)\cdot\nabla_w f\,\mathrm{d}^3w = -n\nabla_w\cdot(\overline{FQ}) \qquad (12.11)$$

We arrive at a kinetic equation for the distribution function f by the very trivial statement that, if there are no changes, there is no explicit dependence on time, or – in other words – the total derivative

$$\frac{\mathrm{d}}{\mathrm{d}t}f = 0 \qquad (12.12)$$

has to be zero. Taking into account all seven variables of f, (12.5), we find by partial differentiation

$$\frac{\mathrm{d}}{\mathrm{d}t}f = \frac{\partial}{\partial t}f + \frac{\partial}{\partial x_1}f\frac{\partial x_1}{\partial t} + \frac{\partial}{\partial x_2}f\frac{\partial x_2}{\partial t} + \frac{\partial}{\partial x_3}f\frac{\partial x_3}{\partial t}$$

$$+ \frac{\partial}{\partial w_1}f\frac{\partial w_1}{\partial t} + \frac{\partial}{\partial w_2}f\frac{\partial w_2}{\partial t} + \frac{\partial}{\partial w_3}f\frac{\partial w_3}{\partial t} \qquad (12.13)$$

or using the velocity vector $\mathbf{w} = (\partial x/\partial t; \partial y/\partial t; \partial z/\partial t)$, and the force density $\mathbf{F}$ where

$$\mathbf{F} = m\frac{\partial}{\partial t}\mathbf{w} \qquad (12.14)$$

if all particles have the same (or averaged) mass m, we arrive at

$$\frac{\partial f}{\partial t} + \mathbf{w}\cdot\nabla f + \frac{\mathbf{F}}{m}\cdot\nabla_w f = 0 \qquad (12.15)$$

This equation is a *kinetic equation* and is called the *Vlasov equation*. If the force is due to electrical quantities ($\mathbf{E} \approx$ electric $\mathbf{H} \approx$ magnetic field strength, $c \approx$ velocity of light)

$$\mathbf{F} = e\left(\mathbf{E} + \frac{1}{c}\mathbf{w} \times \mathbf{H}\right)$$

then

$$\frac{\partial f}{\partial t} + \mathbf{w}\cdot\nabla f + \frac{e}{m}\left(\mathbf{E} + \frac{1}{c}\mathbf{w} \times \mathbf{H}\right)\cdot\nabla_w f = 0 \qquad (12.16)$$

If f does depend explicitly on the time, the total differentiation in (12.12) is not zero but equal to the change due to collisions (subscript c):

$$\frac{\partial f}{\partial t} + \mathbf{w} \cdot \nabla f + \frac{\mathbf{F}}{m} \nabla_w f = \left(\frac{\partial f}{\partial t} \right)_c \tag{12.17}$$

which is the Boltzmann equation. The whole problem is then concentrated in the collision term on the right hand side. An approximation for collisions with neutral atoms is the Krook collision term

$$\left(\frac{\partial f}{\partial t} \right)_c = \frac{f_n - f}{\tau} \tag{12.18}$$

where f_n is the distribution function of the neutral atoms and τ is the averaged collision time.

For Coulomb collisions, (12.17) can be approximated using binary collisions by the Fokker–Planck equation

$$\frac{df}{dt} = \sum_{n=1}^{\infty} \frac{(-1)^n \partial^n}{n! \partial_{w_{i_1}} \partial_{w_{i_2}} \cdots \partial_{w_{i_n}}} (\bar{\alpha}_{(m)} f) \quad (i_n = 1, 2, 3, \ldots, n) \tag{12.19}$$

where the $\bar{\alpha}$s are the Fokker–Planck coefficients (Ray and Hora 1977).

Without discussion we mention Liouville's theorem: for a conservative system, f is constant along a dynamic trajectory. This means that the volume $d^3r d^3w$ which is taken by a number of particles, does not change with time.

The kinetic equations provide the possibility of treating a plasma without having reached thermal equilibrium. The limiting case of equilibrium is expressed by a special distribution function. If one normalizes f at each point of certain particle density $n(\mathbf{r}, t)$, the factor f_M in

$$f(\mathbf{r}, \mathbf{w}, t) = n(\mathbf{r}, t) \hat{f}_M(\mathbf{r}, \mathbf{w}, t) \tag{12.20}$$

is called the Maxwellian distribution (or Maxwell–Boltzmann distribution)

$$\hat{f}_M = \left(\frac{m}{2\pi kT} \right)^{3/2} \exp(-\mathbf{w}^2 / v_{th}^2) \tag{12.21}$$

using Boltzmann's constant k and the average thermal velocity $v_{th} = (2kT/m)^{1/2}$. Equation (12.21) takes into account positive and negative velocity components. Based on the scalar magnitude of the velocity, w, one defines another distribution $g(w)$ by

$$\int_0^{\infty} g(w) dw = \int_{-\infty}^{+\infty} f(\mathbf{w}) d^3w \tag{12.22}$$

where

$$g(w) = 4\pi n(\mathbf{r}, t)\left(\frac{m}{2\pi kT}\right)^{3/2} w^3 \exp(-w^2/v_{\text{th}}^2) \tag{12.23}$$

to be distinguished from f.

12.3 Loss of information

The description of the kinetic equations in the preceding section is sufficient to understand the derivation of the macroscopic equations in the next section. In this section, some basic problems of the kinetic theory should be mentioned. Boltzmann's basic problem was the description of the collisions, even in the most simplified way where the collision time can be neglected. The derivation of the Boltzmann equation can be based on the more general Liouville distribution function for N particles

$$F(\mathbf{r}_1, \mathbf{r}_2, \ldots, \mathbf{r}_N, \mathbf{w}_1, \mathbf{w}_2, \ldots, \mathbf{w}_N, t) \tag{12.24}$$

which gives the combined probability of finding particle 1 at $\mathbf{r}_1$ with velocity $\mathbf{w}_1$, and particle 2 at $\mathbf{r}_2$ with velocity $\mathbf{w}_2$, and particle N at $\mathbf{r}_N$ with $\mathbf{w}_N$. A one-particle distribution function f defines the probability of finding a particle at x, with a velocity $\mathbf{w}$, at a time t from integrating (12.24) over all but the first particle coordinate:

$$f(\mathbf{r}_1, \mathbf{w}_1, t) = \int d^3 r_2, \ldots d^3 r_N d^3 w_2, \ldots d^3 w_N F(\mathbf{r}_1 \ldots \mathbf{r}_N, w_1 \ldots w_N) \tag{12.25}$$

If the particles move independently, an expression by factors is possible:

$$F(\mathbf{r}_1 \ldots \mathbf{r}_N, \mathbf{w}_1, \ldots, \mathbf{w}_N, t) = f(\mathbf{r}_1, \mathbf{w}_1, t) \ldots f(\mathbf{r}_N, \mathbf{w}_N, t) = f^N(\mathbf{r}, \mathbf{w}, t) \tag{12.26}$$

F satisfies the Liouville equation

$$\frac{\partial F}{\partial t} + [F, H] = 0 \tag{12.27}$$

where the Poisson bracket of F with the Hamiltonian H of the complete system

$$[F, H] = \sum_{i=1}^{N} \frac{\partial H}{\partial p_i}\frac{\partial F}{\partial q_i} - \frac{\partial H}{\partial q_i}\frac{\partial F}{\partial p_i} \tag{12.28}$$

remembering Hamilton's equations

$$\dot{q}_i = \frac{\partial H}{\partial p_i}; \quad \dot{p}_i = -\frac{\partial H}{\partial q_i}$$

and converting from the generalized coordinates q_i to the Cartesian x_i and from the generalized momenta p_i to velocities $\mathbf{w}$; equation (12.28) can be written

$$[F, H] = \sum_{i=1}^{N} \left(\mathbf{w}_i \cdot \nabla F + \frac{\mathbf{F}}{m} \cdot \nabla_{w_i} F \right) \tag{12.29}$$

rewriting (12.27) into

$$\frac{\partial}{\partial t} F + \sum_{i=1}^{N} \left(\mathbf{w}_i \cdot \nabla F + \frac{\mathbf{F}}{m} \cdot \nabla_{w_i} F \right) = 0 \tag{12.30}$$

Integration over $N - 1$ coordinates as in (12.25), results in the collisionless Boltzmann equation

$$\frac{\partial}{\partial t} f + \mathbf{w} \cdot \nabla f + \frac{\mathbf{F}}{m} \cdot \nabla_w f = 0 \tag{12.31}$$

Following the results of Blatt and Opie (1974), there are fundamental deficiencies concerning the loss of information about the initial state when the system is described by the kinetic equations. The approximation (12.26) particularly, is a severe restriction. But there are even considerable doubts whether the Liouville equation (12.27) with a general F represents any real thermal system adequately (Blatt, 1959). The careful derivation of the kinetic model (Blatt and Opie 1979) is also made on the basis of minor approximations appropriate to a very dilute gas, ignoring effects of ternary and higher order collisions.

These arguments should be taken into account if criticisms of the macroscopic hydrodynamic models are discussed: the kinetic models also are still not free of deficiencies, especially for the high densities of laser produced plasmas. A return to the single particle description of the microscopic models based on equations (12.1) is difficult in relation to computer capacity, the necessary neglect of low range interactions, the neglect of the Coulomb collisions especially with respect to small angle scattering, and the approximation of collective effects.

12.4   Derivation of macroscopic equations

In this section the macroscopic equations of continuity and motion will be derived from the integration of the Boltzmann equation (12.17)

$$\frac{\partial f}{\partial t} + \mathbf{w} \cdot \nabla f + \frac{\mathbf{F}}{m} \nabla_w f = \left(\frac{\partial f}{\partial t} \right)_c \tag{12.32}$$

after multiplication by a factor. Each term will be integrated over the whole velocity space d^3w, if we assume that there are no velocity dependent forces. The forces are then holonomial following the definition of H. Hertz.

12.4.1 The equation of continuity (mass conservation)

Multiplying (12.32) by the trivial factor 1 and performing the integration by d^3w from $-\infty$ to $+\infty$ of the first term results in

$$\int \frac{\partial f}{\partial t} d^3w = \frac{\partial}{\partial t} \int f d^3w = \frac{\partial}{\partial t} n \tag{12.33}$$

following (12.6) and (12.9). Integration of the second term of (12.32) based on (12.10) with a quantity $Q = 1$ results in

$$\int \mathbf{w} \cdot \nabla f \, dw = \nabla \cdot (\overline{n\mathbf{w}}) = \nabla \cdot \overline{n\mathbf{u}} \tag{12.34}$$

where the velocity of the particles was split into the drift velocity $\mathbf{u}$ and the random (thermal) velocity $\mathbf{v}$

$$\mathbf{w} = \mathbf{v} + \mathbf{u} \tag{12.35}$$

with

$$\mathbf{v} = 0 \tag{12.36}$$

Integration of the third term of (12.17) using (12.11) with $Q = 1$ produces

$$\int \frac{\mathbf{F}}{m} \cdot \nabla_w f \, d^3w = -\int f \nabla_w \cdot \mathbf{F} \, d^3w = -n \overline{\nabla_w \cdot \mathbf{F}} = 0$$

which is zero because we assume velocity independent (holonomic) forces $\mathbf{F}$. The right hand side of (12.32) results in

$$\int_{\omega}^{+\infty} \left(\frac{\partial f}{\partial t} \right)_c d^3w = 0 \tag{12.37}$$

because collisions cannot change the average total number n of particles per cm^3.

Summarizing (12.33)–(12.37) results in the velocity integral of the Boltzmann equations

$$\frac{\partial}{\partial t} n + \nabla \cdot n\mathbf{v} = 0 \tag{12.38}$$

which is the hydrodynamic equation of continuity (11.3), the equation of conservation of mass, remembering that $\rho = mn$.

12.4.2 The equation of motion (momentum conservation)

Another integration of the Boltzmann equation over the velocity space, after multiplying by a factor $(m\mathbf{w})$ of momentum and integration will lead to the hydrodynamic equation of motion (equation of conservation of momentum). From (12.32) we obtain

$$\int m\mathbf{w}\frac{\partial}{\partial t}f\,\mathrm{d}^3w + \int m\mathbf{w}\mathbf{w}\cdot\nabla f\,\mathrm{d}^3w + \int m\mathbf{w}\frac{\mathbf{F}}{m}\cdot\nabla_w f\,\mathrm{d}^3w$$

$$= \int m\mathbf{w}\left(\frac{\partial f}{\partial t}\right)_c \mathrm{d}^3w \tag{12.39}$$

We use, as before, (12.8)–(12.1) where

$$Q(\mathbf{w}) = m\mathbf{w} \tag{12.40}$$

The first term A is from (12.8)

$$A = \int m\mathbf{w}\frac{\partial}{\partial t}f\,\mathrm{d}^3w = \frac{\partial}{\partial t}(\overline{nm\mathbf{w}}) \tag{12.41}$$

where the separation into drift velocity $\mathbf{u}$ and random velocity $\mathbf{v}$, (11.35), leads to

$$A = mn\frac{\partial}{\partial t}\overline{(\mathbf{v}+\mathbf{u})} + \overline{(\mathbf{v}+\mathbf{u})}\frac{\partial}{\partial t}mn \tag{12.42}$$

and using $\bar{\mathbf{v}} = 0$ to

$$A = mn\frac{\partial}{\partial t}\mathbf{u} + \mathbf{u}\frac{\partial}{\partial t}mn \tag{12.43}$$

The second term B from (12.9), (12.35) and (12.40) becomes

$$B = \int m\mathbf{w}\mathbf{w}\cdot\nabla f\,\mathrm{d}^3w$$

$$= \nabla\cdot \overline{n\mathbf{w}m\mathbf{w}} = \nabla\cdot nm\overline{(\mathbf{v}+\mathbf{u})(\mathbf{v}+\mathbf{u})}$$

$$= \nabla\cdot nm\overline{(\mathbf{v}\mathbf{v}+\mathbf{u}\mathbf{u}+\mathbf{v}\mathbf{u}+\mathbf{u}\mathbf{v})} \tag{12.44}$$

where the last two dyadic terms are zero ($\mathbf{u} = 0$).

It is furthermore

$$B = \bar{\mathbf{u}}\bar{\mathbf{u}}\cdot\nabla mn + mn\overline{\mathbf{u}\cdot\nabla\mathbf{v}} + nm\overline{\mathbf{v}\nabla\cdot\mathbf{u}} + \nabla\cdot mn\overline{\mathbf{v}\mathbf{v}} \tag{12.45}$$

The third term of (12.45) r.h.s. is obtained by use of the equation of continuity

$$mn\mathbf{u}\nabla\cdot\mathbf{u} = -\mathbf{u}\frac{\partial}{\partial t}mn - m\mathbf{u}\mathbf{u}\nabla n$$

of which the last term cancels with the first in (12.45). The other terms on the r.h.s. of (12.45) cancels with the second term r.h.s. of (12.43) when adding A and B, resulting in

$$A + B = mn\frac{\partial}{\partial t}\mathbf{u} + mn\mathbf{u}\cdot\nabla\mathbf{u} + \nabla\cdot mn\overline{\mathbf{vv}} \tag{12.46}$$

Here the first two terms on the r.h.s. can be written by using the equation of continuity

$$mn\frac{\partial}{\partial t}\mathbf{u} + mn\mathbf{u}\cdot\nabla\mathbf{u} = \frac{\partial mn\mathbf{u}}{\partial t} + \frac{\partial mn\mathbf{u}^2}{\partial x} \tag{12.47}$$

using a rearrangement of terms and using the temporal derivative on the l.h.s. and the spatial derivative on the r.h.s.

There remains the interpretation of the last term in (12.45). Writing

$$mn\mathbf{vv} = 2mn\tfrac{1}{2}\mathbf{v}^2\mathbf{1} \tag{12.48}$$

using the unity tensor $\mathbf{1} = \mathbf{i}_1\mathbf{i}_1 + \mathbf{i}_2\mathbf{i}_2 + \mathbf{i}_3\mathbf{i}_3$ and remembering that

$$\frac{m}{2}\mathbf{v}^2 = \tfrac{3}{2}kT \tag{12.49}$$

we find, for isotrope geometry with respect to the unity tensor,

$$\nabla\cdot mn\overline{\mathbf{v}}\,\overline{\mathbf{v}} = \nabla\, 3nkT \tag{12.50}$$

In (12.49), the equation of state for the ideal gas of single atoms or fully ionized plasma particles was derived from the random velocity $\mathbf{u}$ of the kinetic theory.

The macroscopic assumption is that the pressure is related to the macroscopic adiabatic constant

$$\gamma = (2 + f)/f \tag{12.51}$$

with the degree of freedom f, this has to be interpreted as a one-dimensional pressure, mentioned by Denisse and Delcroix (1963) as being 'somewhat open to criticism'. The subsequent derivation of these relations from the kinetic theory (Boltzmann equation) was shown by Lalousis and Hora (1983).

At this stage, it should be noted where the modification of the derivation of the macroscopic equation of motion includes the general equations of state. The velocities $\mathbf{w}$ of bound or oscillating particles in degenerate states have then to be formulated from the beginning in the Boltzmann equation and have to be identified and interpreted at the level of (12.48).

It should be mentioned that for the case of 'strong coupling' in the sense of a plasma parameter Λ of the Coulomb logarithm to have values much below two, Hansen (1980) has reviewed the derivation and arrived at the transport coefficients for these special cases of the equation of state.

The third term C in (12.39) is shown in (12.10), (12.11) and (12.40).

$$C = \int m\mathbf{w}\frac{\mathbf{F}}{m}\cdot\nabla_w f\, d^3w = -n_w \mathbf{F} m\mathbf{w} = -n\mathbf{F}\cdot\nabla_w\mathbf{w} - mn\mathbf{w}_w\mathbf{F}$$

(12.52)

The second term in the last expression vanishes because the forces $\mathbf{F}$ should not depend on the velocity $\mathbf{w}$.

$$\nabla_w\cdot\mathbf{F} = 0$$

(12.53)

The tensor of the first term in the last expression of (12.52) is

$$\nabla_w\mathbf{w} = \mathbf{i}_1\mathbf{i}_1\frac{\partial}{\partial w_1}w_1 + \mathbf{i}_1\mathbf{i}_2\frac{\partial}{\partial w_1}w_2 + \mathbf{i}_1\mathbf{i}_3\frac{\partial}{\partial w_1}w_3$$

$$+ \mathbf{i}_2\mathbf{i}_1\frac{\partial}{\partial w_2}w_1 + \mathbf{i}_2\mathbf{i}_1\frac{\partial}{\partial w_2}w_2 + \mathbf{i}_2\mathbf{i}_3\frac{\partial}{\partial w_2}w_3$$

$$+ \mathbf{i}_3\mathbf{i}_1\frac{\partial}{\partial w_3}w_1 + \mathbf{i}_3\mathbf{i}_1\frac{\partial}{\partial w_3}w_2 + \mathbf{i}_3\mathbf{i}_3\frac{\partial}{\partial w_3}w_3$$

where all non-diagonal terms vanish for Cartesian coordinates; therefore, using

$$\nabla_w\mathbf{w} = 1$$

and from (12.52) we arrive at

$$C = mn\mathbf{F}\cdot\mathbf{1} = -mn\mathbf{F}$$

(12.54)

the forces in plasmas should be

$$\mathbf{F} = Ze\mathbf{E} + \frac{Ze}{c}\mathbf{u}\times\mathbf{H} + \mathbf{F}_g$$

(12.55)

where Z is the number of charges for the equation of ions, while $Z = 1$ for electrons, and $\mathbf{F}_g$ are forces of gravitation, Coriolis forces and others.

The last term D in (12.39)

$$D = \int m\mathbf{w}\left(\frac{\partial f}{\partial t}\right)_c d^3w$$

(12.56)

expresses the net momentum per volume transferred to the ions by collisions with the electrons

$$D = \mathbf{P}_{ie}$$

(12.57)

If there are no asymmetric velocity distributions of the electrons,

$$\mathbf{P}_{ie} = 0 \tag{12.58}$$

This is not the case e.g. if an electron beam is fired into the plasma.

Putting together the results of A, B, C and D, we arrive at

$$mn\frac{\partial}{\partial t}\mathbf{u} + mn\mathbf{u}\cdot\nabla\mathbf{u} = \nabla - (nkT)$$

$$+ nmZe\left[\mathbf{E} + \frac{1}{c}\mathbf{u} \times \mathbf{H}\right] + mn\mathbf{F}_g \tag{12.59}$$

where the second term of A, (12.43) and (12.46), and the first term of B, (12.45) cancelled because of (12.46), and the first and second terms of B (12.45), using a vortex force motion (12.45) led to the second term on the right hand side of (12.59). This is the kinetic pressure, whose gradient is usually neglected in comparison with the static pressure nkT. The macroscopic (hydrodynamic) equation of motion is then

$$mn\frac{\partial}{\partial t}\mathbf{u} + mn\mathbf{u}\cdot\nabla\mathbf{u} = -\nabla nkT + mnZe\left[\mathbf{E} + \frac{1}{c}\mathbf{u} \times \mathbf{H}\right] + mn\mathbf{F}_g$$

$$\tag{12.60}$$

where again we emphasize that the first term on the r.h.s. is correct only for the equation of state of an ideal gas.

$$mn\,\mathrm{d}\mathbf{u}/\mathrm{d}t = \mathbf{f} \tag{12.61}$$

where $\mathbf{f}$ is the force density in agreement with (11.10).

From (12.59) we immediately derive the Bernoulli equation for a vortex-free stationary $(\mathrm{d}/\mathrm{d}t = 0)$ motion without external forces $\mathbf{F}_g$ and fields $\mathbf{E}$ and $\mathbf{H}$

$$\nabla\left(nkT + \frac{mn}{2}\mathbf{u}^2\right) = 0 \tag{12.62}$$

or using the (static) pressure $p = nkT$ and the density $\rho = nm$, integration of (12.62) results in Bernoulli's equation

$$p + \frac{\rho}{2}u^2 = \text{constant} \tag{12.63}$$

It should be noted that the macroscopic equation of energy conservation is achieved by multiplying the Boltzmann equation (12.32) by a factor $m\mathbf{w}\mathbf{w}$ and integrating over the whole velocity space. Again, the interpretation of the kinetic velocity integrals leads then to the various formulations of the equation of motion similar to the case we described earlier from the correct interpretation of (12.48) which led to the equation of state of the ideal gas as the most simplified example. As the procedures for the integration of the Boltzmann equation for the energy equation, reproduce the equation of state in the same way as in the case of the equation of motion, we are not presenting this case here.

13

Studies of equations of state from high pressure shock waves in solids

13.1 Introduction

The laws governing the propagation of shock waves through condensed matter are related to the physics of the equations of state (see, e.g., Altshuler (1965), Davison and Graham (1979), McQueen *et al.* (1970), Rice *et al.* (1985), Zeldovich and Raizer (1967)). Particles in a solid state are attracted to each other at large distances while they repel each other at short distances (as a result of the interpenetration of the electronic shell of the atoms). The atoms can be separated to a large distance by supplying the binding energy ($\approx 1\text{eV}/\text{atom}$) in order to overcome the binding forces, while in order to compress a condensed matter it is also necessary to supply energy to overcome the repulsive forces. These forces increase very rapidly as the distance between the atoms is decreased. At zero temperature and zero pressure mechanical equilibrium of a solid is reached by the balance of the attraction and repulsion of the interatomic forces, implying a minimum of elastic potential energy. This minimum of energy is taken as $E_c = 0$ and the specific volume at this 'zero point' is V_{0c}. The value of V_{0c} is smaller than the volume of the solid at standard conditions, $P = 0$ or $1\,\text{atm}$ and $T_0 \approx 300\,\text{K}$. For a volume V larger than V_{0c} the atoms are attracted to each other, while for values of V smaller than V_{0c} the repulsive forces are dominant. The cold pressure increases rapidly with compression and becomes negative with expansion. The negative sign of the pressure denotes the fact that the expansion of a body from its equilibrium at $T = 0$ and $P = 0$ is carried out by a tensile force. This tension force is balanced by the binding forces which react to return the solid to its equilibrium volume V_{0c}.

Small pressures only are needed to compress a gas highly by a shock wave, for example a shock pressure of the order of 100–$1000\,\text{atm}$ can compress a gas to few times its initial density. However, in order to compress a cold metal by 10% only, an external pressure of the order of 10^5 atm is needed. This fact can be seen from the values of the compressibility, κ_0

$$\kappa_0 = -\frac{1}{V}\left(\frac{\partial V}{\partial P}\right)_S \qquad (13.1)$$

which is of the order of 10^{-6} atm^{-1} for metals at standard conditions. To compress a solid by a factor of 2 one has to apply pressures of the order of several million atmospheres. At these high pressures, the elastic compression energy is of the order of $10\,\mathrm{eV/atom}$ and it rises rapidly with further increase of compression. These energies (or the appropriate high pressures) are sufficient to change the crystal structure into a close packing type of liquid. Thus the high pressures of the shock waves cause phase transitions that are related to changes of the crystal lattice symmetry. We will ignore these phase transitions and will describe physical situations below or above the phase transition points.

There are a number of basic techniques for achieving high pressure shock waves, in particular the following are used:

(a) high explosive driven shock waves
(b) shock waves generated by the impact of accelerated projectiles
(c) laser generated shock waves (see, e.g, Salzmann, Eliezer, Krumbein and Gitter (1983), F. Anisimov, Prokhorov and Fortov (1984))
(d) shock waves generated by nuclear explosives (see, e.g., Ragan 1980)

Using the first two techniques (a) and (b) pressures of up to a few megabars (1 megabar $= 10^6$ atm) were achieved in the laboratory, while using big laser facilities or sophisticated guns to accelerate projectiles, pressures higher by an order of magnitude are expected. In all these experiments, high energies are released in a short period of time (in the region domain of picoseconds to microseconds), and therefore very fast measurements are necessary. The experimental measurements are usually taken from a remote location and the diagnostic system must be as complete as possible, since the system cannot be returned to its original state for verifying and checking the measured quantities. It is beyond the scope of this book to discuss the experimental techniques and the equipment used in high pressure shock wave experiments. Rather it is assumed that a strong shock wave is created and the necessary parameters are measured in the laboratory, without discussing how this is achieved (for this purpose see, e.g., Altshuler (1965), Davison and Graham (1979), Rice *et al.* (1958)). We assume a planar one-dimensional steady state shock wave which satisfies the Rankine–

Hugoniot relations as obtained from the conservation of mass, momentum and energy,

$$\frac{V_0}{V} \equiv \frac{\rho}{\rho_0} = \frac{D}{D-u} \tag{13.2}$$

$$P_H - P_0 \approx P_H = \rho_0 D u \tag{13.3}$$

$$\rho_0 D(E_H - E_0 + \tfrac{1}{2}u^2) = P_H u \tag{13.4}$$

In (13.2)–(13.4) the subscript 0 refers to the stationary material ahead of the shock (initial conditions for the shock wave experiment). The material behind the shock wave front reaches a pressure P_H, a density $\rho \equiv 1/V$ and a specific internal energy E_H, while the shock wave is moving with a velocity D and the particles behind the shock surface are moving with velocity u. Using (13.2)–(13.4) a Hugoniot curve is obtained

$$E_H - E_0 = \tfrac{1}{2}(P_H + P_0)(V_0 - V) \tag{13.5}$$

For a known initial state (P_0, V_0) it is necessary to measure any two quantities from (P_H, V, E_H, D, u) in order to fix a point on the Hugoniot curve, usually described by P_H as a function of V (or ρ).

The P–V diagram of Fig. 13.1 describes the Hugoniot curve $P_H(V)$ for the initial conditions V_0, T_0 and $P_0 = 0$. Note that the atmospheric pressure is negligibly small in comparison with pressures related even to very small changes in the specific volume V_0, therefore $P_0 = 0$ or 1 atm are equivalent initial conditions for shock wave experiments in solids. The difference in

Fig. 13.1. The variation of P_H, P_S and P_c with V.

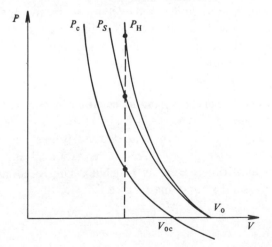

Fig. 13.1 at any particular value of V between P_H and the cold pressure P_0 is equal to the thermal pressure P_T that is caused by the shock wave. Since P_T is positive, the cold pressure at room temperature turns out to be negative, $P_c(V_0) < 0$. Also in Fig. 13.1 the isentrope P_S passing through the foot of the Hugoniot is plotted, so that the qualitatively mutual positions of the curves P_c, P_S and P_H are illustrated.

The existence of an extremely large non-thermal pressure P_c (which does not exist in gases) in condensed matter determines the basic physical behaviour of solids and liquids compressed by shock waves. Besides this cold pressure P_c caused by the repulsive forces, a thermal pressure P_T, caused by the atomic and electronic motions inside the shocked material, is also created. For pressures up to the order of 0.5×10^6 atm, the P_c term is the dominant one, while for shock waves causing pressures of the order of a few millions of atmospheres the P_c and the P_T terms are comparable in magnitude. As the shock wave pressure increases further, the thermal pressure becomes dominant.

In Section 13.2 a Grüneisen type equation of state is assumed and an integro-differential equation is developed for the elastic pressure $P_c(V)$. In Section 13.3 the equations for the initial conditions necessary to solve the equation for P_c are stated. It is shown how a polynominal fit to the shock wave experimental data is used in order to calculate the parameters of the equation of state. Next it is shown how the Hugoniot data together with the knowledge of the Grüneisen coefficient enables one to calculate the isentropic process near the Hugoniot curve. This is associated with the velocity of the end surface of the target, a knowledge of which is needed for the measurement of the second parameter, besides the directly measured shock wave velocity. In Section 13.5 the equations of state from Hugoniot data are calculated for aluminum, copper and lead. An alternative possible calculation is given in Section 13.6, where the cold pressure is calculated using an interpolation method from room temperature empirical data to the Thomas–Fermi–Dirac model.

13.2   The Grüneisen coefficient $\gamma(V)$ and an equation for the cold pressure P_c

In order to study the equation of state deduced from shock wave data it is necessary to assume a model for the equation of state. In particular, one can use the Grüneisen model, which takes into account the lattice oscillations, for a solid state of matter (see Chapter 10)

$$P_T = \frac{\gamma}{V} E_T \tag{13.6}$$

The thermal pressure (P_T) and the cold pressure (P_c) are added to give the total pressure (P)

$$P = P_c + P_T \tag{13.7}$$

and similarly for the appropriate energies

$$E = E_c + E_T \tag{13.8}$$

using (13.6)–(13.8), the following equation of state is obtained

$$P(V, E) = P_c(V) + \frac{\gamma(V)}{V}[E - E_c(V)] \tag{13.9}$$

which has been found experimentally to be valid near the Hugoniot curve (McQueen *et al.* 1970). Since the temperature, the entropy and other thermodynamic functions cannot be calculated explicitly from (13.9), this equation of state is incomplete. A complete equation of state can be obtained, for example, by combining the Grüneisen pressure–energy relation given in (13.9) with the Debye relation between thermal energy and temperature. In the classical limit

$$E_T = c_v T \tag{13.10}$$

And the specific heat c_v has the value (Dulong and Petit law)

$$c_v = 3R \,[\text{erg mol}^{-1}\,\text{deg}^{-1}] \tag{13.11a}$$

where R is the gas constant, $R = 8.314 \times 10^7\,\text{erg mol}^{-1}\,\text{deg}^{-1}$ and the energy in (13.10) is measured in erg mol^{-1}, or equivalently

$$c_v = \frac{3R}{\langle A \rangle} \,[\text{erg g}^{-1}\,\text{deg}^{-1}] \tag{13.11b}$$

where $\langle A \rangle$ is the average atomic weight (g mol^{-1}). In this case the energy in (13.10) is measured in erg g^{-1}, which is in general our standard convention. The good agreement between (13.11) and experiment is demonstrated in Table 13.1. Using (13.6), (13.10) and (13.11) the thermal pressure is given in

Table 13.1.

	$\langle A \rangle$ g mol^{-1}	$3R/\langle A \rangle$ 10^6 erg g^{-1} deg^{-1}	$c_v(\exp)$ 10^6 erg g^{-1} deg^{-1}
Al (aluminum)	27	9.24	8.96
Cu (copper)	63.55	3.92	3.83
Pb (lead)	207.2	1.20	1.29

this case by

$$P_T = 3R\gamma T/V \tag{13.12}$$

From this equation it is evident that an accurate knowledge of γ is needed in order to calculate the temperature T from the pressure measurements.

In the three-dimensional space of $P-V-T$ (or $P-V-S$, etc.), the Hugoniot curve is a single line, while the equation of state is a surface. Therefore, it is evident that one cannot obtain the complete equation of state from shock wave data, in particular taking into account that for practical reasons the known initial conditions are usually the standard values of pressure (P_0), specific volume (V_0) and initial temperature (T_0). However, the experimental Hugoniot data is useful to obtain limits on the parameters of the equation of state, similarly to obtain relations between measurable quantities. In particular, we shall show how a Grüneisen-like equation of state can be calculated using experimental Hugoniot shock wave data.

Several models have been suggested to relate the γ function in the Grüneisen model to the curvature of the cold compression curve. As an example we use the Slater–Landau model in order to relate $\gamma(V)$ to P_c (see Chapter 10)

$$\gamma = -\frac{2}{3} - \frac{V}{2}\left(\frac{d^2 P_c}{dV^2}\right) \bigg/ \left(\frac{dP_c}{dV}\right) \tag{13.13}$$

Equation (13.13) is not the only relation suggested in the literature between the Grüneisen coefficient $\gamma(V)$ and the derivatives of the cold pressure, $P_c(V)$. The models that have been proposed in the literature can be summarized by the following equation (Altshuler 1965).

$$\gamma(V) = -\left(\frac{2}{3} - \frac{t}{3}\right) - \frac{V}{2}\frac{d^2}{dV^2}[P_c V^{2t/3}] \bigg/ \frac{d}{dV}[P_c V^{2t/3}] \tag{13.14}$$

The most popular values for t are $t = 0$ or 1 or 2, where the value $t = 0$ corresponds to (13.13) and $t = 1$ was suggested by Dugdall and McDonald (1953). These three models for γ give almost equal values as the pressure is increased. Therefore, one can choose the model that gives the best value for low pressure data without significantly changing the high pressure results. However, it is important to note that the fact that all three models of (13.14) (i.e., $t = 0, 1, 2$) converge to the same value, does not necessarily imply that any of these models is correct. As mentioned above, we choose $t = 0$, i.e., (13.13) to pursue our development of the equation of state from shock wave data.

Using (13.9), γ is related to the experimental Hugoniot pressure (P_H) and energy (E_H) by

$$\gamma = V(P_H - P_c)/(E_H - E_c) \tag{13.15}$$

From (13.13) and (13.15) and the thermodynamic relation

$$P_c(V) = -\frac{dE_c(V)}{dV} \tag{13.16}$$

one obtains an integro-differential equation for $P_c(V)$, or a third order differential equation for $E_c(V)$, which can be integrated numerically.

Equation (13.13) and (13.15) give

$$\gamma = V(P_H - P_c)/(E_H - E_c) = -\frac{2}{3} - \frac{V}{2}\left(\frac{d^2 P_c}{dV^2}\right) \Big/ \left(\frac{dP_c}{dV}\right) \tag{13.17}$$

Substituting E_H from the Hugoniot relation, (13.5), into (13.17), one obtains the following integro-differential equation for $P_c(V)$,

$$\frac{V(P_H - P_c)}{E_0 + \frac{1}{2}P_H(V_0 - V) + \int_{V_{0c}}^{V} P_c dV} = -\frac{2}{3} - \frac{V}{2}\left(\frac{d^2 P_c}{dV^2}\right) \Big/ \left(\frac{dP_c}{dV}\right)$$

$$\tag{13.18}$$

where V_{0c} satisfies the relations:

$$P_c(V_{0c}) = 0 \quad E_c(V_{0c}) = 0 \tag{13.19}$$

The initial conditions required for the numerical integration of (13.18) are the specific volume V_{0c} and the slope $(dP_c/dV)_{V_{0c}}$, as well as the internal energy E_0 for the experimental initial conditions (V_0, T_0). E_0 can be obtained by integrating the Debye specific heat curve from $T = 0$ to T_0,

$$E_0 = \int_0^{T_0} c_v(T) dT \tag{13.20}$$

The experimental initial data (V_0, T_0) before the arrival of the shock wave should be known. The initial conditions for solving (13.18) can be derived from extrapolations of available data at standard conditions. The solution of (13.18) for $P_c(V)$ and subsequently the knowledge of $\gamma(V)$ from (13.17) give the desired equation of state of the model under consideration as given by (13.9). However, in order to reach this goal, the initial conditions necessary to solve (13.18) numerically are needed. For this purpose we now derive an equation from which the value of V_{0c} can be obtained.

13.3   The specific volume V_{0c} of the 'zero point' and the initial conditions for the P_c equation

First, we use the first order Taylor expansion for $\gamma(V)$ in terms of the initial experimental value $V_0(=1/\rho_0)$.

$$\gamma(V_{0c}) = \gamma_0 + \left(\frac{d\gamma}{dV}\right)_0 (V_{0c} - V_0) \tag{13.21}$$

where $\gamma_0 \equiv \gamma(V_0)$ and $(d\gamma/dV)_0 \equiv (d\gamma/dV)_{V=V_0}$, γ_0 is given in terms of experimental quantities measured in the laboratory (see Chapter 10)

$$\gamma_0 = \frac{3V_0\alpha}{c_v\kappa_0} \tag{13.22}$$

where α is the coefficient of (linear) thermal expansion, κ_0 is the isothermal compressibility and c_v is the specific heat at constant value of $T = T_0$ (usually room temperature). From (13.9) and (13.5) one gets

$$V(P_H - P_c) = \gamma[E_0 + \tfrac{1}{2}P_H(V_0 - V) - E_c] \tag{13.23}$$

Evaluating P_H at $V = V_{0c}$, i.e., using $E_c(V_{0c}) = 0$ and $P_c(V_{0c}) = 0$, (13.23) can be rewritten as

$$V_{0c}P_H(V_{0c}) = \gamma(V_{0c})[E_0 + \tfrac{1}{2}P_H(V_{0c})(V_0 - V_{0c})] \tag{13.24}$$

Equations (13.21) and (13.24) yield the relation that the initial specific volume should satisfy

$$V_{0c}P_H(V_{0c}) = \left[\gamma_0 + \left(\frac{d\gamma}{dV}\right)_0 (V_{0c} - V_0)\right]$$
$$\times [E_0 + \tfrac{1}{2}P_H(V_{0c})(V_0 - V_{0c})] \tag{13.25}$$

In order to solve (13.25) for V_{0c}, the value of $(d\gamma/dV)_0$ is needed. It is shown in the next few paragraphs how to reach this goal.

The Hugoniot curve, P_H as a function of V, can be fitted experimentally by the following polynomial

$$P_H = \sum_{k=1}^{N} A_k^H(\sigma - 1)^k \qquad \sigma \equiv \frac{\rho}{\rho_0} = \frac{V_0}{V} \tag{13.26}$$

where as usual, ρ is the density behind the shock wave front and ρ_0 is the density of the initial state ahead of the shock wave front. For example, Altshuler *et al.* (1960) make an experimental fit with $N = 7$ for shock wave pressures in aluminum, copper and lead up to 5×10^6 atm ($= 5$ Mb); the coefficients are given in Table 13.2.

A similar expansion to (13.26) can be made for an isentropic curve

$(dS = 0)$ *near* the Hugoniot curve,

$$P_S = \sum_{k=1}^{N} a_k{}^S(\sigma - 1)^k, \quad \sigma \equiv \frac{\rho}{\rho_0} = \frac{V_0}{V} \tag{13.27}$$

There is a second order tangency at the origin of the coordinate between the Hugoniot and the isentropic curve (see, e.g., Zeldovich and Raizer (1967), p. 56 and pp. 63–7), i.e.,

$$\left(\frac{dP_S}{dV}\right)_{V_0} = \left(\frac{dP_H}{dV}\right)_{V_0} \tag{13.28}$$

$$\left(\frac{d^2 P_S}{dV^2}\right)_{V_0} = \left(\frac{d^2 P_H}{dV^2}\right)_{V_0} \tag{13.29}$$

These two relations for the polynomial expansions, (13.26) and (13.27), imply

$$a_1{}^S = A_1{}^H, \quad a_2{}^S = A_2{}^H \tag{13.30}$$

A third relation can be obtained by using the Grüneisen equation of state $\gamma = V(P_S - P_H)/(E_S - E_H)$ in the limit $V \to V_0$, where

$$E_S = -\int_{V_0}^{V} P_S dV \tag{13.31}$$

and the Hugoniot relation $E_H = E_0 + \frac{1}{2}P_H(V_0 - V)$ is used. This relation, obtained in Appendix 6, is

$$a_3{}^S = A_3{}^H - \tfrac{1}{6}\gamma_0 A_2{}^H \tag{13.32}$$

The isotherm $T = 0$ is an isentropic process as well, since for $T = 0$ the

Table 13.2. *The coefficients $A_k{}^H$ appearing in the expansion of P_H (see (13.26))*

	Values of $A_k{}^H$ in 10^{10} dyne cm^{-2} (≈ 10 kbars)		
k	Al (aluminum)	Cu (copper)	Pb (lead)
1	73.1	137.0	41.4
2	152.7	271.7	101.7
3	143.5	224.0	120.0
4	−887	1078	−43
5	2862	−2967	547
6	−3192	3674	−801
7	1183	−1346	312

Note: Table adapted from Altshuler *et al.* (1960).

entropy is $S = 0$ (Nernst's theorem), thus $P_c(V)$ describes an isentropic curve. Assuming that (13.13) for $\gamma(V)$ is also true for other isentropic processes not too far from $T = 0$, in particular that (13.13) is also valid for the isentropic process passing through the foot of the Hugoniot at $V = V_0$, we can use the polynomial expansion for P_S at $V = V_0$ to calculate γ_0 and $(d\gamma/dV)_0$

$$\gamma_0 \equiv \gamma(V_0) = -\frac{2}{3} - \frac{V_0}{2}\left(\frac{d^2 P_S}{dV^2}\right)_{V_0} \bigg/ \left(\frac{dP_S}{dV}\right)_{V_0} \tag{13.33}$$

and

$$\left(\frac{d\gamma}{dV}\right)_0 \equiv \left(\frac{d\gamma}{dV}\right)_{V_0} = \frac{1}{2}\left[\left(\frac{d^2 P_S}{dV^2}\right)_{V_0} \bigg/ \left(\frac{dP_S}{dV}\right)_{V_0}\right.$$
$$\left. + V_0\left(\frac{d^3 P_S}{dV^3}\right)_{V_0} \bigg/ \left(\frac{dP_S}{dV}\right)_{V_0} - V_0\left(\frac{d^2 P_S}{dV^2}\right)_{V_0}^2 \bigg/ \left(\frac{dP_S}{dV}\right)_{V_0}^2\right] \tag{13.34}$$

Substituting (13.27) in the last two equations and taking the limit $V \to V_0$ one gets, using (13.30) and (13.32) also [see Appendix 6]

$$\gamma_0 = \frac{1}{3} + \frac{a_2^S}{a_1^S} = \frac{1}{3} + \frac{A_2^H}{A_1^H} \tag{13.35}$$

$$V_0\left(\frac{d\gamma}{dV}\right)_0 = 4 + 7\frac{a_2^S}{a_1^S} - 3\frac{a_3^S}{a_1^S} + 2\left(1 + \frac{a_2^S}{a_1^S}\right)^2 \tag{13.36}$$

or using (13.30) and (13.32) to relate to Hugoniot parameters

$$V_0\left(\frac{d\gamma}{dV}\right)_0 = 4 + 7\frac{A_2^H}{A_1^H} + 2\left(1 + \frac{A_2^H}{A_1^H}\right)^2 - \frac{3A_3^H}{A_1^H} + \tfrac{1}{2}\gamma_0\frac{A_2^H}{A_1^H} \tag{13.37}$$

A_1^H, γ_0 and thus A_2^H are measured experimentally at room temperature, since a_1^S is given by the isentropic compressibility coefficient (κ_0^{-1})

$$A_1^H = a_1^S = -V_0\left(\frac{\partial P_S}{\partial V}\right)_{V_0} = \frac{1}{\kappa_0} \tag{13.38}$$

and γ_0 is given by measurable quantities in (13.22). Thus, A_1^H and A_2^H are measured experimentally at standard conditions, while A_3^H is taken from the experimental fit of the Hugoniot. In this way one can calculate V_{0c} from (13.25). The standard volume of metals V_0 is usually only about 2% larger than V_{0c}.

Finally, in order to solve the integro-differential equation (13.18), the initial slope $(dP_c/dV)(V_{0c})$ is needed. For this purpose, once again the

Grüneisen equation of state is our starting point. Substituting $P = P_H$ in (13.9),

$$P_H = P_c + \frac{\gamma}{V}(E_H - E_c) \tag{13.39}$$

and subtracting (13.39) from (13.9), the Grüneisen equation of state can be rewritten as

$$P - P_H = \frac{\gamma}{V}(E - E_H) \tag{13.40}$$

Taking the derivative of this equation at $V = V_{0c}$ we get

$$V_{0c}\left(\frac{dP_c}{dV}\right)_{V_{0c}} = V_{0c}\left(\frac{dP_H}{dV}\right)_{V_{0c}} + P_H(V_{0c})$$

$$- E_H(V_{0c})\left(\frac{d\gamma}{dV}\right)_{V_{0c}} - \gamma(V_{0c})\left(\frac{dE_H}{dV}\right)_{V_{0c}} \tag{13.41}$$

where E_H is given in (13.5) with $P_0 = 0$ for all practical purposes. V_{0c} is known from the solution of (13.25), $\gamma(V_{0c})$ and $(d\gamma/dV)_{V_{0c}}$ are obtained from (13.21), while γ_0 and $(d\gamma/dV)_0$ are derived in (13.35) and (13.37) in terms of the parameters $A_1{}^H$, $A_2{}^H$ and $A_3{}^H$ from the experimental fit of the Hugoniot curve. Taking all these into consideration, the last of the initial conditions, $(dP_c/dV)_{V_{0c}}$, necessary for the numerical solution of (13.18), is obtained.

At this stage, a short summary, of the long procedure carried out so far to calculate the equation of state, might be useful. First, a model given by (13.9) is assumed for the equation of state. In this equation, $P_c(V)$ and $\gamma(V)$ are not known. The next step is to obtain a relation between $\gamma(V)$ and $P_c(V)$. This is given in our example by the differential equation (13.13). (13.9) and (13.13) are used together with the Hugoniot relation, (13.5), to obtain an integro-differential equation (13.18) for $P_c(V)$. In this equation P_H is given by the shock wave experimental data, fitted, for example, by the polynomial of (13.26). From the solution of $P_c(V)$ the function $\gamma(V)$ is obtained from (13.13). However, before being able to calculate $\gamma(V)$ and $P_c(V)$ from (13.18), one needs the initial conditions V_{0c} and $(dP_c/dV)_{V_{0c}}$ in addition to (13.19). For the calculation of V_{0c} (13.25) was introduced. This equation needs as input the values of γ_0 and $(d\gamma/dV)_0$, which were obtained from the fit of the experimental Hugoniot in (13.35) and (13.37). For the initial condition $(dP_c/dV)_{V_{0c}}$ one has to solve (13.41), so that knowledge of the initial conditions is complete. Therefore, simultaneous solution, as described above, of (13.5), (13.9), (13.13), (13.18), (13.19), (13.25), (13.26), (13.35), (13.37) and (13.41) enables us to calculate numerically the Grüneisen equation of state.

13.4   Isentropic processes near the Hugoniot curve and the free surface velocity

We will next show how the Hugoniot data together with the knowledge of $\gamma(V)$ enables one to calculate the isentropic $(dS = 0)$ process near the Hugoniot curve. After the determination of the volume dependence of $\gamma(V)$, one can differentiate (13.40) with respect to V for an isentropic process (i.e., $P = P_S$ and $E = E_S$ in (13.40)) to obtain

$$P_S = -\frac{dE_H}{dV} + (P_H - P_S)\frac{d}{dV}\left(\frac{V}{\gamma}\right) + \frac{V}{\gamma}\left(\frac{dP_H}{dV} - \frac{dP_S}{dV}\right) \tag{13.42}$$

where the relation

$$\frac{dE_S}{dV} = -P_S \tag{13.43}$$

for a $dS = 0$ process has been used. Equation (13.42) is a first order differential equation for an isentropic change in the pressure $P_S(V)$ in terms of the known $P_H(V)$ (experimental data), $E_H(V)$ (13.5) and $\gamma(V)$ (13.17). Numerical integration of (13.42) gives the pressure P_S as a function of the specific volume. The temperature change in this transition is obtained using the thermodynamic identity

$$T dS = c_v dT + T\left(\frac{\partial P}{\partial T}\right)_V dV = 0 \tag{13.44}$$

and the relation

$$\gamma = V\left(\frac{\partial P}{\partial T}\right)_V \Big/ c_v \tag{13.45}$$

The solution of (13.44) and (13.45) is given by

$$T = T_0 \exp\left(-\int_{V_0}^{V}\frac{\gamma dV}{V}\right) \tag{13.46}$$

where T_0 and V_0 are the initial conditions for the isentropic process. Equation (13.46) together with the solution of (13.42) can be used to obtain the temperature at a (P, V) point near the Hugoniot curve.

One of the objects of calculating isentropes is for the evaluation of the Riemann integral which is associated with the particle velocity in the rarefaction wave. Such a wave results for example when a shock wave is reflected from a free surface. When the shock wave reaches the end surface of the solid bounded by the vacuum (or the atmosphere), the free surface develops in a very short time (≈ 10 psec) a velocity U_{fs} given by

$$U_{fs} = u + u_r \tag{13.47}$$

where u is the particle velocity relative to the shock wave and appears in (13.2)–(13.4) (the Rankine–Hugoniot equations). The material velocity increase, u_r, due to the rarefaction wave from some point on the Hugoniot to zero pressure is given by the Riemann integral along an isentropic

$$u_r = \int_{V(P_H)}^{V(P=0)} \left(-\frac{dP_S}{dV} \right)^{1/2} dV \tag{13.48}$$

The derivative in (13.48) is obtained from (13.42), so that a knowledge of P_S gives the value of u_r. Layers adjacent to the free surface go into motion under the influence of the shock transition from P_0, V_0 to P, V and subsequent isentropic expansion in the reflected rarefraction wave from P, V to P_0, V_0^1 ($> V_0$). Although these two processes are not the same, it turns out that for $u \ll D$ (shock velocity) one has

$$u \approx u_r \tag{13.49}$$

From (13.47) and (13.49) the following practical relation is obtained:

$$U_{fs} \approx 2u \tag{13.50}$$

Therefore, from free surface velocity measurements, one can calculate the particle velocity of the shock wave compressed material. This free surface velocity together with the experimental measurement of the shock wave velocity might serve as the two necessary quantities (out of five: P_H, V, E_H, D, u) to fix a point on the Hugoniot.

We conclude this section with the important remarks that the procedure described here to obtain information about the equation of state from Hugoniot data is not unique. For example the Hugoniot polynomial fit in (13.26) is not the only possible experimental fit. The shock wave data can be fitted very well by (Altshuler (1965), Davison and Graham (1979), McQueen et al. (1970), Rice et al. (1958))

$$D = A + Bu \tag{13.51}$$

Table 13.3. *The coefficients A and B (see (13.51)) obtained from experimental fit*

	A (km s^{-1})	B
Al	5.328	1.338
Cu	3.958	1.497
Pb	2.028	1.517

Note: Table adapted from McQueen & Marsh (1960).

where D and u are defined in (13.2)–(13.4) and A and B are constants. Typical values of the constants A and B (obtained from experimental best fit) for aluminium, copper and lead are given in Table 13.3. Sometimes a higher order polynomial (in u) is needed in (13.51). Algebraic solution of (13.2), (13.3) and (13.51) yield the following fit for the Hugoniot curve

$$P_H = \frac{A^2(V_0 - V)}{(B-1)^2[AV/(A-1) - V_0]^2} \tag{13.52}$$

Although the difference in the Hugoniot curve between P_H as given by (13.52) and P_H from (13.26) (the polynomial fit) is negligible, less than a fraction of 1% in V/V_0, the differences in the solution of $\gamma(V)$ might be significant! Therefore, taking this fact into consideration, the difficulties in calculating P_c in a well defined and unique form (e.g., (13.13) is not unique!) and also the uncertainties in estimating the thermal pressure and energy, it can by said that the study of equations of state from shock wave data is not yet complete and more research is needed.

13.5   Equations of state for aluminium, copper and lead

The pressure and the internal energy of a material can be considered as divided into two contributions: (a) the elastic term related to the interaction forces between the atoms of the material, which is therefore independent of temperature; the appropriate pressure P_c and the energy E_c are defined as the 'cold pressure' and 'cold energy' respectively and (b) the thermal term related to the temperature T of the body. This latter contribution is usually divided into two parts, one part describing the motion of the atoms in the lattice and the second part related to the thermal motion of the electrons. The thermal lattice pressure and energy are denoted by P_{TL} and E_{TL} respectively, and the appropriate thermal contribution to the pressure and to the energy by the electrons is P_{TE} and E_{TE}. Therefore we can write the equation of state as follows:

$$P = P_c + P_{TL} + P_{TE} \tag{13.53}$$
$$E = E_c + E_{TL} + E_{TE} \tag{13.54}$$

The cold pressure and energy are related by the thermodynamic relation (see (13.16))

$$E_C = -\int_{V_{0c}}^{V} P_c dV \tag{13.55}$$

where V_{0c} is the 'zero point' specific volume which was discussed in Section 13.3. The lattice thermal energy and pressure are related through the Grüneisen equation of state (see (13.6)), and the energy is described by the (Debye) solid state theory. Therefore in this approximation one has

$$E_{TL} = c_v(T - T_0) + E_0 \tag{13.56}$$
$$P_{TL} = \gamma(V)E_{TL}/V \tag{13.57}$$

where c_v is the lattice specific heat at constant volume and is given, to a good approximation, by the Dulong and Petit law (13.11b), T_0 is the initial temperature in the shock wave experiment, usually the room temperature, E_0 is the internal energy at T_0 defined by

$$E_0 = \int_0^{T_0} c_v(T)\mathrm{d}T \tag{13.58}$$

For a constant value of c_v Eqs (13.56) and (13.58) imply the previously mentioned (13.10). The values of E_0 can be obtained from tables given in the solid state literature (see Table 13.4). For high temperatures $(T \gg T_0)$ one can write (13.10) as a very good approximation even if c_v is not a constant since the difference between E_0 and $c_v T_0$ is negligible in comparison with $c_v T$. $\gamma(V)$ is the Grüneisen coefficient and it is assumed to be a function only of the specific volume V. This coefficient is related to the cold pressure derivatives through (13.14).

Assuming for the electron excitations an equation of state similar to the Grüneisen equation of state for the lattice, namely

$$P_{TE} = \gamma_E E_{TE}/V \tag{13.59}$$

Table 13.4. *The normal density* ρ_0 $(= 1/V_0)$, *the lattice specific heat at constant volume* c_v, *the internal energy at normal conditions* E_0, *the coefficient of compressibility at constant temperature* κ_C, *the coefficient of linear thermal expansion* α, *the Grüneisen coefficient at normal conditions* γ_0 *(using (13.22)) and the electron specific heat coefficient* β_0 *for Al, Cu and Pb*

	Al	Cu	Pb
$\rho_0 (\text{g cm}^{-3}) \equiv \dfrac{1}{V_0}$	2.71	8.93	11.34
$c_v(10^6 \text{ erg g}^{-1} \text{deg}^{-1})$	8.96	3.83	1.29
$E_0(10^7 \text{ erg g}^{-1})$	161.0	77.1	32.3
$\kappa_0(10^{-12} \text{ cm}^2 \text{ dyne}^{-1})$	1.37	0.73	2.42
$\alpha(10^{-6} \text{ deg}^{-1})$	23.1	16.5	29.0
γ_0	2.09	1.98	2.46
$\beta_0(\text{erg g}^{-1} \text{deg}^{-2})$	500	110	144

and a relation between the energy E_{TE} and the temperature as suggested from the free electron model ($c_{vE} \approx T, E_{TE} \approx T^2$)

$$E_{TE} = \tfrac{1}{2}\beta T^2 \tag{13.60}$$

one gets

$$P_{TE} = \tfrac{1}{2}\gamma_E \beta T^2 / V \tag{13.61}$$

In order to find a relation between β and the specific volume V one can use the thermodynamic relation

$$\left(\frac{\partial E}{\partial V}\right)_T = T\left(\frac{\partial P}{\partial T}\right)_V - P \tag{13.62}$$

Substituting (13.60) and (13.61) in (13.62) one has

$$\frac{\partial \beta}{\partial V} = \frac{\gamma_E}{V}\beta \tag{13.63}$$

or after integration

$$\beta = \beta_0 \exp\left(\int_{V_0}^{V} \gamma_E \frac{dV}{V}\right) \tag{13.64}$$

It was found (Altshuler *et al.*, 1960) that it is accurate and consistent with shock wave data up to a few megabars pressure to take

$$\gamma_E = \tfrac{1}{2} \tag{13.65}$$

so that (13.64) has the solution

$$\beta = \beta_0 \left(\frac{V}{V_0}\right)^{1/2} \tag{13.66}$$

At extremely high pressures, where the TF model is applicable, the value found for the electronic Grüneisen coefficient $\gamma_E = \tfrac{2}{3}$. The value of β_0 is measured experimentally at very low temperatures since for $T \to 0$ the lattice specific heat is dominated by quantum effects which imply $c_v \approx T^3$ for lattice, while the electron specific heat is proportional to T only. At

Table 13.5. *The 'zero-point' specific volume V_{0c} and the initial condition for the cold pressure $(dP_c/dV)_{V_{0c}}$ for Al, Cu and Pb*

	Al	Cu	Pb
V_{0c}/V_0	0.988	0.990	0.979
$V_0(dP_c/dV)_{V_{0c}} (10^{10}\,\text{dyne cm}^{-2})$	−77.1	−142.8	−45.9

Note: Table adapted from Altshuler *et al.* (1960).

room temperature, on the other hand, the electronic specific heat is usually smaller by a factor of 10–100 relative to the lattice specific heat. The measured experimental values of β_0 are given in Table 13.4. Substituting (13.65) and (13.66) in (13.60) and (13.61) and using them together with (13.55), (13.56) and (13.60), the equations of state (13.53) and (13.54) can be written in the following form,

$$P = P_c + \frac{\gamma}{V} [c_v(T - T_0) + E_0] + \frac{1}{4} \frac{\beta_0}{V_0} \left(\frac{V_0}{V}\right)^{1/2} T^2 \tag{13.67}$$

$$E = - \int_{V_{0c}}^{V} P_c dV + c_v(T - T_0) + E_0 + \tfrac{1}{4} \beta_0 \left(\frac{V}{V_0}\right)^{1/2} T^2 \tag{13.68}$$

Substituting in the left hand side of (13.67) the Hugoniot data as suggested by the best fit of (13.26) (the polynomial fit of P_H), and for the left hand side of (13.68) we take the Hugoniot relation (13.5) (again taking (13.26) into account for P_H), one has two equations with three unknowns: P_c, γ and T. The third equation is the differential equation for P_c as suggested in the previous sections (see (13.14) or (13.18)). The initial conditions are found as suggested in Section 13.3 and the derived values of V_{0c} and $(dP_c/dV)_{V_{0c}}$ are given in Table 13.5 for the shock wave experiments carried out with aluminum, copper and lead. The experimental input data necessary to solve the set of the integro-differential equations ((13.14)[†], (13.67) and (13.68)) is given in Tables 13.4 and 13.2 for aluminum, copper and lead. The numerical solutions of the equations of state for these metals are given in Tables 13.6, 13.7 and 13.8 (Altshuler *et al.* (1960)).

Table 13.6 illustrates shock wave experiments for aluminum up to about 2 Mbars. One can see that in this region the thermal electronic contribution to the equation of state is negligible. However, the thermal contribution of the lattice is about 25% at a pressure of 2 Mbars and about 10% for a pressure of 0.5 Mbars. From the table one can see that the temperature increases with the strength of the shock wave. However, for a 2 Mbar shock the temperature is still less than one electron volt. The value of the Grüneisen coefficient decreases with increase of compression.

Table 13.7 for copper and Table 13.8 for lead illustrate shock wave experiments up to about 4 Mbars. One can see from these tables the trend of the thermal contributions relative to the cold contributions. It is important to note that for strong shock waves with high compression the electronic thermal contribution starts to be significant. In all three metals the decrease of $\gamma(V)$ with density was observed.

[†] Altshuler *et al.* (1960) use $t = 1$. However, a similar solution with $t = 0$ changes the given results very little (less than 1–2%).

Table 13.6. *EOS for Al. The Hugoniot experimental fit for the pressure P_H, the Grüneisen coefficient γ, the temperature T_H for the appropriate points on the Hugoniot, the calculated cold pressure P_c and cold energy E_c, the lattice thermal contribution to the Hugoniot pressure P_{TL}/P_H and the electron thermal contribution to the Hugoniot pressure P_{TE}/P_H*

$V_0/V = \rho/\rho_0$	P_H* (10^{12} dyne cm^{-2})	γ	T_H*** (K)	P_c* (10^{12} dyne cm^{-2})	E_c** (10^8 erg g^{-1})	P_{TL}/P_H	P_{TE}/P_H
1.05	0.042	2.00	315	0.031	2.0	0.26	0.00
1.10	0.090	1.81	348	0.078	10.5	0.13	0.00
1.15	0.145	1.56	401	0.134	26.0	0.08	0.00
1.20	0.211	1.37	488	0.197	47.0	0.07	0.00
1.25	0.288	1.28	625	0.266	76.0	0.08	0.00
1.30	0.374	1.28	818	0.345	111.0	0.08	0.00
1.35	0.472	1.31	1097	0.429	151.0	0.09	0.00
1.40	0.582	1.37	1476	0.519	198.0	0.11	0.00
1.45	0.713	1.41	1980	0.619	249.0	0.13	0.00
1.50	0.861	1.44	2640	0.727	306.0	0.16	0.00
1.55	1.030	1.44	3440	0.845	368.0	0.18	0.00
1.60	1.217	1.43	4410	0.970	435.0	0.19	0.01
1.65	1.427	1.41	5530	1.100	508.0	0.22	0.01
1.70	1.652	1.39	6790	1.250	585.0	0.23	0.01
1.75	1.897	1.34	8180	1.400	667.0	0.24	0.02
1.80	2.160	1.30	9670	1.570	754.0	0.25	0.02

Notes: Table adapted from Altshuler et al. (1960)
*10^{12} dyne cm^{-2} ≈ 1Mb ≈ 10^6 atm. **10^8 erg g^{-1} ≈ 7.2 × 10^{-3} eV/atom. ***11 600 K ≈ 1 eV.

Table 13.7. EOS for Cu. notation as in Table 13.6

$V_0/V = \rho/\rho_0$	$P_H (10^{12} \text{ dyne cm}^{-2})$	γ	$T_H (K)$	$P_c (10^{12} \text{ dyne cm}^{-2})$	$E_c^{**} (10^8 \text{ erg g}^{-1})$	P_{TL}/P_H	P_{TE}/P_H
1.05	0.075	1.89	317	0.061	1.3	0.19	0.00
1.10	0.167	1.85	360	0.151	6.3	0.10	0.00
1.15	0.280	1.87	438	0.254	15.1	0.09	0.00
1.20	0.413	1.88	577	0.375	27.9	0.09	0.00
1.25	0.566	1.84	802	0.516	44.6	0.09	0.00
1.30	0.755	1.77	1 150	0.674	64.7	0.11	0.00
1.35	0.971	1.70	1 630	0.853	89.1	0.12	0.00
1.40	1.225	1.63	2 300	1.052	117.0	0.14	0.00
1.45	1.522	1.59	3 180	1.272	148.7	0.16	0.00
1.50	1.858	1.55	4 350	1.518	184.8	0.18	0.00
1.55	2.252	1.53	5 760	1.783	224.0	0.21	0.00
1.60	2.714	1.53	7 530	2.075	267.2	0.23	0.01
1.65	3.247	1.53	9 710	2.387	314.5	0.25	0.01
1.70	3.880	1.54	12 425	2.730	366.0	0.29	0.01

Note: $**10^8 \text{ erg g}^{-1} \approx 17.0 \times 10^{-3} \text{ eV/atom}$

Table 13.8. *EOS for Pb. Notation as in Table 13.6.*

$V_0/V = \rho/\sigma_0$	P_H (dyne cm^{-2})	γ	T_H K	P_c (10^{12} dyne cm^{-2})	E_c** (10^8 erg g^{-1})	P_{TL}/P_H	P_{TE}/P_H
1.10	0.053	2.20	364	0.042	1.2	0.21	0.00
1.20	0.134	2.00	563	0.116	6.2	0.13	0.00
1.30	0.250	1.90	1045	0.216	15.3	0.14	0.00
1.40	0.423	1.84	2000	0.346	28.8	0.18	0.00
1.50	0.655	1.77	3550	0.510	46.7	0.21	0.01
1.60	0.955	1.69	5730	0.713	69.0	0.23	0.02
1.70	1.330	1.60	8485	0.953	95.8	0.25	0.03
1.80	1.765	1.48	11590	1.237	127.3	0.26	0.04
1.90	2.255	1.35	15000	1.560	163.2	0.25	0.06
2.00	2.776	1.21	18470	1.926	203.5	0.24	0.07
2.10	3.355	1.07	22150	2.330	248.2	0.22	0.09
2.20	4.010	0.98	26230	2.770	297.0	0.21	0.10

Note: ** 10^8 erg g$^{-1} \approx 55 \times 10^{-3}$ eV/atom.

We conclude this section with Fig. 13.2 in which the scaled TF electronic zero pressure (P_c) is plotted against the atomic scaled volume given by

$$ZV = \frac{Z}{\rho_0}\left(\frac{A}{N_A}\right)$$

where A is the atomic mass and N_A the Avogadro's number and ρ_0 is the experimentally measured normal density. In the figure, the calculated values are compared with the shock wave experimental data of Altshuler *et al.* (1958, 1960), Krupnikov *et al.* (1963) and McQueen and Marsh (1960).

13.6 Semi-empirical interpolation equation of state

As was mentioned in Section 13.4, the experimental shock wave data can be fitted to within 1% to a polynomial of $P_H(\rho/\rho_0)$ as given in (13.26) or by a simple polynomial of shock wave velocity $D(u)$ as given by (13.51). In the second case the Hugoniot pressure is given by (13.52), which is analytically completely different from (13.26), although numerically both

Fig. 13.2. The TF electronic zero pressure (P) plotted against the atomic scaled volume compared with experimental cold isotherms derived from the shock wave experiments of Altshuler *et al.* (1958, 1960), Krupnikov *et al.* (1963), McQueen and Marsh (1960). (The authors would like to thank Dr H Szichman for plotting this figure and carrying out the calculations.)

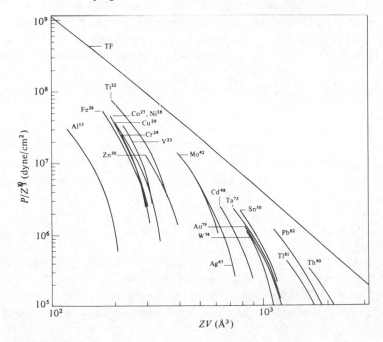

functions (13.26) and (13.52) are to within 1% equal! However, both approaches (13.26) and (13.52) can give different values for $\gamma(V)$ not only in magnitude (e.g., 30%–40%) but also in the nature of the solutions. This is due to the fact that $\gamma(V)$ is dependent not only on the values of the Hugoniot pressure P_H but also on the derivatives of P_H (see (13.14)) which are not determined from experimental data.

This difficulty can be avoided if one can calculate P_c from principles other than the Hugoniot data, and use the shock wave experiments for a consistency check and to calculate the thermal contribution of the Hugoniot. (see, e.g., Kormer and Urlin (1960)). For example, assuming that we know the function $P_c(\rho)$ and use it in (13.67) and (13.68) where $P = P_H$ and $E = E_H$ in these equations. In this case we have to solve two equations with two unknowns, namely $\gamma(\rho)$ and $T(\rho)$ for every Hugoniot point P_H, $V_H(= 1/\rho)$. For this purpose we will now develop an interpolation formula for the cold compression P_c and cold energy E_c.

The cold compression is described by the equation (Kormer and Urlin 1960).

$$P_c(\rho) = \sum_{i=1}^{6} a_i \delta^{(i/3+1)} \tag{13.69}$$

$$\delta \equiv \rho/\rho_c \equiv V_{0c}/V \tag{13.70}$$

The six coefficients $a_1 \ldots a_6$ are found from the properties of the material under normal conditions (room temperature) and by using the Thomas–Fermi–Dirac model at high compressions. ρ_c is the density of the material at zero point ($P = 0$, $T = 0$) as was discussed in Section 13.3. However, since

$$\frac{\rho_c - \rho_0}{\rho_0} \approx \frac{V_0 - V_{0c}}{V_0} \approx 1\% \tag{13.71}$$

for simplicity, we take for the expansion of the cold pressure the ratio $\sigma \equiv \rho/\rho_0$, where ρ_0 is the density of the material at room temperature, instead of the ratio δ which one really should take in the expansion of (13.69). That (13.71) is a good estimate can be seen from the following Taylor expansion,

$$\rho_c(T = 0) = \rho_0(T_0) + \left(\frac{\partial \rho}{\partial T}\right)_P (0 - T_0) = \rho_0(1 + 3\alpha T_0) \tag{13.72}$$

where $\partial/\partial\rho = -(1/V^2)(\partial/\partial V)$ and the definition of α as the thermal expansion: $3\alpha = (1/V_0)(\partial V/\partial T)_P$, was used. At ρ_0 and ρ_c one has $P = 0$, so that equation (13.72) is justified. Substituting the values of α from Table 13.4 one gets the approximation (13.71). A much better and reliable approach was discussed in Section 13.3.

The six coefficients $a_1, a_2 \ldots a_6$ are calculated in the following way For $\delta = 1$ the cold pressure is zero, thus

$$\sum_{i=1}^{6} a_i = 0 \tag{13.73}$$

The cold energy E_C is given by

$$E_C = -\int_{V_{0c}}^{V} P_c(V) dV = \int_{\rho_c}^{\rho} P_c(\rho) \frac{d\rho}{\rho^2} \tag{13.74}$$

so that the binding energy is given by

$$E_B = \int_{\rho_0}^{0} \frac{P_c d\rho}{\rho^2} = -\frac{3}{\rho_c} \sum_{i=1}^{6} \frac{a_i}{i} \tag{13.75}$$

and E_B is knwon experimentally and for ρ_c one can take ρ_0 as mentioned above. The compressibility κ is given by

$$\kappa = -\frac{1}{V}\left(\frac{\partial V}{\partial P}\right)_T = \delta\left(\frac{\partial P_c}{\partial \delta}\right)^{-1} \tag{13.76}$$

so that using the approximation $\delta \approx \rho/\rho_0 = 1$, one has

$$\kappa_0 = \sum_{i=1}^{6}\left(\frac{i}{3}+1\right)a_i = \sum_{i=1}^{6} \frac{ia_i}{3} \tag{13.77}$$

where κ_0 is known experimentally from room temperature measurements; see Table 13.4 and (13.73) which were used for the right hand side of equation (13.77). Now using (13.13) to relate the cold compression to the

Table 13.9. *The coefficients $a_i (i = 1, \ldots, 6)$ describing P_c in (13.69), the binding energy E_B, the chosen values for the high compression δ_1 and the appropriate TFD pressures for Al, Cu and Pb*

	Al	Cu	Pb
$E_B (10^{10} \, \mathrm{erg\,g^{-1}})$	11.64	5.37	0.90
δ_1	5.0	6.0	14.67
$P_{TFD}(\delta_1) (\mathrm{Mbar})$	24.6	244	659.7
$a_1 \times 10^{-10}$	-192.28	$1\,291.48$	-49.83
$a_2 \times 10^{-10}$	1823.01	$-10\,971.62$	440.13
$a_3 \times 10^{-10}$	-4927.19	$27\,771.97$	988.1
$a_4 \times 10^{-10}$	5347.01	$-31\,202.77$	728.67
$a_5 \times 10^{-10}$	-2473.48	$16\,090.45$	-145.12
$a_6 \times 10^{-10}$	422.93	-2979.51	14.25

Grüneisen coefficient

$$\gamma = \frac{1}{3} + \frac{1}{2}\rho\frac{d^2 P_c}{d\rho^2}\frac{dP_c}{d\rho} = \frac{1}{3} + \frac{\delta}{2}\frac{d^2 P_c}{d\delta}\frac{dP_c}{d\delta} \tag{13.78}$$

and again assuming this relation for $\delta = 1$ and $\gamma = \gamma_0$ (for a more accurate expression see (13.21) and the following calculation)

$$\gamma_0 = \frac{1}{3} + \frac{1}{2}\sum_{i=1}^{6}\frac{i}{3}\left(\frac{i}{3}+1\right)a_i\left[\sum_{i=1}^{6}\frac{i}{3}a_i\right] \tag{13.79}$$

where γ_0 is known from experimental measurements at room temperature (see (13.22) and Table 13.4). So far we have obtained four equations: (13.73), (13.75), (13.77) and (13.79) for the six coefficients in (13.69). All these expressions are based on experimental input from 'room temperature' measurements. The other two expressions are obtained by using the Thomas–Fermi–Dirac (TFD) theory at high compression, e.g., for $\delta = 10$. Let us assume that P_c satisfies the TFD model at $\delta = \delta_1$, then

$$\sum_{i=1}^{6} a_i\delta_1^{(i/3 + 1)} = P_{\text{TFD}}(\delta_1) \tag{13.80}$$

$$\sum_{i=1}^{6} a_i\left(\frac{i}{3}+1\right)\delta_1^{i/3} = \left(\frac{dP_{\text{TFD}}}{d\delta}\right)_{\delta = \delta_1} \tag{13.81}$$

so that (13.80) and (13.81) complete the set of equations necessary to calculate P_c. The numerical solution from Kormer and Urlin (1960) is given in Table 13.9. The values for P_{TFD} and δ_1 in the table were chosen at

$$(P_{\text{TF}} - P_{\text{TFD}})/P_{\text{TF}} < 0.2 \tag{13.82}$$

where P_{TF} is the pressure in the TF model, since in this region the TFD model seems to be justified. The coefficients from Table 13.9 for $a_i (i = 1\ldots6)$, together with (13.67) and (13.68), fit the experimental shock wave data with an accuracy better than 10% for pressure regimes up to 5 Mbars.

We end this section with the comment that the interpolation formula given by (13.69) is not unique. Moreover this interpolation does not have the correct asymptotic value for the TF model, so this expansion is limited to pressures of a few tens of Megabars and cannot be extended to the regime of the TF model.

In conclusion we may say that the study of equations of state from shock wave data is not yet complete and more research is needed. It is necessary to increase the strength of the shock waves by using sophisticated technological devices and also to improve our theoretical and computational knowledge in order further to advance this important field of physics.

14

Equation of state and inertial confinement fusion

Having guided the reader through the details of the equations of state, the application to the field of high density, high temperature plasmas will be explained. Though this is a very important field and activities are on a huge scale, examples will be given showing the poor or modest level the field has reached with respect to the use of the correct equation of state.

14.1 Pellet fusion

Nuclear fusion refers to the phenomenon in which two or more light atomic nuclei combine to form a heavier atomic nucleus. These reactions for nuclei with low atomic numbers are exothermic. Fusion is the energy source of the Sun and other stars. In 1952 a large amount of nuclear fusion energy was achieved for the first time on our planet by a thermonuclear explosion, by utilizing fusion reactions in a mixture of deuterium (D) and tritium (T). Since then, research work has sought ways of using controlled nuclear fusion energy for peaceful needs. As water in the oceans contains one deutron for every 6000 protons the oceans are an inexhaustible source of nuclear energy.

In order to realize nuclear fusion, the two light nuclei must come close enough to interact. Since each nucleus carries a positive electrical charge, the nuclei must have enough initial energy to overcome the electrostatic repulsion forces. The repelling force increases with the electrical charge, therefore it is desirable that the interacting nuclei should have the lowest possible charge. Thus the hydrogen isotopes are the best candidates for controlled fusion (Oliphant *et al.* 1934),

$$D + D \rightarrow {}^3He + n + 3.2\,MeV \tag{14.1}$$

$$D + D \rightarrow T + H + 4.0\,MeV \tag{14.2}$$

$$D + T \rightarrow {}^4He + n + 17.6\,MeV \tag{14.3}$$

where ^{3}He and ^{4}He are isotopes of helium, n denotes the neutron and the released energy per reaction is given in millions of electron volts (MeV).

The electrostatic repulsion is overcome by raising the temperature of the fusion fuel. It is necessary to obtain a sufficiently high temperature for fusion reaction to occur. The temperature required for efficient fusion depends on the reactions that are being employed. From calculations based on measured cross-sections, it has been determined that the D–T fusion fuel is efficiently burned at 10 keV (about 10^8 K) (or less if reheat and ignition is involved (Hora 1981)). The fusion of deuterium alone, as described in reactions (14.1) and (14.2), would require an even higher temperature of about 30 keV. Therefore, it is believed that controlled fusion will be realized through the D–T reaction before other possible reactions such as D–D can be used.

At these very high temperatures all the atoms will be stripped of their electrons and create a highly (or fully) ionized plasma. At the high temperature necessary for fusion, a plasma loses a considerable amount of its energy in the form of radiation. The main radiation results from decelerating electrons interacting with ions (bremsstrahlung). The system is self-sustaining when the rate of energy produced by fusion exceeds the rate of losses from the plasma by radiation. This requirement determines the critical *ignition temperature* of a nuclear fusion reaction. After the ignition temperature is reached, the plasma must be confined long enough to compensate for the supplied initial energy to heat the plasma. This criterion is due to J.D. Lawson (1957) and is expressed by the condition (see Appendix 7)

$$n\tau \gtrsim 10^{14}\,\mathrm{s\,cm^{-3}}\quad \text{(for D–T)} \tag{14.4}$$

when n is the plasma density in units of particles per cm^3 and τ is the time in seconds for which the plasma of that density is confined.

There are two basic approaches to achieve the desired controlled thermonuclear fusion: (a) magnetic confinement (b) inertial confinement. In the first approach the plasma is confined (i.e., τ in (14.4) is increased) with the help of magnetic fields. Many different magnetic field arrangements have been proposed for the confinement of the high temperature plasmas required for fusion reactions. In this chapter we do not discuss magnetic confinement though it has to be mentioned that inertial confinement fusion research by lasers has revealed the basic mechanism of internal electric fields in inhomogeneous plasmas, causing a rotation within the magnetic fields as a basic (and not second order) property of these plasmas which was previously overlooked (Hora *et al.* 1983; Hora *et al.* 1984; Hora *et al.* 1985). In the second approach, the basic idea is to heat a pellet of deuterium–

tritium mixture very rapidly to thermonuclear temperature and the confinement is due to the inertia of free expansion. The Lawson criterion is obtained by maximizing n through the implosion and compression of the fusion fuel, while the confinement time τ is fixed by the inertia of the imploded material. The pellet compression can be reached in principle by using the following drivers: (i) lasers, (ii) electron beams, (iii) ion beams (Yonas 1984).

When the incident pulse of the driver is absorbed by a pellet, a plasma is created in the outer surface (corona) of the target. The expanding corona causes a recoil by transferring momentum to the inner pellet mass to move it rapidly inward, i.e., an implosion takes place. Converging shock wave pressure compresses and heats the D–T fuel very rapidly causing nuclear fusion interactions. The pellet compression is *essential* for such a scheme to work (Nuckolls *et al.* 1972).

14.2 The limiting case of isentropic (shock-free) volume ignition (self-similarity model)

In order to clarify the need for compression we will explain the ideal model of self-similarity compression which includes the ideal equation of state. This can be illustrated straightforwardly by following the mathematical behaviour of an idealized spherical plasma. Suppose that at an initial time $t = 0$, a plasma of density $n_e = Zn_i$, consisting of Z-times

Fig. 14.1. Ideal adiabatic compression of a Gaussian density profile with compression velocities at times $t_1 < t_2$ for spatially constant plasma temperatures $T(t_1) < T(t_2)$.

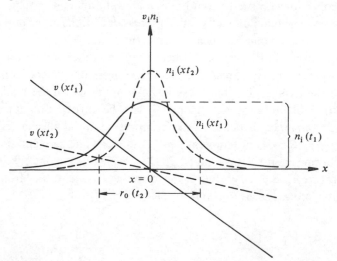

charged ions (subscript i) and electrons (subscript e), has a Gaussian density profile in radius, a spatially constant temperature, and a linear velocity profile in radius (see Fig. 14.1). Results originally obtained in astrophysical studies show that a 'self-similarity' solution exists to describe the time evolution of the system. According to this solution, the Gaussian density shape is preserved during compression (only the central density $n_c(t)$ and the plasma radius $r_0(t)$ vary in time), and the linearity of the velocity profile is retained; only the angle of slope of $v(x, t)$ in Fig. 14.1 varies with time:

$$v(r, t) = \frac{v_0(t)r}{R_0} \tag{14.5}$$

The temperature $T(t)$ follows an adiabatic law for ideal gases if there is no degeneracy

$$T = T_0[n_c(t)/n_c(0)]^{2/3} \tag{14.6}$$

This has been shown on the basis of the Boltzmann equation (Hora and Pfirsch 1972). If a compression motion initially occurs (i.e., negative slope of v at $t = 0$), the compression and kinetic energies of the imploding plasma are converted into increased temperature until (at a time t^*, where $r_0(t^*) \approx 0$) all the kinetic energy of motion has been thermalized with a maximum value of $T_{max} = T(t^*)$. Thereafter, the velocity slope switches to a positive value, describing expansion and adiabatic cooling. Final transfer of all thermal energy into kinetic energy of expansion occurs at $T(t = \infty) = 0$.

This represents an ideal process (as entropy remains constant) for conversion of the low energy density into high density at $t = t^*$ with conversion factors of 100 and more. In practice, however, the necessary initial conditions for such a compression are very difficult to achieve in the laboratory. The main difficulty is to suppress deviations of the initial linear velocity at $t = 0$ while producing a uniform radial velocity profile with a small deviation in the various directions. The model is (like the Carnot process for thermodynamics) the ideal case of isentropic compression.

With the assumption that these initial conditions can be realized, the calculation of nuclear-fusion gains (G) can be carried out numerically, starting from an equivalent box-like compressed initial plasma density n_0, input energy E_0, kinetic temperature T_0, radius R_0 and zero initial velocity. The resulting energy gains for a 50:50 DT plasma for the optimized kinetic temperature $T_0 \approx 10.3$ keV follow the formula (derived in 1964 from a fit to computer plots; see Hora (1964) or Hora and Pfirsch (1972))

$$G = \left(\frac{E_0}{E_{BE}}\right)^{1/3}\left(\frac{n_0}{n_s}\right)^{2/3} \tag{14.7}$$

Here n_s is the solid-state DT density, and the break-even energy $E_{BE} = 1.6\,MJ$. This formula is algebraically identical with the formula for the gain $G \approx n_0 R_0$ derived independently by Kidder (1974) in order to substitute the Lawson criterion. The substitution can be seen from the use of the input energy into the pellet

$$E_0 = 2\pi R_0{}^3 n_0 (1 + Z)kT_0 \tag{14.8}$$

with an initial radius R_0 and an initial temperature T_0 of about $10\,keV$ derived from the optimum calculation of (14.7). With this value, (14.7) becomes

$$G = \text{const}\, n_0 R_0 \quad \text{const} = 1.66 \times 10^{-22}\,cm^2 \tag{14.9}$$

where the agreement with Kidder's (1974) constant was within a factor of 2, which has to be considered a good approximation.

Equation (14.7) shows immediately how an increase of the initial density n_0, e.g., by a factor of ten, reduces the necessary input energy E_0 quadratically (to 1/100 in the example). *This is a straightforward illustration of the need for compression of the plasma.*

Fusion then occurs by a nearly homogeneous ignition of the whole plasma volume, i.e., 'volume ignition'. An improvement of the model was possible through inclusion of added effects, i.e., fuel burn-up, bremsstrahlung losses (with self-absorption), and reheat of the pellet by the energy loss in the fuel of the high-energy alphas created by fusion reactions (Hora 1981).

Despite the identity of Kidder's (1974) result, (14.9) and despite it being based on the very simplified assumptions of neglecting alpha reheat, bremsstrahlung losses and fuel depletion, (as for the Hora model (1964), its reproduction with an alternative derivation is given here. This simple model assumes: (a) isentropic compression, (b) homogeneous compression and (c) spherical symmetry. In an isentropic compression the hydrodynamic work is preferable to the application of a shock wave sequence to compress the pellet. In order to describe the homogeneous compression (at least approximately with respect to internal friction) in a one-dimensional spherical coordinate system it is convenient to work in Lagrangian coordinates.

The fluid equations with the variables (velocity, pressure, energy, entropy, etc.) described by functions of space coordinates and time are called the Euler equations, or equivalently the fluid equations in Euler coordinates. In contrast to the Eulerian coordinates which determine the fluid variables at a given point in space (for all times), the Lagrangian coordinates describe (in time) the fluid variables of a given fluid particle.

Mathematically the time derivative in Lagrangian coordinates is equal to the total derivative D/Dt which is related to the usual (Eulerian) time derivative $\partial/\partial t$ by the equation

$$\frac{D}{Dt} = \frac{\partial}{\partial t} + \mathbf{u} \cdot \nabla \tag{14.10}$$

where $\mathbf{u}$ is the fluid particle velocity and ∇ is the usual (Eulerian) space derivative. The Lagrangian coordinates are useful in general for one-dimensional problems with plane, cylindrical or spherical symmetry. The particle is described either by the fluid's particle mass or by the position of the particle at $t = 0$. We summarize the Lagrangian coordinates in

Table 14.1. *A one-dimensional plane geometry*

	Eulerian coordinates	Lagrangian coordinates
The flow variables	u, ρ, P, etc., are functions of (x, t)	u, ρ, P, etc., for a given fluid particle are functions of t
The coordinates	x	$a \text{ or } m \equiv \displaystyle\int_{x_1}^{x} \rho \, dx \equiv \rho_0 a \quad \text{(A2)}$
		$dm = \rho \, dx \qquad \dfrac{\partial}{\partial x} = \rho \dfrac{\partial}{\partial m} \quad \text{(A3)}$
		or
		$da = \dfrac{\rho}{\rho_0} dx \qquad \dfrac{\partial}{\partial x} = \dfrac{\rho}{\rho_0} \dfrac{\partial}{\partial a} \quad \text{(A4)}$
		$(\rho_0 \equiv \rho(x, t = 0))$
Time derivative	$\dfrac{\partial}{\partial t}$	$\dfrac{D}{Dt}$
Continuity equation (mass conservation)	$\dfrac{\partial \rho}{\partial t} + \dfrac{\partial}{\partial x}(\rho \mu) = 0$	$\dfrac{\partial V}{\partial t} = \dfrac{\partial u}{\partial m} \text{ or } \dfrac{1}{V_0} \dfrac{\partial V}{\partial t} = \dfrac{\partial u}{\partial a} \quad \text{(A5)}$
		$\left(V \equiv \dfrac{1}{\rho} \right)$
Momentum conservation (Newton's law)	$\dfrac{\partial u}{\partial t} + u \dfrac{\partial u}{\partial x} = -\dfrac{1}{\rho} \dfrac{\partial P}{\partial x}$	$\dfrac{\partial u}{\partial t} = -\dfrac{\partial P}{\partial m} \text{ or }$
		$\dfrac{\partial u}{\partial t} = -V_0 \dfrac{\partial P}{\partial a} \quad \text{(A6)}$

Note: In (A2)–(A6), the x coordinate does not enter explicitly in the Lagrangian coordinates. After the Lagrangian equation (A5)–(A6) are solved for $V = V(m, t)$ the dependence of the flow variables is on m and t.

comparison with the Eulerian coordinates in two tables, one for a one-dimensional plane geometry and the second for a one-dimensional spherical geometry. We denote the fluid velocity by u, the density by ρ, the related specific volume by $V = 1/\rho$, the pressure is P, m is the mass of a column of fluid of unit cross-section in plane geometry and the mass of a fluid sphere of a radius r in spherical geometry. The comparison, made only for mass and momentum equations on the Eulerian coordinate x, is obtained by the integration of (A3) (A' indicates equation numbers in Tables 1 and 2), derived from (14.10) and (A2)

$$\mathrm{d}x = V(m, t)\mathrm{d}m \qquad (14.11)$$

$$x(m, t) = \int_0^m V(m, t)\mathrm{d}m + x_1(t) \qquad (14.12)$$

where x_1 is the reference particle (or the point of reference necessary to measure the coordinate of the particles).

It is convenient to describe the equation of motion using (A5) and (A11) by

$$\frac{\partial u}{\partial t} = -\frac{1}{\rho}\frac{\partial P}{\partial r} \qquad (14.13)$$

Table 14.2. *Fluid equations in one-dimensional spherical Lagrangian coordinates*

The coordinates m	or	$r_0 \equiv r(t = 0)$	
$m = \int_0^r \rho 4\pi r^2 \mathrm{d}r$		$\frac{4}{3}\pi r_0^3 \rho_0 - \int_0^r 4\pi r^2 p \mathrm{d}r$	(A7)
$\mathrm{d}m = \rho 4\pi r^2 \mathrm{d}r$		$\mathrm{d}r_0 = \frac{r^2}{r_0^2}\cdot\frac{\rho}{\rho_0}\mathrm{d}r$	(A8)
$\frac{\partial}{\partial r} = \rho 4\pi r^2 \frac{\partial}{\partial m}$		$\frac{\partial}{\partial r} = \frac{r^2}{r_0^2}\frac{\rho}{\rho_0}\frac{\partial}{\partial r_0}$	(A9)
Mass conservation			
$\frac{\partial V}{\partial t} = \frac{\partial}{\partial m}(4\pi r^2 u)$		$\frac{1}{V_0}\frac{\partial V}{\partial t} = \frac{1}{r_0^2}\frac{\partial}{\partial r_0}(r^2 u)$	(A10)
Momentum conservation			
$\frac{\partial u}{\partial t} = -4\pi r^2 \frac{\partial P}{\partial m}$		$\frac{\partial u}{\partial t} = -\frac{1}{\rho_0}\frac{r^2}{r_0^2}\frac{\partial P}{\partial r_0}$	(A11)

and the mass conservation by using (A8) (instead of (A10)) in the following form

$$\frac{\rho}{\rho_0} = \frac{r_0^2}{r^2} \frac{\partial r_0}{\partial r} \tag{14.14}$$

Equations (14.13) and (14.14) are solved together with (A7) and the appropriate boundary conditions. For realistic problems, the energy equation (or the entropy equation) has to be considered as well.

In the Lagrangian coordinate systems, $r(t)$ is the instantaneous position of a mass point particle which has an initial position $r_0 = r(t = 0)$. The volume element in a homogeneous compression changes with time everywhere at the same rate, so that one can write

$$d^3 r = \underline{h}^3(t) d^3 r_0 \tag{14.15}$$

where in Lagrangian coordinates the position r of the fluid particle is a function of the initial position r_0 and time,

$$r = r(r_0, t) \tag{14.16}$$

and from (14.15) one can write with the separation function $\underline{h}$

$$r(r_0, t) = r_0 \cdot \underline{h}(t) \tag{14.17}$$

Equations (14.15) and (14.17) are the actual description and the definition of the homogeneous motion of the fluid. The fluid velocity is derived from (14.17)

$$u(r_0, t) = \frac{dr}{dt} = r_0 \frac{d\underline{h}}{dt} \equiv r_0 \underline{\dot{h}} \tag{14.18}$$

where the dot denotes the time differentiation. Now using the mass conservation equation (14.14) and (14.17) one gets the compression

$$\frac{\rho}{\rho_0} = \frac{r_0^2}{r^2} \frac{\partial r_0}{\partial r} = \underline{h}^{-3} \tag{14.19}$$

The momentum conservation of (A11) can be written and compared with the time derivative of (14.18)

$$\dot{u} = -\frac{r^2}{\rho_0 r_0^2} \frac{\partial P}{\partial r_0} = r_0 \underline{\ddot{h}} \tag{14.20}$$

r^2/r_0^2 is given in (14.17) while in order to calculate $\partial P/\partial r$ the equation of state has to be used. For simplicity we use the EOS, of an adiabatic process of a fully degenerate Fermi–Dirac electron gas or a non-degenerate fully ionized plasma (see Chapter 6)

$$P(r_0, t) = \left(\frac{P_0}{\rho_0^\gamma}\right)\rho^\gamma(r_0, t) \tag{14.21}$$

where $P_0 \equiv P(r_0, 0)$ is the initial profile of the pressure and $\rho_0 \equiv \rho(r_0, 0)$ is the initial density distribution of the pellet. Substituting (14.20) in (14.21) one has,

$$P = P_0 \underline{h}^{-3\gamma} \tag{14.22}$$

$$\frac{\partial P}{\partial r_0} = \frac{\partial P_0}{\partial r_0} h(t)^{-3\gamma} \tag{14.23}$$

which can be inserted into the equation of motion (14.20) to yield

$$\underline{h}^{3\gamma-2}\underline{\ddot{h}} = -\frac{1}{\rho_0 r_0}\frac{\partial P_0}{\partial r_0} \equiv -\frac{1}{t_c^2} \tag{14.24}$$

The left hand side of (14.24) is a function of time only, while the right hand side of this equation depends on r_0, therefore each side should be equal to a constant since t and r_0 are independent variables. The constant of (14.24) is defined by $-1/t_c^2$, where t_c has the dimension of time. Now assuming an equation of state with $\gamma = \frac{5}{3}$, the differential equation of (14.24),

$$\underline{h}^3\underline{\ddot{h}} = -\frac{1}{t_c^2} \tag{14.25}$$

with the initial conditions $\underline{h}(0) = 1, \underline{\dot{h}}(0) = 0$ has the solution (i.e., $\underline{h}(0) = 1$ means $r = r_0$ while $\underline{\dot{h}}(0)$ implies a zero initial velocity; see (14.17), (14.18))

$$\underline{h} = (1 - \tau^2)^{1/2} \quad \tau \equiv t/t_c \tag{14.26}$$

From this solution one can see that the time t_c is the time of collapse, i.e., the time that the pellet collapses to zero radius. (It should be clear that we are discussing an idealization; in reality a pellet cannot collapse to zero radius.)

For a pellet of radius R, the pressure at the surface necessary to cause the collapse of the pellet is given from (14.22) by

$$P(R, t) = P(R, 0)(1 - \tau^2)^{-5/2}; \quad \tau = \frac{t}{t_c} \tag{14.27}$$

Next, one has to solve the spatial differential equation (14.24)

$$\frac{1}{\rho_0 r_0}\frac{\partial P_0}{\partial r_0} = \frac{1}{t_c^2} \tag{14.28}$$

We take the initial condition for this problem in a rather artificial way modifying Kidder's (1974) derivation by assuming that the initial fluid state has a uniform entropy. For an ideal gas the entropy is proportional to $\ln(P/\rho^\gamma)$, therefore Kidder's assumption[†] is

$$\frac{\partial}{\partial r_0}\left(\frac{P_0}{\rho_0{}^\gamma}\right) = 0 \tag{14.29}$$

or equivalently

$$\frac{P_0}{\rho_0{}^\gamma} = \text{const} = \frac{P_{00}}{\rho_{00}{}^\gamma} \tag{14.30}$$

where $P_{00} = P_0(0)$ and $\rho_{00} = \rho_0(0)$. The solution of (14.28) and (14.30) is for $\gamma = \frac{5}{3}$

$$\frac{P_0(r_0)}{P_0(0)} = (1 + \beta x_0{}^2)^{5/2} \tag{14.31}$$

$$\beta \equiv \frac{R_0{}^2}{3c_0{}^2 t_c{}^2} \tag{14.32}$$

$$x_0 \equiv \frac{r_0}{R_0} \quad 0 < x_0 < 1 \tag{14.33}$$

$$c_0{}^2 = \frac{\gamma P_0(0)}{\rho_0(0)} \quad (\gamma = \tfrac{5}{3}) \tag{14.34}$$

where c_0 is the sound velocity at the center of the pellet at $t = 0$. The radial solutions for density and the temperature (assuming an ideal equation of state) are

$$\frac{\rho_0(r_0)}{\rho_0(0)} = (1 + \beta x_0{}^2)^{3/2} \tag{14.35}$$

$$\frac{T_0(r_0)}{T_0(0)} = a + \beta x_0{}^2 \tag{14.36}$$

The time dependence of the temperature, density and pressure at every point inside the pellet is proportional to $\underline{h}^{-2}$, $\underline{h}^{-3}$ and $\underline{h}^{-5}$ respectively (see (14.19) and (14.22) and the ideal EOS, $P/\rho \approx T$). Therefore, using (14.26) one has

$$\frac{T(r_0, t)}{T_0(0)} = \frac{1 + \beta x_0{}^2}{1 - \tau^2} \tag{14.37}$$

[†] Kidder compares his results with detailed numerical calculations and the 'good' agreement might be an *a posteriori* justification of the whole model discussed in this section.

$$\frac{\rho(r,t)}{\rho_0(0)} = \left(\frac{1 + \beta x_0^2}{1 - \tau^2}\right)^{3/2} \tag{14.38}$$

$$\frac{P(r,t)}{P_0(0)} = \left(\frac{1 + \beta x_0^2}{1 - \tau^2}\right)^{5/2} \tag{14.39}$$

The mechanical power that does the work to compress the pellet is given by

$$P_M = -4\pi R^2 P(R,t)u(r,t) \tag{14.40}$$

where $R = R_0 h(t)$ is the external boundary of the pellet and R_0 is the initial radius of the pellet. Using for $P(R,t)$, (14.27), and for $u(r,t)$, (14.18) and (14.26) one gets

$$P_M = \frac{4\pi R_0^3 P_0(0)(1 + \beta)^{5/2}\tau}{t_c(1 - \tau^2)^2} \quad [\text{W}] \tag{14.41}$$

The inertial confinement parameter $\langle \rho R \rangle$ (which is equivalent to the Lawson parameter $n\tau$) is given by

$$\langle \rho R \rangle \equiv \int \rho \, dr = \frac{R_0 \rho_0(0) f(\beta)}{(1 - \tau^2)} \tag{14.42}$$

$$f(\beta) = \int_0^1 (1 + \beta x_0^2)^{3/2} \, dx_0 \tag{14.43}$$

$$\approx 1 + \tfrac{1}{2}\beta + \tfrac{3}{40}\beta^2 + \cdots \quad (\beta \ll 1) \tag{14.44}$$

$$\approx \frac{\beta^{3/2}}{4} \quad (\beta \gg 1) \tag{14.45}$$

The relation between the hydrodynamic power P_M and the laser power P_L depends on the interaction physics between the laser and the plasma surrounding the pellet. Denoting by ε the efficiency of the absorbed laser energy to convert into useful work to compress the pellet (see Linhart, 1984), i.e.,

$$\varepsilon = P_M/P_L \tag{14.46}$$

one can obtain the necessary laser pulse shape $P_L(t)$ (by using (14.41)). However, the value of ε is a result of laser–plasma interactions and it is not yet a very well known parameter.

We end this section with the comment that the necessary energy (or power) to compress the fuel during the process of a 'cold' isentrope can increase significantly due to fast electrons or hard X-ray preheating. This can happen if these particles or photons have a mean free path of the order of the pellet size or the shell thickness for shell structure pellets. The preheat can also be caused by undesired shock waves which might arise during the interaction processes. If we assume that there is some heating prior to the desired compression and that this heating is proportional to the compression energy:

$$dQ = -\xi P dV \tag{14.47}$$

with ξ as a parameter. Then using the first law of thermodynamics (i.e., energy conservation)

$$dE = dQ - P dV = -(1 + \xi)P dV \tag{14.48}$$

together with the equation of state for a Fermi–Dirac gas or an ideal gas

$$E = \tfrac{3}{2} PV \tag{14.49}$$

we obtain for the energy requirement to compress the fuel,

$$E = E_0 \left(\frac{\rho}{\rho_0} \right)^{2(1 + \xi)/3} \tag{14.50}$$

where as usual $\rho = 1/V$. The preheating increases the value of E_0 as well as the ρ/ρ_0 contribution to the required energy. For example, if $\xi \approx 10\%$ then for a 10^4 ($= \rho/\rho_0$) compression, E is increased by a factor of 1.85 relative to the case of no preheating ($\xi = 0$).

14.3   Central core ignition with minimized entropy production

The simple hydrodynamic transfer of laser energy into an ideal self-similarity compression profile (i.e., volume ignition) is prevented by the fact that the laser initially deposits its thermal and mechanical energy into a shallow volume of plasma at the pellet surface. However, the momentum of ablation of this hot plasma corona causes a compressing recoil to the interior. Then, including the thermal conduction from the hot corona, a thermalized, dense, thick, compressed plasma is generated in the interior of the target. The central core in this high-density and high-temperature region can undergo ignition. This, in turn, can cause a self-sustaining fusion combustion wave that travels outward, burning D–T fuel in the process. The energy-flux density (intensity) near the core required to cause this process is 10^{19} W cm^{-2} (Emmett *et al.* 1974). This central ignition process (Meldner 1981) has been studied extensively following the early 1974 results at KMS (Glass 1983) where a nearly 360° symmetric irradiation of DT-filled (10 atmospheres) glass balloons demonstrated that core compression (seen from X-ray pinhole pictures) was possible. The nearly symmetric irradiation was achieved using only two beams together with a unique 'clamshell' focusing mirror system.

An additional advantage of the high core density is the absorption of the $\alpha\,(=\mathrm{He^4})$ particles by the dense material. Since the mean free path (l) of the α-particles is inversely proportional to the density, one has

$$l_\alpha \approx R^3 \tag{14.51}$$

Thus, if for example the radius is reduced by a factor of 10, the mean free path of α-particles is reduced by a factor of 1000. In this case the energy carried by the α-particles, which is 3.52 MeV per reaction, would be deposited within the pellet and a self-sustaining thermonuclear burn wave will propagate outward causing fusion to the rest of the pellet. In this case it is necessary to compress only the centre of the pellet in order to start the ignition process there. Moreover the rest of the compressed pellet should be as cold as possible in order to minimize the overall energy input into the pellet.

The most effective compression is isentropic as shown in the preceding section and this might be achieved approximately if the ablation pressure could be increased according to an ideal time profile. Besides the time-shaping of the driver pulse, structured targets have to be used to vary the time-evolution of pressure on the compressed material and to avoid preheating of the pellet core in order to achieve the desired isentropic compression. These considerations have shown that high compression can be achieved by (a) shaping in time the input energy of the driver, and (b) 'clever' pellet design. For example, Livermore designers (Ahlstrom 1982) suggested for a Nd:Yag laser driver, with maximum irradiance of $10^{14}\,\mathrm{W\,cm^{-2}}$, a 'double shell' pellet with a profile such as that shown in Fig. 14.2.

Fig. 14.2. Pellet concept for high gain laser induced fusion.

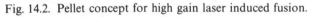
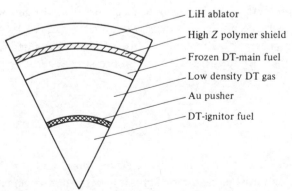

The main fuel region in Fig. 14.2 in the outer shell is physically distinct from the ignition fuel of D–T in the inner pellet. This structure makes it possible for the main fuel to be on a cold isentrope in order to maximize the pellet gain in energy.

The interaction physics between the high power driver and the pellet, leading to high compression causing thermonuclear burn, can be summarized schematically by the following intercorrelated process:

> Energy absorption (of driver) → Energy Transport
> → Compression → Nuclear Fusion.

The irradiated matter (by lasers, or ions, etc.) consists of a dense inner core, surrounded by a less dense and hot corona. The energy is absorbed by the outer periphery of the plasma and it is transported inwards to the pellet. Due to the strong heating of matter in the absorption region, high pressure is exerted on the surrounding material which leads to the formation of intense shock waves, moving into the interior of the target. The idea of compressing the pellet by convergent compression waves has significantly reduced the required energy for inertial confinement fusion. The Lawson criterion, (14.4), for the break-even condition may be substituted by the form (see (14.7) and (14.8))

$$\rho R \gtrsim 1 \, \text{g cm}^{-2} \tag{14.52}$$

where ρ is the compressed fuel density for a core radius R.

The properties of matter at high density and high temperature are important to explain and to calculate the compression process. In particular the equation of state data for different materials (see, e.g., Fig. 14.2) are necessary to calculate the shock wave propagation into the pellet. For inertial confinement fusion a knowledge of the properties of matter is needed for temperatures up to 100 keV and for densities up to 10^4 times solid density. The corresponding pressures are enormous. For example, the pressure of the degenerate electrons of hydrogen with 10^4 times liquid density (i.e. an electron density of about $n_e \approx 5 \times 10^{26} \, \text{cm}^{-3}$) is about 10^{12} *atmospheres* ($\approx 10^6$ Mbars). This estimate is obtained using the expression for the Fermi degenerate electron pressure at zero temperature

$$P = \tfrac{2}{5} n_e \varepsilon_F \tag{14.53}$$

$$\varepsilon_F = \frac{h^2}{8m} \left(\frac{3}{\pi} n_e \right)^{2/3} \tag{14.53a}$$

where ε_F is the Fermi energy, m is the electron mass and h is Planck's constant. For comparison, the thermal pressure of the non-degenerate ions $P = n_e kT$ with $n_e = 5 \times 10^{26} \, \text{cm}^{-3}$ and a temperature kT of 10 keV is about

5×10^{12} atmospheres. Thus, it can be summarized that for inertial confinement purposes, one needs the data of equations of state for many materials in the domain of

$$0 \leqslant T \leqslant 100 \, \text{keV} \quad 10^{-4} \leqslant \frac{\rho}{\rho_0} \leqslant 10^4 \quad 0 \leqslant P \leqslant 10^{13} \, \text{atm} \qquad (14.54)$$

where ρ_0 is the initial liquid or solid density of the material under consideration. The numerical values in (14.54) are only orders of magnitude, as it might be possible in one scheme or another to go above these values.

As mentioned above, the main idea of inertial confinement fusion is the aim of achieving very high compression using laboratory facilities, up to $\rho/\rho_0 = 10\,000$! This concept (Nucksolls *et al.* 1972), can be easily understood by using a 'realistic' equation of state. Using for example the Thomas–Fermi model for the hydrogen isotopes; a D–T mixture with initial (liquid) density of $\rho_0 = 0.2 \, \text{cm}^3$, one needs an energy of 3.0 keV per atom to increase the temperature of the fuel to 1 keV ($\approx 10^7$ K) without changing its density. However, for an extra energy of 0.2 keV/atom (i.e., an extra energy of about 7%!) at 1 keV temperature one gets a compression of $\rho/\rho_0 = 20$! Now using elementary knowledge of nuclear reaction cross-sections it is evident that it is necessary to use the driver's energy in order to compress the material as much as possible instead of heating it by 'brute force'. Although *the idea seems to be self evident from the equation of state data*, the way to put it into practice is still unresolved.

The physics of inertial confinement fusion is based on the hydrodynamics of one or more fluids, or equivalently on transport (e.g., Boltzmann) equations. In order to solve these equations a knowledge of the equation of state and transport coefficients (such as thermal conductivity, electrical conductivity, radiation capacities, etc.) is necessary. We write down explicitly an example of a typical set of hydrodynamic equations one has to solve numerically for *laser induced* inertial confinement fusion in order to see where the equation of state data are necessary.

The physical model to be described (Christiansen *et al.* 1974; Tahir *et al.* 1984; Meyer-ter-Vehn *et al.* 1984) treats the electrons and the ions as distinct species with separate thermodynamic variables. The two systems of electrons and ions have one hydrodynamic velocity and are coupled by energy exchange. The basic equations of the model are the mass, momentum and energy conservations for the various species involved in the process. In particular the momentum equation is

$$\rho \frac{\mathrm{d}u}{\mathrm{d}t} = -\nabla P \qquad (14.55)$$

where P is the total pressure and u is the hydrodynamic velocity

$$u(r, t) = \frac{dr}{dt} \qquad (14.56)$$

The total pressure is made up of the electron pressure plus the ion pressure. The viscous terms are not included explicitly in (14.55), but are accounted for in the finite difference expression by the inclusion of 'artificial' dissipative terms. There are also, in the system of equations, two energy equations, one for the electrons and one for the ions. For example, the electron energy equation is

$$\left(\frac{\partial E_e}{\partial T_e}\right)_V \frac{dT_e}{dt} + \left(\frac{\partial E_e}{\partial V}\right)_T \left(-\frac{1}{\rho^2}\right)\frac{d\rho}{dt} + P_e \frac{dV}{dt} = \Phi_e \left[\frac{\text{erg}}{\text{s cm}^2 \text{ g}}\right]$$
$$(14.57)$$

where the quantities with subscript e refer to the electrons, E_e is the energy, e.g. $E = E(T, V)$, the specific volume V is related to the 'fluid' density $\rho = 1/V$, P_e is the electron pressure and Φ_e is the energy source term. A similar equation to (14.57) is written for the ions.

The source term Φ includes:

(a) *Energy absorption from the laser* (in the domain of laser absorption only)

$$\Phi_L = \frac{I(t)}{m_e}\left\{1 - \exp\left(-\int k_{ib} dl\right)\right\} \quad [\text{erg s}^{-1} \text{cm}^{-2} \text{g}^{-1}] \quad (14.58)$$

Where the integral is over the optical path of the laser light, $I(t)$ is the laser intensity (erg s^{-1} cm^{-2}) on an element of mass m_e which is equal to the total electron mass per cell. k_{ib} is the absorption per unit optical path length l for inverse bremsstrahlung and is given by

$$k_{ib} = \frac{8\pi e^2 \ln \Lambda n_e^2 \langle z^2 \rangle}{3c(\omega_L/2\pi)^2 (2\pi m k T_e)^{3/2} \langle z \rangle}\left(1 - \frac{\omega_p^2}{\omega^2}\right)^{-1/2} \quad [\text{cm}^{-1}] \quad (14.59)$$

where c and $(\omega_L/2\pi)$ are the speed of light and the laser frequency respectively, $\ln \Lambda$ is the Coulomb logarithm (Spitzer 1962) which is of the order of 10, ω_p is the plasma frequency,

$$\omega_p^2 = \frac{4\pi e^2 n_e}{m_e} \qquad (14.60)$$

$\langle z^2 \rangle$ is the average of the squared charge number and is given by

$$\frac{1}{n_i}\sum_{j=0}^{z} z_j^2 n_{ij} = \langle z^2 \rangle \qquad (14.61)$$

where z_j is the charge of the ions ionized j times and n_{ij} is their partial density, so that

$$n_i = \sum_{j=1}^{z} n_{ij} \tag{14.62}$$

(b) Thermal conduction of energy – Φ_T

In order to transport energy between the absorption region and the interior of the target, good thermal conduction is necessary. The classical value for the electron thermal conductivity is given by

$$\Phi_T = \frac{1}{\rho} \nabla(k_e \nabla T_e) [\text{erg s}^{-1} \text{cm}^{-2} \text{g}^{-1}] \tag{14.63}$$

where the electron thermal conductivity coefficient k_e is a rapidly increasing function of T_e (proportional to $T_e^{5/2}$)

$$k_e = 20 \left(\frac{2}{\pi}\right)^{3/2} \frac{(kT_e)^{5/2} k \langle z \rangle}{m^{1/2} e^4 \ln \Lambda \langle z^2 \rangle} [\text{erg s}^{-1} \text{cm}^{-1} \text{K}^{-1}] \tag{14.64}$$

where k is the Boltzmann constant, m is the electron mass, etc. $\langle z \rangle$ is the average charge number of the ions and $\langle z^2 \rangle$ is defined in (14.61). The large reduction of thermal conductivity (Hora and Miley 1984) is due to the strong electric fields in the nonlinear force produced double layers (Hora *et al.* 1984a) where a strong a disturbance can occur also by a new type of resonance for perpendicular incidence (Hora and Ghatak 1985). Other field effects are found by Destler *et al.* (1984, 1984a).

(c) Electron–Ion energy exchange – Φ_{ei}

This is given as a function of the local temperature difference by

$$\Phi_{ei} = \left(\frac{\partial E_j}{\partial T}\right)_V v_{ei}(T_i - T_e) \tag{14.65}$$

where the quantities with subscript i correspond to ionic functions,

$$\left(\frac{\partial E_i}{\partial T}\right)_V = c_{vi} \tag{14.66}$$

is the specific heat for the ionic system, and v_{ei} is the electron–ion collision frequency given by (for the correction at high laser intensities see Hora (1981) Chapter 6; and a quantum correction: p. 37)

$$v_{ei} = \left(\frac{\pi}{2}\right)^{3/2} \frac{m^{1/2} e^4 n_i \ln \Lambda \langle z^2 \rangle}{M_i (kT_e)^{3/2}} [\text{s}^{-1}] \tag{14.67}$$

where M_i is the ion mass and n_i is the ion density.

(d) Radiation losses – Φ_R

For a Maxwellian distribution of electrons these losses can be described by the bremsstrahlung radiation

$$\Phi_R = -\frac{8\pi^2 n_e T_e^{1/2} \langle z^2 \rangle}{3hc^3 m^{3/2} M_i} \, [\mathrm{erg\,s^{-1}\,cm^{-2}\,g^{-1}}] \tag{14.68}$$

where the symbols were defined above. The negative sign is taken since it is considered here that the bremsstrahlung radiation is lost from the plasma.

(e) Thermonuclear energy absorption – Φ_N

For the D–T process this can be expressed by

$$\Phi_N = \langle \sigma u \rangle_{DT} n_D n_T E_{DT} \, [\mathrm{erg\,s^{-1}\,cm^{-2}\,g^{-1}}] \tag{14.69}$$

Where E_{DT} is the energy released during the reaction, i.e., 17.6 MeV per reaction (in c.g.s. units) if the neutrons are absorbed, and 3.6 MeV per reaction if only the charged particles are absorbed. n_D and n_T are the densities of the deuterium and the tritium and the $\langle \sigma u \rangle_{DT}$ is a function of the ion temperature T_i and is given by the approximation (see Clark *et al.* 1978)

$$\langle \sigma u \rangle_{DT} \approx 3.7 \times 10^{-12} \, T_i^{-2/3} \exp(-20\, T_i^{-1/3}) \, [\mathrm{cm^3\,s^{-1}}] \tag{14.70}$$

Equation (14.69) is correct only if the particles created in the D–T reaction deposit their energy locally. If, however, the neutrons created in the D–T process leave the pellet and the charged α-particles do not deposit their energy locally, an α-particle transport equation is necessary so that nuclear burn wave propagation can be described. For large reactor size pellets, the neutrons may deposit some of their energy in the pellet, and in this case a proper transport description for the neutrons is necessary. Concerning the controversy about the α-stopping power, see Cicchitelli *et al.* (1984).

Taking into account the energy deposition and loss mechanisms described above, the energy source term Φ_e in (14.57) can be written as

$$\Phi_e = \Phi_L + \Phi_T + \Phi_{ei} + \Phi_R + \Phi_N \tag{14.71}$$

where Φ_L, Φ_T, Φ_{ei}, Φ_R and Φ_N are given in (14.58), (14.63), (14.65), (14.68) and (14.69) respectively. Although (14.71) seems to be complicated enough, it should be mentioned that many approximations and assumptions were made in writing down the energy source terms. For example, in the energy absorption term for the laser – Φ_L, it was assumed that only inverse bremsstrahlung is the dominant mechanism. However, from high irradiance laser–plasma interactions, it is known (Hora 1981) that other mechanisms, such as resonance absorption and parametric instabilities, are also responsible for the absorption of the laser light. These last mentioned processes introduce many difficulties and nonlinear effects into the expression for Φ_L (see Section 14.4.1). For the thermal conduction term Φ_T the diffusion approximation was used (see (14.63)) for highly ionized plasmas. However, it is known from laser–plasma research that steep temperature gradients exist in the plasma domain so that the diffusion flux

$$F_e = k_e \nabla T_e \tag{14.72}$$

becomes impossibly large. In order to restrict the thermal flux a phenomenological approach was introduced by imposing the thermal 'inhibited free stream flux' F_{ef}

$$F_{ef} = \frac{f n_e (kT_e)^{3/2}}{m} \tag{14.73}$$

where f is an adjustable parameter known as the flux limiter parameter. The physics which describes the parameter f can be sophisticated and complicated, e.g., strong magnetic fields of the order of megagauss or turbulence, or the high electric 'inhomogeneity fields' (Eliezer and Ludmirsky 1983; Hora *et al.* 1983; Hora and Miley 1984) may reduce the value of f.

For practical calculations one may obtain an effective $\bar{F}$ from

$$\frac{1}{\bar{F}} = \frac{1}{F_e} + \frac{1}{F_{ef}} \tag{14.74}$$

where F_e and F_{ef} are given in (14.72) and (14.73). Therefore, the expression for Φ_T as given by the diffusion approximation (14.63) is simplified and in many cases it does not describe the real physical process.

Even the electron–ion exchange term as given by (14.65) and (14.67) is not complete. The electron–ion collision frequency may be dominated by nonlinear and collective collisions phenomena. In the radiation energy term (14.68), one might also need to introduce the effects of photon absorption (and appropriate atomic physics) by the plasma and the pellet material. In cases where a significant percentage of the input energy is transformed into soft X-rays, the interaction between this radiation and the pellet might be crucial in explaining the inertial confinement fusion process.

In order to calculate $\langle z \rangle$ and $\langle z^2 \rangle$, necessary for (14.64), (14.67) and (14.68), the rate coefficients for ionization and recombination for ions with different degrees of ionization are needed. The problem is more complicated for multi-layered targets as shown, for example, in Fig. 14.2. A Saha type of equation is necessary in order to simplify the calculations. However, for this purpose a knowledge of the equation of state is necessary.

The equation of state is of fundamental importance in order to understand the physics of inertial confinement fusion and to be able to design the appropriate pellet to achieve high gain. The equation of state determines how much energy is needed to compress the fuel to a desired high density. For example, it is easy to compress an ideal gas because there are no repulsive forces. Therefore, by using an ideal gas in the formulations described by (14.55)–(14.59), an unrealistically large value for the fuel (ρR) is obtained which incorrectly implies high gain pellets. Since it is necessary to reach a high compression of at least 1000 times solid density for the D–T fuel, the compression in the later stages is dominated by the degeneracy pressure described by Fermi–Dirac statistics for electrons. The Fermi–Dirac equation of state:

$$P_e = n_e \varepsilon_F \left[\frac{2}{5} + \frac{\pi^2}{6} \left(\frac{KT_e}{\varepsilon_F} \right)^2 + \cdots \right] \tag{14.75}$$

where ε_F is the Fermi energy (see (14.53a) may be a good approximation at high temperature and high pressures, but it is definitely a poor approximation at solid density because it gives too high a pressure for low temperature solid densities. For example, (14.75) gives for $T = 0$ and aluminum solid density a pressure of 10 Mbars ($\approx 10^7$ atm!) instead of a zero pressure! Moreover, the Fermi–Dirac model does not take into account Coulomb interactions between the electrons and the ions. A more accurate model which includes degeneracy pressure and the nucleon–electron interactions is the Thomas–Fermi model. In this model the effective one-electron potential is calculated and the electrons are then distributed in the atomic (spherical) potential according to Fermi–Dirac statistics. In this model the pressure is much lower at solid density, although not low enough to fit the experimental data. While the Fermi–Dirac model neglects completely the possibility of ionization of the atoms, the Thomas–Fermi model can allow for calculating quasi-free electrons representing the ionization state of the ions. The Thomas–Fermi model is improved by taking exchange and quantum corrections into account. These corrections imply binding forces so that the overall pressure could be zero at zero temperature for solid density. The immediate compensation of the quantum pressure by the electrostatic energy of attraction has been formulated (Hora 1981, p. 29).

Moreover, hydrodynamic codes which perform simulations of inertial confinement fusion require the use of equation of state models which describe the thermodynamic functions of both electrons and ions. (Szichman *et al.* 1983, Section 14.5). The main values of interest are the pressure and internal energy, as well as their respective temperature and volume derivatives as a function of material density and temperature. The electronic equation of state can be taken, to a good approximation, from the corrected Thomas–Fermi–Dirac model. The ionic contribution to the equation of state can be described by the Debye–Grüneisen equation of state, with the appropriate density variations of the Debye temperature and the Grüneisen coefficient. At sufficiently high temperature and/or low densities the ion contribution can be described by the ideal gas equation of state. To join these two limiting cases a semi-empirical interpolation method can be used. The extent to which the inclusion of different models into the computer codes will influence the inertial confinement fusion results is a subject mainly for study and research. In general, since the heating mechanisms, the energy transport and the effects occurring in the corona are not very dependent on the ion parameters, one would expect not much dependence of these phenomena on the ion equation of state model. However, for processes which may be more sensitive to the ion parameters,

Fig. 14.3. The problems for the central ignition model for pellet fusion due to the equation of state:
(1) The Corona.
    2 (or 3) Temperature EOS
    Ideal gases
    Weak Coulomb Interaction (Debye–Hückel)
    Non-Local Thermal-Equilibrium
(2) Fusion.
    Strong Coulomb Interactions ⎫
                                ⎬ – (Thomas–Fermi)
    Fermi Statistics   ⎭

(3) On the Way to Fusion. Most difficult
(4) 'Easy' Experiments Region.

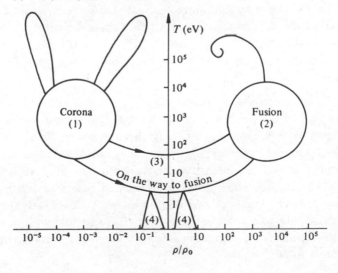

such as those taking place in the compressed solid, e.g. shock wave phenomena, one may expect variations in the calculated results due to the use of different models in computing the ion contribution to the equation of state. The equation of state also determines the velocity of shock waves and therefore the timescale of the whole implosion occurring in inertial confinement fusion phenomena.

For a summary and conclusion see Fig. 14.3 – with rather negative consequences for the central ignition scheme.

14.4   Alternative driving schemes: nonlinear force, cannon ball

If the central ignition scheme proves not to be successful at present, because of insufficient knowledge of radiation transport on internal electric fields and double layers causing the reduction of thermal transport, on the stopping power etc., there are two alternative schemes to solve the laser-fusion in the future:

(a)   the nonlinear-force scheme (Hora, 1981) which may provide the conditions of ideal adiabatic compression, and

(b)   the cannon ball scheme (Yamanaka 1981, Hora *et al.* 1984) of pellet compression without most of the disadvantages of the scheme outlined in the preceding section.

It should be mentioned, as a further possibility, that a special kind of volume ignition (Section 14.2) may have been realized by linear gas dynamics despite the fact that it seemed before to be impossible. The specially time-shaped laser irradiation, *by carefully avoiding shocks and central compression and bouncing,* was achieved (Yamanaka-compression) leading to the record fusion gain of 1.2×10^{12} fusion neutrons at 8 kJ second harmonics of neodymium glass laser input on a 800 μm diameter glass microballoon (Yamanaka *et al.* 1986). Yamanaka-compression is the simplest solution without any nonlinear effect and without the complications of central ignition. Its success is based on the nearly ideal adiabatic conditions.

14.4.1   *The nonlinear-force pushing*

One of the difficulties of the central ignition scheme is the fact that the laser radiation is not simply thermalized in the pellet corona for the subsequent hydromechanical ablation–compression dynamics; at neodynium–glass laser intensities above a threshold of 10^{14} W cm^{-2}, parametric instabilities and absorption anomalies appear and nonlinear (ponderomotive) forces become predominant. To avoid all this, some concepts remain to use lower intensities. The technologically expensive decrease of the laser wavelength by a factor of three increases the threshold moderately to 10^{15} W cm^{-2}.

Rather than trying to avoid nonlinear effects during laser–plasma interaction, one can try to find ways of actively using these effects, especially the nonlinear force. This involves the domain of very high laser intensities. The principal advantage is that thermalization (and entropy generation) for the transfer of optical energy into plasma motion is relatively small and produces only slight perturbations. At the same time, direct mechanical action by the electric and magnetic laser fields E, H and the subsequent current density j arising from the nonlinear, intensity-dependent refractive index $\tilde{n}$, results in a nonlinear force density f_{NL} which is predominant over the thermokinetic force (given by the pressure gradient), (see Hora 1981)

$$f_{NL} = \frac{1}{c} jxH + 4\frac{1}{\pi} E\mathbf{V} \cdot E - 4\frac{1}{\pi}\mathbf{V} \cdot (1 - \tilde{n})EE \qquad (14.76)$$

This equation includes all the necessary terms, as demonstrated by the fact that it does not result in shear motion at obliquely incident plane waves. For perpendicular incidence, the force f_{NL} reduces to the component in the x-direction along the wave propagation direction and can be expressed in terms of the variation of the refractive index with plasma density:

$$\overline{f_{NL}} = -\frac{\partial}{\partial x}\overline{(E^2 + H^2)}/8\pi = -\frac{F_v^2}{16\pi}\frac{\omega_p^2}{\omega^2}\frac{\partial}{\partial x}\frac{1}{|\tilde{n}|} = (\tilde{n}^2 - 1)\frac{\partial}{\partial x}\frac{E^2}{8\pi}$$

$$(14.77)$$

These specific results are restricted to monochromatic radiation. The treatment of ultra-short-wave packets and very fast amplitude changes is in progress. Note that oblique incidence requires further terms.

Results from detailed numerical calculations for high-intensity ($I > 10^{16}\,\mathrm{W\,cm^{-2}}$) neodymium laser interactions show the predominance of the nonlinear force in plasma dynamics while other parametric and anomalous effects decrease with increasing intensity. The inclusion of nonlinear optical properties confirms the relatively small effect of collision heating. Solutions of the wave field with respect to travelling- and standing-wave components indicate the appearance of Brillouin-like density ripples and subsequent high reflection.

The positive result is that firing of laser pulses onto pre-irradiated (moderately heated) plasma layers of soft density profiles does result in thick compression blocks of plasma possessing nearly linear velocity profiles as required for the ideal self-similarity compression. An ablation layer similar to this has been seen from an almost perfectly symmetric pellet irradiation where the ablation front of 60 keV ion energy consumed $\approx 50\%$ of the laser energy (Meyer *et al.* 1978; Slater 1977).

The results of extensive computations (Hora *et al.* 1979; Hora 1981) demonstrated that compressing the thick (cold) plasma blocks following their interior motion by nonlinear-force acceleration, produced approximately the ideal conditions of the self-similarity model such that the following purely kinematic (force-free) compression and reexpansion could be computed from the volume ignition model for deuterium–tritium fuel, including alpha reheat, radiation loss and fuel depletion. One example based on irradiation of a (profile shaping) pre-irradiated hollow sphere of DT plasma with a laser input energy of 500 kJ and compression to 300 times solid state density produced a net gain (= fusion energy/input laser energy) of 50 (Hora *et al.* 1975; Hora 1981).

The parameters were chosen in such a way that the electron and the ion gas fuel filled the ideal adiabatic non-degenerate equation of state neglecting the ionization processes at pre-irradiation. Any modifications of parameters have to take into account possible changes of the equation of state. The alternative suggested identity of this nonlinear force compression scheme with 'hot electron' driving was confirmed by Yonas (1984).

14.4.2   The cannon ball scheme

A method of avoiding all anomalies, instabilities and difficulties due to nonlinearities of back scattering, preventing laser light interacting with the plasma itself, and to provide the lowest degree of interaction, absorption and spherical compression symmetry, is to use a medium- or high-Z mantle around the pellet and to shine the laser light into the space (hohlraum) between pellet and mantle through prefabricated (or laser self-generated) holes in the mantle (Fig. 14.4).

This filling of the hohlraum with monochromatic radiation has been achieved (Yamanaka 1986) providing reassurance that this approach can

Fig. 14.4. The cannonball with laser irradiation through holes H into an empty space between an outer high-Z mantle and the DT fusion pellet.

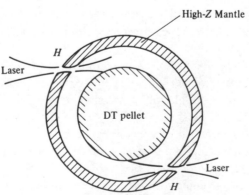

meet the requirements. At laser intensities near 10^{15} W cm^{-2} the laser radiation is quickly transformed into a wide spectrum of X-rays. For producing fusion energy from lasers, higher intensities are of interest where the radiation stays monochromatic. Experimental indications for slowing down the transfer to a continuous spectrum were given by Alexandrov *et al.* (1984). In this respect, it should be noted that a relativistic stabilization does exist to prevent deviations from the spherical shape of the surfaces of the hohlraum.

The monochromatic hohlraum radiation acts as an explosive similar to the gunpowder in a grenade: it drives both the mantle and the pellet by the brute force of the radiation pressure even if energy transfer is shut off by any of the effects mentioned. If the compression velocity (radiation pressure) of the pellet is always larger than the plasma thermal velocities at the interface (supersonic), the pellet will be compressed through a snowplough process.

In a very preliminary, nonoptimized (and perhaps pessimistic) calculation, one arrives at the following snowplough compression result. We assume a nearly constant inner radius $R_2 = 300\,\mu$m of the mantle (initial radius of a solid state pellet is $0.95\,R_2$) and a compression of 7 keV DT to: (i) 10^{25} ions cm^{-3}, or (ii) 10^{24} ions cm^{-3}, at a speed greater than thermal ion velocity. This produces a static confinement of ≈ 500 ps and requires a laser energy respectively for cases (i) and (ii) of

$$E = 3.2 \times 10^6 \text{ J} \quad \text{and} \quad 3.2 \times 10^5 \text{ J}$$

and average laser power of

$$P = 2.7 \times 10^{16} \text{ W} \quad \text{and} \quad 2.7 \times 10^{15} \text{ W}.$$

with a laser-pulse duration in both cases of 120 ps, we predict a fusion gain (Hora *et al.* 1984)

$$G = 50.4 \quad 5.04 \tag{14.78}$$

The laser intensity in the hohlraum is

$$I = 2.6 \times 10^{20} \text{ W cm}^{-2} \quad 8.2 \times 10^{19} \text{ W cm}^{-2} \tag{14.79}$$

while the average focused laser intensity in the entrance holes (3×10^{-3} cm diameter) is:

$$I_{\rm H} = 2.7 \times 10^{21} \text{ W cm}^{-2} \quad 2.7 \times 10^{20} \text{ W cm}^{-2}. \tag{14.80}$$

This is sufficient to produce relativistic hole drilling based on the Yamanaka effect for relativistically self-focused beam propagation (Hora 1981) which reduces the diffraction limitation by total reflection as in fibre optics.

These estimates were made on the basis of a simplified snowplough model. The detailed model, again, will have to include the highly turbulant plasma state at compression and a version of the equation of state with highly non-equilibrium thermodynamic conditions.

14.5 ˙ The two-temperature equation of state

Having explained the complexity and incompleteness of the use of the correct equation of state in the field of inertial confinement fusion we give here an example of how a first step has been made to take into account various systems (degenerate and non-degenerate) of equations of state for the practical case of two fluids (electron and ion) with different temperatures.

Hydrodynamic codes which perform simulations of inertial confinement interactions (laser–target or electron–target or ion–target interactions) require the use of EOS models which describe the thermodynamic functions of both electrons and ions. These codes usually contain a quasi-independent system of electrons at some temperature T_e and a second quasi-independent system of ions at a temperature T_i. The main values of interest are the pressure and internal energy as well as their respective temperature and volume derivatives as a function of material density and temperature (see e.g., (14.102)).

One of the models widely used in the literature (Trainor *et al.* 1978) is the statistical Thomas–Fermi (TF) model of the atom for the electronic contribution to the EOS and ideal gas behaviour for the ions. The TF model predicts pressure values which are considerably higher than those obtained experimentally. (See Fig. 13.2 of Chapter 13.) In particular the TF model gives a few megabars of pressure at normal conditions ($T \approx 0$, $\rho \simeq \rho_0$)! instead of a zero pressure, so that this model is definitely wrong at low temperature and normal densities.

A more realistic model can be found in the SESAME EOS library (Cooper 1979, 1983) which is a collection of tabular data giving the pressure and internal energy per unit mass as a function of temperature and density for a variety of materials. The values in the tables were obtained from a variety of sources, including phenomenological and theoretical models and fits to experimental data. This library represents the best approach to realistic EOS, although they are *single temperature* EOS. Cooper (1983) made the first attempts to formulate two-temperature (electron and ions) tables of EOS, however, the authors caution that these tables are at present very crude, at best. Therefore, for nearly all materials, only the single temperature tables are considered to be at all reliable. Since for laser–plasma or particle-beam–plasma calculations one requires separate tables for electrons and for ions, one has the task of either recalculating these separate values independently or, alternatively, of finding some scheme of separating the existing SESAME tables into their component parts (Szichman *et al.* 1983). In general, according to SESAME authors (Cooper 1983) the theoretical portions of the EOS were constructed from three distinct contributions: (i) a cold curve due to electronic forces at $T = 0$, (ii) thermal electronic excitations and (iii) thermal (nuclear) or ionic excitations. The total pressure and energy can be written as

$$P(\rho, T) = P_c(\rho) + P_e^T(\rho, T) + P_i^T(\rho, T) \equiv P_e + P_i \qquad (14.81)$$

$$E(\rho, T) = E_c(\rho) + E_e^T(\rho, T) + E_i^T(\rho, T) \equiv E_e + E_i \qquad (14.82)$$

where P_c and E_c are the cold (electron) terms while P^T and E^T are the appropriate thermal contributions. In general, for very high compression the pressure and the energy do not separate into two additive contributions. However, for practical purposes it is convenient to assume (14.81) and (14.82) until a better and more fundamental approach is satisfactorily resolved. For related treatments see More (1981) and DeWitt (1984).

14.5.1 Electronic contributions to the EOS

The Thomas–Fermi EOS was extended by taking into account electron-exchange effects and other quantum corrections. This model (Kirshnits 1959; Kalitkin 1960; McCarthy 1965), known as the corrected Thomas–Fermi–Dirac (TFC) EOS, is accepted today as being a reasonably accurate, available theory in the range of density, temperature and pressure of the SESAME tables and of interest to plasma-beam (laser or particle) interactions. The solution of the TFC equation involves considerable difficulties and requires extensive numerical calculations. Following McCarthy (1965) one can find numerical tables for TFC equations of state. In order to use these tables, one has to know the quantities ZV and $Z^{-4/3}T$, where

$$V = \frac{A}{N_A \rho} = \frac{1}{N_0 \rho} \qquad (14.83)$$

where A is the atomic mass, N_A is the Avogadro number, and N_0 is the number of particles per unit mass. From the tables one can read off the values of P_e and E_e in the form $Z^{-10/3}P$ and $Z^{-7/3}E$. These are the uncorrected (TF) values, while the tabulated electron and quantum exchange corrections δP and δE are in the form $Z^{-8/3}\delta P$ and $Z^{-5/3}\delta E$. The required TFC values for P_e and E_e are

$$P_e = \left[\left(\frac{P_{TF}}{Z^{10/3}} \right) + Z^{-2/3} \left(\frac{\delta P}{Z^{8/3}} \right) \right] Z^{10/3} \qquad (14.84)$$

$$E_{\mathrm{e}} = \left[\left(\frac{E_{\mathrm{TF}}}{Z^{7/3}} \right) + Z^{-2/3} \left(\frac{\delta E}{Z^{5/3}} \right) \right] Z^{7/3} \qquad (14.85)$$

where the values in round brackets ( ) are read from the tabulated values of McCarthy (1965). The values for the zero temperature isotherms are obtained in a similar manner from the tables of Kalitkin (1960).

Using McCarthy (1965) and Kalitkin's (1960) tables we plot (Szichman *et al.* 1983) in Figs. 14.5 and 14.6 the isotherms of the internal energy for electrons E_{e} and the electron pressure P_{e} versus the compression ρ/ρ_0 for aluminum ($Z = 13$, $A = 26.982$). The dotted lines are particular isotherms from the TFC model to be used in (14.81) for comparison with the SESAME EOS. (See Figs. 14.7 and 14.8.)

14.5.2   *The ion contributions to the EOS*

The nuclear (or ion) contributions to the EOS are known explicitly for the two extreme cases: the gas phase and the solid phase. The thermo-

Fig. 14.5. Isotherms of electronic internal energy per unit mass as a function of compression for aluminum ($10^{12}\mathrm{erg/g} = 1\,\mathrm{Mbar\,cm^3/g}$)

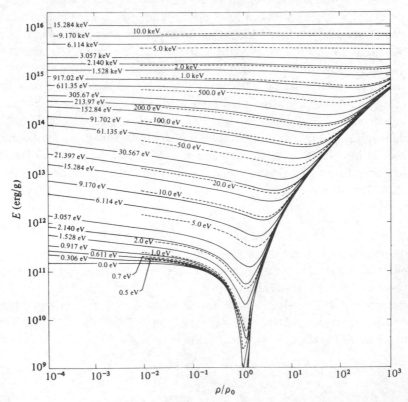

dynamics of simple solids are well described by the Grüneisen equation
of state (see Chapter 10) while for high temperatures (or low densities)
they are described by the ideal gas EOS. To join the two limiting cases
a semi-empirical interpolation method has been used (Kormer *et al.* 1962).
This simple picture does not include phase transitions. To deal with this
problem knowledge of a great amount of empirical data in a 'hard domain'
of experiments is needed. Moreover, since in beam–plasma interactions
for fusion research we are interested mainly in the domain of high pressure
($\gtrsim$ Mbar) and high temperature, the importance of the phase transitions
might be small (because the phase transitions have already occurred). We
now discuss briefly the gas, Grüneisen and the interpolation methods.

Fig. 14.6. Isotherms of electronic pressure as a function of
compression for aluminum (10^{12} dynes/cm^2 = 1 Mbar)

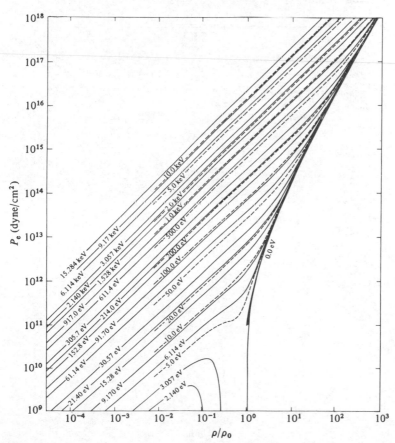

(a) Ideal gas (Model 1)
The equations of state in this case are (see Chapter 3)

$$E_{IG} = \tfrac{3}{2} N_0 kT \tag{14.86}$$

$$P_{IG} = N_0 \rho kT \tag{14.87}$$

where the subscript IG denotes the ideal gas phase and N_0 is defined in (14.83).

Fig. 14.7. Isotherms of the total internal energy per unit mass as a function of compression for aluminum. (———) Taken from SESAME tables (Cooper 1983). (– – –) Calculated using the TFC model for the electrons and the perfect gas model for the ions (When this line does not appear, it overlaps that of Model 2). (······) Calculated using the TFC model for the electrons and the perfect gas – solid interpolation for the ions (Model 2). (—·—) Calculated using the TFC model for the electrons. The ion contribution is deduced from the SESAME tables (Model 3).

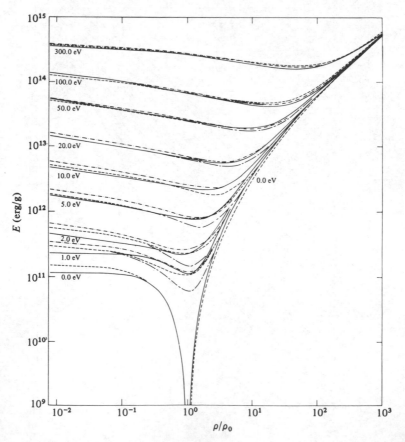

(b) Debye-Grüneisen EOS for solids

The equations of state in this case are (see Chapter 10)

$$E_{DG} = 3N_0 kTD(T_D/T) \tag{14.88}$$

$$P_{DG} = 3\Gamma N_0 \rho kTD(T_D/T) \tag{14.89}$$

where the subscript DG denotes the Debye–Grüneisen solid phase, Γ is the Grüneisen coefficient, T_D is the Debye temperature and the Debye

Fig. 14.8. Isotherms of the total pressure as a function of compression for aluminum. Curves as in Fig. 14.7.

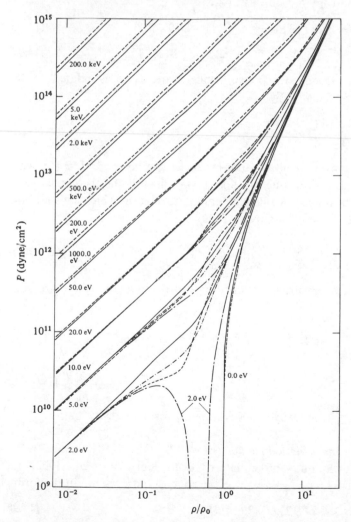

integral D(x) is given by

$$D(x) = \frac{3}{x^3} \int_0^x \frac{y^3 dy}{e^y - 1} \tag{14.90}$$

In the limit of $T \gg T_D$ (i.e., $x \to 0$) the Debye integral goes to the limit $D \to 1$, so that the EOS reduces to

$$E_{DG} = 3N_0 kT \tag{14.91}$$

$$P_{DG} = 3\Gamma N_0 \rho kT \tag{14.92}$$

For completeness we shall also give the expression for the free energy F_{DG} and for the entropy S_{DG} in this model:

$$F_{DG} = N_0 kT [3 \ln(1 - e^{-T_D/T}) - D(T_D/T)] \tag{14.93}$$

$$S_{DG} = -N_0 k [3 \ln(1 - e^{-T_D/T}) - 4D(T_D/T)] \tag{14.94}$$

The (density dependent) Debye temperature $T_D(\rho)$ is related to the Grüneisen coefficient Γ by

$$\Gamma = \frac{\rho}{T_D} \frac{dT_D}{d\rho} \tag{14.95}$$

It is observed experimentally that for small compressions $\Gamma \approx 1/\rho$, while for large compressions one can take $\Gamma \approx \frac{2}{3}$ which is the theoretical value for a free electron gas. A phenomenological relation which satisfies both limits can be written (Thompson 1972):

$$\Gamma = \frac{\Gamma_0 \rho_0}{\rho} + \frac{2}{3}\left(1 - \frac{\rho_0}{\rho}\right)^2 \qquad \frac{\rho}{\rho_0} > 1 \tag{14.96}$$

The Debye temperature is obtained by substituting (14.96) in (14.95) and integrating

$$\frac{T_D(\rho)}{T_D(\rho_0)} = \left(\frac{\rho}{\rho_0}\right)^{2/3} \exp\left\{\Gamma_0\left(1 - \frac{\rho_0}{\rho}\right) - \frac{1}{2}\left[3 - 4\frac{\rho_0}{\rho} + \left(\frac{\rho_0}{\rho}\right)^2\right]\right\}$$

$$\text{for } \frac{\rho}{\rho_0} > 1 \quad (14.97)$$

where $T_D(\rho_0)$ is the Debye temperature at normal density.

(c) *Solid–gas interpolation method (Model 2)*
Using the technique and equations from Kormer *et al.* (1962) and Thompson (1972) one can express the energy and pressure in the form

$$E_i = 3N_0 kT_D(T_D/T)(2 + \xi)/(1 + \xi) \tag{14.98}$$

$$P_i = N_0 \rho k T_D (T_D/T)(3\Gamma + \xi)/(1 + \xi) \qquad (14.99)$$

Here ξ is called the interpolation parameter and is expressed by Thompson (1972)

$$\xi = \frac{N_0^{5/3} h^2 \rho^{2/3} T}{2\pi k T_D^2} \qquad (14.100)$$

where h is Planck's constant, k is Boltzmann's constant and the Debye temperature T_D is a function of the density ρ (see, e.g., (14.97)). For $T \to 0$ one obtains from (14.100) that $\xi \to 0$, so that (14.98) and (14.99) reduce to the Debye–Grüneisen equations of state (14.88) and (14.89). For $T \to \infty$ one gets $\xi \to \infty$ and $D \to 1$ so that (14.98) and (14.99) reduce to the ideal gas (14.86) and (14.87).

In order to use (14.98) and (14.99) for the domain $\rho/\rho_0 < 1$ as well as for $\rho/\rho_0 \geqslant 1$ one has to know the functions ξ T_D and Γ in the whole reason. For $\rho/\rho_0 > 1$ we know T_D from (14.97), Γ is given in (14.96) and ξ therefore is defined in (14.100). However, for $\rho/\rho_0 < 1$, the Debye temperature and the Grüneisen theory of solids, (14.96) is not at all appropriate for $\rho < \rho_0$ since Γ diverges for $\rho \to 0$. Therefore, since we don't have any theoretical basis to calculate Γ for $\rho < \rho_0$, one assumes (Thompson 1972) an artificial extrapolation for Γ. Taking a Taylor expansion in the form of a second order polynomial in ρ/ρ_0, and assuming Γ and $d\Gamma/d\rho$ continuous at $\rho = \rho_0$, one gets

$$\Gamma = (1 - 2\Gamma_0)\left(\frac{\rho}{\rho_0}\right)^2 + (3\Gamma_0 - 2)\left(\frac{\rho}{\rho_0}\right) + 1 \quad \frac{\rho}{\rho_0} < 1 \qquad (14.101)$$

where $\Gamma(0) = 1$ has been assumed. The form of (14.101) has no physical basis but it is *a posteriori* justified to use it because it works well. (Thompson 1972). Solving (14.95) for Γ given in (14.101) one gets,

$$T_D(\rho) = T_D(\rho_0)\left(\frac{\rho}{\rho_0}\right) \exp\left\{\tfrac{1}{2}(1 - 2\Gamma_0)\left(\frac{\rho}{\rho_0}\right)^2 + (3\Gamma_0 - 2)\left(\frac{\rho}{\rho_0}\right)\right.$$
$$\left. + (\tfrac{3}{2} - 2\Gamma_0)\right\}; \qquad \text{for } \rho/\rho_0 < 1 \qquad (14.102)$$

Equations (14.98) and (14.99) are now defined for $\rho/\rho_0 \geqslant 1$ as well as for $\rho/\rho_0 \leqslant 1$ by using (14.96), (14.97), (14.91), (14.92) and (14.100). For completeness we now write the ion free energy F_i and the ion entropy S_i in the solid–gas interpolation method,

$$F_i = N_0 k T \{3 \ln (T_D/T) - 1 + \tfrac{3}{2} \ln (1 + \xi)\} \qquad (14.103)$$

$$S_i = -N_0 k \{3 \ln (T_D/T) - 4 + \tfrac{3}{2} \ln (1 + \xi) + \tfrac{3}{2}(\xi/(1 + \xi))\} \qquad (14.104)$$

Once again we mention that the EOS given by (14.98), (14.99), (14.103) and (14.104) do not describe the 'true' phase transition (i.e. melting) between the solid and the gas phase, but in many cases this approach is acceptable for hydrodynamic numerical efforts. The derivatives of the thermodynamic functions, which are usually required for the codes, are

$$c_{vi} = \left(\frac{\partial E_i}{\partial T}\right)_V = \frac{E_i}{T}\left\{1 - \frac{\xi}{(1+\xi)(2+\xi)}\right\} \tag{14.105}$$

$$\left(\frac{\partial P_i}{\partial T}\right)_V = \frac{P_i}{T}\left\{1 - \left(\frac{\xi}{1+\xi}\right)\left(\frac{3\Gamma-1}{3\Gamma+\xi}\right)\right\} \tag{14.106}$$

$$\left(\frac{\partial P_i}{\partial \rho}\right)_T = \frac{P_i}{\rho}\left\{1 + \left(\frac{\xi}{1+\xi}\right)\frac{2(3\Gamma-1)^2}{3(3\Gamma+\xi)}\right\} + \frac{3N_0\rho kT}{1+\xi}\frac{d\Gamma}{d\rho} \tag{14.107}$$

where

$$\frac{d\Gamma}{d\rho} = \frac{\rho_0}{\rho^2}\left\{\Gamma_0 - \tfrac{4}{3}\left(1 - \frac{\rho_0}{\rho}\right)\right\}; \quad \text{for } \rho \geqslant \rho_0 \tag{14.108}$$

$$\frac{d\Gamma}{d\rho} = 2(1 - 2\Gamma_0)\left(\frac{\rho}{\rho_0}\right) + (3\Gamma_0 - 2); \quad \text{for } \rho \leqslant \rho_0 \tag{14.109}$$

(d) E_i and P_i deduced from the SESAME tables (Model 3)

For E_i, P_i, E_e and P_e one can write the following phenomenological expressions, which are taken by analogy to the ideal equations of state for a neutral plasma. We denote by $\langle Z^* \rangle$ the degree of (average) ionization of the atoms in the plasma, so that for the thermal part of the EOS one can write

$$E_i = \frac{E_{SE}(T,\rho) - E_{SE}(0,\rho)}{1 + \langle Z^* \rangle} \tag{14.110}$$

$$P_i = \frac{P_{SE}(T,\rho) - P_{SE}(0,\rho)}{1 + \langle Z^* \rangle} \tag{14.111}$$

where E_{SE} and P_{SE} are the SESAME values of energy and pressure for the appropriate temperature T and density ρ. However, the value of $\langle Z^* \rangle$ is not yet known in (14.110) and (14.111). Similarly to (14.110) and (14.111), we can write for the thermal electron contribution

$$E_e^T = \frac{[E_{SE}(T,\rho) - E_{SE}(0,\rho)]\langle Z^* \rangle}{1 + \langle Z^* \rangle} \tag{14.112}$$

$$P_e^T = \frac{[P_{SE}(T,\rho) - P_{SE}(0,\rho)]\langle Z^* \rangle}{1 + \langle Z^* \rangle} \tag{14.113}$$

By comparing (14.112) and (14.113) with the TFC model, i.e.,

$$E_e^T \equiv E_{TFC}(T, \rho) - E_{TFC}(o, \rho) \tag{14.114}$$

$$P_e^T \equiv P_{TFC}(T, \rho) - P_{TFC}(o, \rho) \tag{14.115}$$

where E_{TFC} and P_{TFC} are the energy and pressure in the TFC model, one can calculate the degree of ionization $\langle Z^* \rangle$. Using this value of $\langle Z^* \rangle$ in (14.110) and (14.111) one gets the ion contribution to the EOS.

14.5.3 Results and discussion

The ion contribution, as calculated using each of the models (1, 2 and 3) in Section 14.5.2, was combined with the TFC electron contribution for the same temperature and the result compared with the values found in the SESAME tables (for one temperature!). These comparisons are shown in Figs. 14.7 and 14.8 (as taken from Szichman *et al.* 1983.) One can see that models 2 and 3 seem to be in general agreement with the SESAME tables. The results using model 3 also give the degree of ionization which is necessary for plasma simulation, and besides this important practical result, this approach is only a self-consistent approach for the ion contribution and the SESAME tables.

The extent to which the inclusion of any model in hydrodynamic codes is necessary is a question for further research and calculations. In general, since the heating mechanisms, the energy transport and the effects generally occurring in the corona are not very dependent on the ion parameters, one would not expect much dependence of these on the EOS ion contribution model used. However, for phenomena which may be more sensitive to the ion parameters, such as shock wave phenomena in compressed solids, one may expect to see changes by using different models in computing the ion contribution to the EOS.

A proper statistical calculation for two fluids (or gases) with a strong interaction between them does not in general give (14.81) and (14.82) which was our starting point for the above discussion. The pressure and energy can not in general be separated into two additive contributions. Therefore, one possible approach is to define a free energy $F_e(\{R_j\}, T_e)$ for a fluid of electrons at temperature T_e, where $\{R_j\}$ is a set of coordinates describing the position of the ions. The free energy F_e includes all the electrostatic energies (including ion–ion!) and the electronic kinetic energy. This free energy is an effective potential for the ion motions. This approach is reasonable in the case where the ions are moving slowly on the collision time-scale of the electrons. The partition function in this case can be written as

$$Q = \int \frac{d^3 R_1 \cdots d^3 R_N d^3 p_1 \cdots d^3 p_N}{N! h^{3N}}$$

$$\times \exp\left(-\frac{1}{kT_I}\left[\sum_{j=1}^{N} \frac{p_j{}^2}{2M_i} + F_e(\{R_i\}, T_e)\right]\right\} \tag{14.116}$$

where N is the number of ions and R_i, p_i are the canonical coordinates and momenta for the ions. The free energy of the electron–ion system is,

$$F(T_e, T_I, V) = -kT_I \ln Q \tag{14.117}$$

The entropy, in this approach, is defined by

$$S_I = -\left(\frac{\partial F}{\partial T_I}\right)_{V, T_e} \qquad S_e = -\left(\frac{\partial F}{\partial T_e}\right)_{V, T_I} \tag{14.118}$$

where S_I and S_e are the ion and electron entropies. The generalized thermodynamic consistency condition is

$$\left(\frac{\partial E}{\partial V}\right)_{T_e, T_I} = T_e\left(\frac{\partial P}{\partial T_e}\right)_{T_I, V} + T_I\left(\frac{\partial P}{\partial T_I}\right)_{T_e, V} - P \tag{14.119}$$

while the basic thermodynamic relation is

$$dE = T_e dS_e + T_I dS_I - P dV \tag{14.120}$$

Although this last approach seems to be on a more fundamental level than the phenomenological approach of Section 14.2, one has in this case to solve very complicated nonlinear equations and to check its applicability to the hydrodynamic codes. This subject is still being researched and poses profound and conceptual difficulties for the thermodynamics of a mixture of two- (or more) temperature fluids with strong interactions.

15

Applications of equations of state in astrophysics

15.1 Overview

In this chapter we will see how the equation of state can be used to determine the structure of stellar interiors. Of course, our treatment will be at a level just adequate to bring out the salient aspects of the subjects; for more rigorous treatments we refer to the numerous books (and edited volumes) written on the subject (see e.g., Aller (1953, 1954), Aller and McLaughlin (1965), Chandrasekhar (1939), Cox and Giuli (1968a, 1968b), Giacconi and Ruffini (1978), Menzel, Bhatnagar and Sen (1963), Schatzman (1958), Schwarzschild (1958), Stein and Cameron (1966), Stodolkiewicz (1973), Zeldovich and Novikov (1971)).

Stars, because of their high temperatures, are gaseous throughout, and quite fortuitously in most stars the matter behaves either as a perfect gas or as a completely degenerate gas; in either case, the equation of state is quite simple. For example, inside the Sun the correction to the pressure derived by assuming the perfect gas law is about 0.43% and for a dense star like σ_2 EriC the correction is about 2% (the numerical values are quoted from Aller (1953), p. 60). In this section we will write the two equations of state in a form which is convenient in studying astrophysics problems and also discuss their domain of applicability. In Section 15.2 we will derive the condition for hydrostatic equilibrium which is one of the basic equations in understanding stellar structure and will be used extensively throughout the chapter. We use the condition for hydrostatic equilibrium [In Section 15.3] to show that if the density variation (inside the star) is known then it is possible to obtain pressure and temperature variations inside the star; the calculation of temperature variation also requires a knowledge of the equation of state which we assume to be given by the perfect gas law. In Section 15.4 we obtain estimates of the pressure and temperature variation by assuming uniform density inside the star and in Section 15.5 we derive some useful theorems which gives us lower and upper bounds for the pressure and temperature variations inside the star. We calculate in Section 15.6 the gravitational potential energy inside the star and derive the virial theorem, which are used in Section 15.7 to obtain a qualitative understanding of the evolution of a star. In Section 15.8 we show that whereas radiation pressure is rather unimportant in stars like the Sun, it plays an important role in very massive stars and indeed, for very massive stars the radiation pressure is so large that the star itself becomes unstable.

In Section 15.9 we discuss in detail the polytropic model of the star in which the pressure–density relationship is assumed to be of the form

$$P(r) = K[\rho(r)]^{1 + 1/n}$$

where the parameters K and n are assumed constant inside the star. Using the above relation and the equation for hydrostatic equilibrium (derived in Section 15.2) we derive explicit expressions for the pressure, density and temperature variations inside the star. The polytropic model has played a very important role in the understanding of the stellar interior. In particular, the case $n = 3$ (which is usually referred to as the *standard model* – see Section 15.10) leads to a fairly good understanding of the interior of the Sun and some other stars. Finally, in Section 15.11 we discuss the white dwarf stars in which the densities are so high that the electrons may be assumed to form a highly degenerate gas.

The values of some physical and astronomical constants (which we will be using throughout the chapter) are given in Table 15.1.

Table 15.1. *Values of some physical and astronomical constants to be used in this chapter*

Quantity	Symbol	Value
Mass of the hydrogen atom	H	1.673×10^{-24} g
Mass of the electron	m	9.110×10^{-28} g
Speed of light in vacuum	c	2.998×10^{10} cm s^{-1}
Planck's constant	h	6.626×10^{-27} erg s
	$\hbar = h/2\pi$	1.055×10^{-27} erg s
Gravitational constant	G	6.672×10^{-8} cm^3 s^{-2} g^{-1}
Boltzmann constant	k	1.381×10^{-16} erg K^{-1}
Radiation pressure constant	$a = \dfrac{\pi^2 \, k^4}{15(\hbar c)^3}$	7.566×10^{-15} erg cm^{-3} K^{-4}
Mass of the sun	$M_\odot$	1.989×10^{33} g
Radius of the sun	$R_\odot$	6.960×10^{10} cm

15.1.1 The equation of state for an ideal gas

For an ideal gas, the equation of state is given by (see Section 13.2)

$$PV = NkT \tag{15.1}$$

where N represents the number of particles of the gas in the volume V. We may write (15.1) in the form

$$P = nkT \tag{15.2}$$

where n represents the number of particles per unit volume. If the gas consists of many types of particles then

$$P = kT \sum_i n_i \tag{15.3}$$

where n_i represents the number of independent constituent particles per unit volume. For example, if the gas consists of only hydrogen and nitrogen molecules (in undissociated forms) then

$$P = kT(n_{H_2} + n_{N_2}) \tag{15.4}$$

where n_{H_2} and n_{N_2} represent respectively the number of hydrogen and nitrogen molecules per unit volume. However, if the gas is at a very high temperature so that the molecules have dissociated into atoms and each atom is completely ionized then

$$P = kT(n_e + n_1 + n_2) \tag{15.5}$$

Table 15.2. *Composition of the Sun from spectroscopic observations*

Element	Atomic weight	Relative abundance	
		by number	by weight
H	1	1000	1000
He	4	80	320
C	12	0.1	1
N	14	0.2	3
O	16	0.5	8
Ne	20	0.5	10
Mg	24	0.06	1
Si	28	0.03	1
S	32	0.02	1
A	40	0.05	2
Fe	56	0.02	1

Note: Table adapted from Schwarzschild (1958); the data are due to L.H. Aller; the original paper by Aller was not available to the authors.

where n_e represents the number of electrons per unit volume, n_1 the number of hydrogen nuclei (i.e. protons) per unit volume and n_2 the number of nitrogen nuclei per unit volume.

We assume that stellar matter is in a completely ionized state and let 1 g of the stellar matter consist of X g of hydrogen, Y g of helium and $(1 - X - Y)$ g of other large atomic weight elements. We may mention here that from the analysis of spectroscopic data it is possible to estimate the composition of stars (see, e.g., Aller 1953). In Table 15.2 we have given the relative abundance of various elements in the Sun. As can be seen, hydrogen and helium are predominant and it is for this reason that hydrogen and helium are being considered separately and the heavier elements are grouped together. From the data given in Table 15.2, we readily obtain

$$X \approx \tfrac{1000}{1348} \approx 0.742$$
$$Y \approx \tfrac{320}{1348} \approx 0.237 \tag{15.6}$$
$$(1 - X - Y) \approx \tfrac{28}{1348} \approx 0.021$$

Spectroscopic data also indicate that most nearby stars have approximately the same composition, indicating similar values for X and Y.

Now, each hydrogen atom will give rise to two particles (one electron and one proton) and therefore X g of hydrogen will give rise to

$$\frac{2X}{H}$$

particles, here H ($\approx 1.67 \times 10^{-24}$ g) represents the mass of the hydrogen atom. On the other hand, each helium atom will give rise to three particles (two electrons and one helium nucleus) and therefore Y g of helium will give rise to

$$\frac{3X}{4H}$$

particles, where in writing the above expression we have assumed that the mass of the helium atom is about 4 times that of the hydrogen atom. Finally, if we consider an atom of atomic number Z and atomic weight A then each atom will give rise to $(Z + 1)$ particles, thus each gram of the substance will give rise to

$$\frac{(Z + 1)}{AH} \approx \frac{1}{2H}$$

particles; in the last step we have assumed $(Z + 1)/A \approx \tfrac{1}{2}$ which is approximately true for most elements included in Table 15.2, except, of course, for hydrogen and helium. For example, for nitrogen $Z = 7$, $A = 14$ giving $(Z + 1)/A \approx 0.57$; for oxygen $Z = 8$, $A = 16$ giving $(Z + 1)/A \approx 0.56$ and for iron $Z = 26$, $A = 56$ giving $(Z + 1)/A \approx 0.48$. Thus, $(1 - X - Y)$ g of large atomic weight elements will give rise to

$$\frac{1}{2H}(1 - X - Y)$$

particles. If the density of the stellar material is ρ (g/cm^3) we would get

$$\sum_i n_i = \left[\frac{2X}{H} + \frac{3Y}{4H} + \frac{1}{2H}(1 - X - Y)\right]\rho$$

or

$$\sum_i n_i = [\tfrac{1}{2} + \tfrac{3}{2}X + \tfrac{1}{4}Y]\frac{\rho}{H}$$

(15.7)

We may therefore write (15.3) as

$$P = \left(\frac{k}{\mu H}\right)\rho T$$

(15.8)

where

$$\mu = \frac{1}{0.5 + 1.5X + 0.25Y}$$

(15.9)

is usually referred to as the mean molecular weight of matter. It can readily be seen that

$$0.5 \leqslant \mu \leqslant 2.0$$

(15.10)

For example, when stellar matter consists only of hydrogen, $X = 1$ and $Y = 0$ giving $\mu = 0.5$ and if the star were to contain only heavier elements, then $X = 0$, $Y = 0$ and $\mu = 2.0$. The perfect gas law is usually written in the form given by (15.8) for use in astrophysics problems.

We may similarly derive an expression for the electron density which we will require in later sections. Since each hydrogen atom gives rise to one electron, each helium atom gives rise to two electrons and an atom with atomic number Z gives rise to Z electrons, the total number of electrons per cm^3 would be given by

$$n_e = \left[\frac{X}{H} + \frac{2Y}{4H} + \frac{Z}{AH}(1 - X - Y)\right]\rho$$

(15.11)

or

$$n_e = \frac{\rho}{\mu' H}$$

(15.12)

where

$$\mu' \approx \frac{2}{1 + X} \tag{15.13}$$

represents the average mass (in atomic weight units) per electron of the (fully ionized) gas and in writing the expression for μ' we have assumed $Z \approx A/2$ for larger atomic weight elements.

We end this section by noting that almost all stellar interiors are at temperatures high enough ($\gtrsim 10^6$ K) for matter to be in a completely ionized state; however, the surface temperatures may not be high enough for matter to be completely ionized. For example, the ionization energy of the hydrogen atom is about 13.6 eV, and for $kT \gtrsim 13.6$ eV, we must have $T \gtrsim 1.6 \times 10^5$ K which may be compared with the surface temperature of the Sun, which is about 5700 K. Nevertheless, temperatures even at the surface are high enough for the perfect gas law to be applicable.

15.1.2   *The equation of state for a degenerate electron gas*

The perfect gas law is applicable to gases at high temperatures and low pressures. Indeed, at any given temperature, if the pressure of a gas is made low enough then the perfect gas law is always applicable. For a given temperature, as the pressure is increased the equation of state becomes quite complicated; however, at very high pressures[†] the electrons form an almost completely degenerate gas and the equation of state becomes simple again. We may point out here that even at low temperatures, if the pressure is high enough ionization will occur – this is known as *pressure ionization*. Furthermore, usually the densities are such that whereas the electron gas is in a degenerate state, the gas consisting of protons and other nuclei is in a non-degenerate state so that perfect gas law is applicable to these nuclei; however, the pressure due to the gas formed by the nuclei is usually negligible in comparison with the pressure due to the degenerate electron gas and therefore we will consider only the pressure due to the degenerate electron gas.

In Section 6.2.1 we showed that in the non-relativistic limit, the pressure of a completely degenerate[‡] electron gas is given by

---

[†] We will quantify these statements later in this section.

[‡] Rigorously speaking, a completely degenerate gas corresponds to $T = 0$ K (see Section 6.2.1); however, if $\varepsilon_F/kT \gg 1$, (where ε_F is the Fermi energy – see Section 6.2.1) we may apply the theory corresponding to a completely degenerate gas. For example, the temperature inside a particular star may be as high as 10^6 K but the matter may be so compressed that the Fermi energy may be as high as 10^6 eV so that $\varepsilon_F/kT \approx 10^3$ and very little error will be involved if we assume $T = 0$.

$$P = \frac{\hbar^2}{5m}(3\pi^2)^{2/3} n_e^{5/3} \tag{15.14}$$

where m represents the electron mass and the n_e the electron density[†]. If we use (15.12) we may write (15.14) in the form

$$P = K_1 \left(\frac{\rho}{\mu'}\right)^{5/3} \tag{15.15}$$

where

$$K_1 = (3\pi^2)^{2/3} \frac{\hbar}{5mH^{5/3}} \tag{15.16}$$

or

$$K_1 \approx (3\pi^2)^{2/3} \frac{(1.05 \times 10^{-27})^2}{5 \times 9.1 \times 10^{-28} \times (1.67 \times 10^{-24})^{5/3}}$$

$$\approx 9.87 \times 10^{12} \text{ c.g.s. units} \tag{15.17}$$

Thus (assuming the validity of the non-relativistic theory) stellar matter can be assumed to form a degenerate gas if the pressure given by (15.15) is much greater than that given by (15.8), i.e., when

$$K_1 \left(\frac{\rho}{\mu'}\right)^{5/3} \gg \rho \left(\frac{k}{\mu H}\right) T \tag{15.18}$$

and conversely. Substituting the numerical values for various parameters, we may rewrite (15.18) in the following form

$$\rho \gg (2.42 \times 10^{-8}) \frac{\mu'^{5/2}}{\mu^{3/2}} T^{3/2} \tag{15.19}$$

where we have used the c.g.s. system of units so that ρ is measured in g cm^{-3} and T in K. If we assume that the only element present in the star is hydrogen then $X = 1$ and $Y = 0$ so that $\mu = 0.5$ and $\mu' = 1.0$, thus (15.19) becomes

$$\rho \gg 6.9 \times 10^{-8} T^{3/2} \tag{15.20}$$

The above expression tells us that the equation of state can be assumed to be given by (15.15) when

$$\left.\begin{array}{ll} \rho \gg 0.07 \text{ g cm}^{-3} & \text{for } T = 10\,000 \text{ K} \\ \rho \gg 2.2 \times 10^3 \text{ g cm}^{-3} & \text{for } T = 10^7 \text{ K} \end{array}\right\} \tag{15.21}$$

[†] It can be seen immediately from (15.14) that even if the proton gas is completely degenerate with the same number of particles per unit volume, the corresponding pressure would be considerably less because of its much larger mass.

In Fig. 15.1 the initial part of the curve is the straight line defined by the equation

$$\log \rho = \log(6.9 \times 10^{-8}) + \tfrac{3}{2} \log T$$

or   (15.22)

$$\log T = 4.8 \qquad\qquad + \tfrac{2}{3} \log \rho$$

Thus the $\log T$–$\log \rho$ curve is a straight line with slope $\tfrac{2}{3}$. Well below the curve, (15.18) will be valid and we may assume the pressure to be given by (15.15). On the other hand, if we are well above the curve then we may assume the validity of the perfect gas law. In the vicinity of the curve, the equation of state is somewhat complicated but as mentioned earlier, in most stars the state of the matter would correspond to either well above or well below the curve.

At higher densities the maximum kinetic energy of the electron becomes comparable to its rest mass energy ($= mc^2$) and the equation of state has to be derived using relativistic formulae (see Section 6.3). We derive below the condition indicating when relativistic formulae should be used. In the non-relativistic theory, the maximum kinetic energy is simply the Fermi energy and is given by (see Section 6.2)

$$\frac{P_{max}^{\,2}}{2m} = \varepsilon_{F_0} = (3\pi^2)^{2/3} \frac{\hbar^2}{2m} n_e^{\,2/3} \qquad\qquad (15.23)$$

Fig. 15.1. Well above the line the equation of state is accurately described by the perfect gas law. Well below the line the equation of state is approximately that of a completely degenerate electron gas. (Figure adapted from Aller (1953) who had adapted from a diagram by G. Wares, *Astrophys. J.* **100**, 159, 1944.)

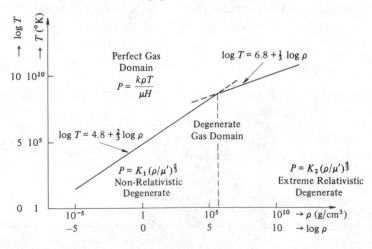

Thus for the non-relativistic theory to be valid we must have

$$(3\pi^2)^{2/3} \frac{h^2}{2m} n_e^{2/3} \ll mc^2 \tag{15.24}$$

If we now use (15.12) and simplify we would obtain

$$\rho \ll \frac{16\sqrt{2\pi}}{3} H\mu' \left[\frac{mc}{h}\right]^3 \tag{15.25}$$

The quantity h/mc is the Compton wavelength of the electron and is given by

$$\frac{h}{mc} \approx 2.43 \times 10^{-10}\,\text{cm} \tag{15.26}$$

If we now substitute it in (15.25), we obtain the following condition for the validity of the non-relativistic theory:

$$\rho \ll 3 \times 10^6\,\text{g cm}^{-3} \quad (\text{for } \mu \approx 1) \tag{15.27}$$

$$\rho \ll 6 \times 10^6\,\text{g cm}^{-3} \quad (\text{for } \mu' \approx 2) \tag{15.28}$$

Indeed inside some of the white dwarf stars (see Section 15.11) the densities are high enough to require the use of the proper relativistic formulae. In Section 6.3 we showed that for a completely degenerate electron gas, if we use proper relativistic formulae, the expressions for the pressure P and the total internal energy per unit volume $(=U/V)$ would be given by

$$P = A f(x) \tag{15.29}$$

$$\frac{U}{V} = A g(x) \tag{15.30}$$

where

$$A = \frac{m^4 c^5}{24\pi^2 h^3} \approx 6.02 \times 10^{22}\,\text{c.g.s. units} \tag{15.31}$$

$$x = \frac{p_{F0}}{mc} = \frac{h}{mc} (3\pi^2 n_e)^{1/3} \approx 1.195 \times 10^{-10} n_e^{1/3}\,\text{c.g.s. units} \tag{15.32}$$

$$f(x) = x(2x^2 - 3)(x^2 + 1)^{1/2} + 3\sinh^{-1} x \tag{15.33}$$

and

$$g(x) = 8x^3 [(x^2 + 1)^{1/2} - 1] - f(x) \tag{15.34}$$

Using (15.12) we obtain, for the density

$$\rho = B x^3 \tag{15.35}$$

where

$$B = \frac{1}{3\pi^2}\left(\frac{mc}{\hbar}\right)^3 \mu'H \approx 9.81 \times 10^5 \mu' \tag{15.36}$$

and *all* numerical values are in c.g.s. units.

In the non-relativistic limit, $x \ll 1$ and we have the following asymptotic forms (see Section 6.3)

$$\left.\begin{array}{l} f(x) \to \frac{8}{5}x^5 \\ g(x) \to \frac{12}{5}x^5 \end{array}\right\} (x \ll 1) \tag{15.37}$$

using which we readily obtain (15.15) and

$$\frac{U}{V} = \frac{3}{2}P \tag{15.38}$$

On the other hand, in the extreme relativistic limit, $x \gg 1$ and the asymptotic forms are

$$\left.\begin{array}{l} f(x) \to 2x^4 \\ g(x) \to 6x^4 \end{array}\right\} x \gg 1 \tag{15.39}$$

using which we obtain

$$\left.\begin{array}{l} \dfrac{U}{V} = 3P \end{array}\right. \tag{15.40}$$

and

$$\left.\begin{array}{c} \\ P = \dfrac{(3\pi^2)^{2/3}}{4} \hbar c n_e^{4/3} \end{array}\right\} x \gg 1 \tag{15.41}$$

If we now use (15.12) we get

$$P = K_2 \left(\frac{\rho}{\mu'}\right)^{4/3} \quad (x \gg 1) \tag{15.42}$$

where

$$K_2 = \frac{(3\pi^2)^{2/3}}{4} \frac{\hbar c}{H^{4/3}} \mu'^{4/3} \approx 1.23 \times 10^{15} \mu'^{4/3} \text{ c.g.s. units} \tag{15.43}$$

Thus in the extreme relativistic limit (i.e., when the densities are very high) stellar matter can be assumed to form a degenerate gas if the pressure given by (15.42) is much greater than that given by (15.8), i.e., when

$$K_2 \left(\frac{\rho}{\mu'}\right)^{4/3} \gg \rho \left(\frac{k}{\mu H}\right) T \tag{15.44}$$

and conversely. Rearranging we get

$$\rho \gg \frac{\mu'^4}{\mu^3} \left(\frac{k}{HK_2}\right)^3 T^3 \tag{15.45a}$$

or,

$$\rho \gg 3 \times 10^{-22} \frac{\mu'^4}{\mu^3} T^3 \qquad (15.45b)$$

where ρ is measured in g cm^{-3}. Once again assuming $X = 1$ and $Y = 0$ (i.e., the only element present in the star is hydrogen) we get

$$\rho \gg (2.4 \times 10^{-21}) T^3 \qquad (15.46)$$

In Fig. 15.1 the second section of the curve is the straight line defined by the equation

$$\log T = 6.8 + \tfrac{1}{3} \log \rho \qquad (15.47)$$

which implies a slope of $\tfrac{1}{3}$.

15.1.3 *The radiation pressure*

We end this section by noting that, in addition to the gas pressure, we also have the pressure due to radiation which plays an important role in some of the stars. The radiation pressure is given by (see (5.19)).

$$P = \tfrac{1}{3} u = \tfrac{1}{3} a T^4 \qquad (15.48)$$

where u represents the radiation energy per unit volume and

$$a = 7.57 \times 10^{-15} \, \text{erg cm}^{-3} \, \text{K}^{-4} \qquad (15.49)$$

represents the radiation pressure constant.

15.2 The equation of hydrostatic equilibrium

We assume that a star is in hydrostatic equilibrium under its own gravitation, i.e. the inward directed gravitational force exactly balances the outward directed force due to the gas and radiation pressure. Since a star changes very slowly during most of its life, the assumption of hydrostatic

Fig. 15.2. A star of radius R. The quantity $dM(r)$ represents the mass of the spherical shell whose inner and outer radii are r and $r + dr$ respectively.

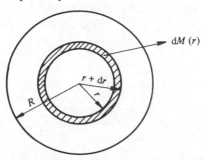

equilibrium is well justified. We will later show that using the basic equation of hydrostatic equilibrium and the equation of state, it is possible to obtain qualitative estimates of the pressure and temperature inside a star.

We assume a spherically symmetric distribution of matter inside the star so that the density $\rho(r)$ depends only on the spherical coordinate r (see Fig. 15.2). Let $dM(r)$ represent the mass of the spherical shell whose inner and outer radii are r and $r + dr$ respectively. Since the volume of the spherical shell is $4\pi r^2 dr$ we have

$$dM(r) = 4\pi r^2 \rho(r)dr \tag{15.50}$$

or

$$\frac{dM(r)}{dr} = 4\pi r^2 \rho(r) \tag{15.51}$$

Thus, if $M(r)$ represents the mass enclosed inside a sphere of radius r then

$$M(r) = \int dM(r) = \int_0^r 4\pi r^2 \rho(r)dr \tag{15.52}$$

Obviously

$$M(0) = 0 \quad \text{and} \quad M(R) = M \tag{15.53}$$

where R represents the radius of the star and M represents its total mass. It should be pointed out that since $M(r)$ is quite different from M, it is extremely important to write the r in parenthesis in writing $M(r)$.

We next consider the gravitational force acting on the spherical shell shown in Fig. 15.2. Obviously, the mass lying outside the shell does not contribute and the (magnitude of the) force due to the sphere of radius r would be given by

$$\frac{GM(r)dM(r)}{r^2} \tag{15.54}$$

where

$$G \approx 6.672 \times 10^{-8} \, \text{cm}^3 \, \text{s}^{-2} \, \text{g}^{-1} \tag{15.55}$$

represents the gravitational constant. The gravitational force will be directed inwards. Let $dP(r)$ represent the increment in pressure over the distance dr; then for the star to be in (hydrostatic) equilibrium we must have[†]

[†] We may mention here that even a small imbalance in the hydrostatic equilibrium will cause catastrophe. Thus a 1% imbalance in the two forces would lead to a 1% change in the solar radius in about 15 minutes. We can easily show this by the following argument: consider the mass $dM(r)$ shown in Fig. 15.2; if a fraction f of the gravitational force is left unbalanced then

$$-4\pi r^2 \mathrm{d}P(r) = \frac{GM(r)\mathrm{d}M(r)}{r^2} \tag{15.56}$$

or,

$$\frac{\mathrm{d}P}{\mathrm{d}r} = -\frac{GM(r)}{r^2}\rho(r) \tag{15.57}$$

where we have used (15.50). Simple manipulation gives

$$\frac{1}{r^2}\frac{\mathrm{d}}{\mathrm{d}r}\left(\frac{r^2}{\rho(r)}\frac{\mathrm{d}P(r)}{\mathrm{d}r}\right) = -\frac{G}{r^2}\frac{\mathrm{d}M(r)}{\mathrm{d}r} = -4\pi G\rho(r) \tag{15.58}$$

where, in the last step, we have used (15.51). The above equation is one of the basic equations in understanding stellar structure and we will be using it throughout the chapter.

15.3 Expressions for pressure and temperature inside a star

In this section we will use (15.50)–(15.58) to derive expressions for pressure and temperature inside a star. We rewrite (15.56) in the form

$$-\frac{\mathrm{d}P}{\mathrm{d}r} = \frac{G}{4\pi r^4}M(r)\frac{\mathrm{d}M(r)}{\mathrm{d}r} \tag{15.59}$$

the mass $\mathrm{d}M(r)$ will be accelerated by $\mathrm{d}^2 r/\mathrm{d}t^2$ where

$$\frac{\mathrm{d}^2 r}{\mathrm{d}t^2} = f\frac{GM(r)}{r^2}$$

Let r move to $r + \alpha r$ in time Δt, then assuming uniform acceleration

$$\alpha r = \frac{1}{2}\frac{\mathrm{d}^2 r}{\mathrm{d}t^2}(\Delta t)^2$$

or

$$\Delta t = \left[\frac{2\alpha r^3}{fGM(r)}\right]^{1/2}$$

If we consider midway between the center and the surface then $r = R/2$ and roughly assuming $M(R/2) \approx M/2$ we obtain

$$\Delta t \approx \left[\frac{\alpha R^3}{2fGM}\right]^{1/2}$$

If we apply the above formula to the Sun for which $R = R_\odot \approx 7 \times 10^{10}\,\mathrm{cm}$, $M = M_\odot \approx 2 \times 10^{33}\,\mathrm{g}$ then for $f \approx 0.01$ and $\alpha \approx 0.01$ we obtain $\Delta t \approx 10^3\,\mathrm{s} \approx 15$ minutes implying that a 1% departure from hydrostatic equilibrium would result in a 1% change in the solar radius in 15 min. Notice that for $f \approx 0.01$ and $\alpha \approx 0.1$, $\Delta t \approx 1$ hr i.e., in 1 hour the solar radius would change by 10%!

and integrate from $r = 0$ to $r = R$ to obtain

$$-\int_0^R \frac{dP(r)}{dr}\,dr = -[P(R) - P_c] = \frac{G}{4\pi}\int_0^R \frac{M(r)}{r^4}\frac{dM(r)}{dr}\,dr$$

where

$$P_c = P(r = 0) \tag{15.60}$$

represents the pressure at the center of the star. Since the pressure at the surface of the star is zero, i.e.,

$$P(R) = 0 \tag{15.61}$$

we obtain

$$P_c = \frac{G}{4\pi}\int_0^R \frac{1}{r^4}M(r)\frac{dM(r)}{dr}\,dr \tag{15.62}$$

Let

$$\bar{\rho}(r) \equiv \frac{M(r)}{(4\pi/3)r^3} \tag{15.63}$$

represent the mean density within r. Thus we may write

$$r = \left(\frac{3}{4\pi}\right)^{1/3}\left[\frac{M(r)}{\bar{\rho}(r)}\right]^{1/3} \tag{15.64}$$

and if we use the above equation in (15.62) we would obtain

$$P_c = \frac{G}{4\pi}\left(\frac{4\pi}{3}\right)^{4/3}\int_0^R [\bar{\rho}(r)]^{4/3}M^{-1/3}(r)\frac{dM(r)}{dr}\,dr \tag{15.65}$$

We next derive an expression for the mean temperature of the star. Now, if we divide a mass into a large number of parts of mass $dm_1, dm_2, \ldots$ having temperatures $T_1, T_2, \ldots$ then the mean temperature of the body $\bar{T}$ is defined as

$$\bar{T} = \frac{\sum_i T_i dm_i}{\sum_i dm_i} \tag{15.66}$$

Thus for a continuously distributed temperature, we have

$$\bar{T} = \frac{1}{M}\int T(r)\,dM(r) \tag{15.67}$$

In order to evaluate the above integral we need to know the equation of state. We assume the radiation pressure (see 15.48) to be negligible and the gas kinetic pressure to be given by the perfect gas law (see 15.8):

$$P(r) = \frac{k}{\mu H} \rho(r) T \tag{15.68}$$

Thus

$$\bar{T} = \frac{1}{M} \frac{\mu H}{k} \int_0^R \frac{P(r)}{\rho(r)} \frac{dM(r)}{dr} dr \tag{15.69}$$

or

$$\bar{T} = \frac{4\pi\mu H}{Mk} \int_0^R P(r) r^2 dr \tag{15.70}$$

where in the last step we have used (15.51). We integrate by parts to obtain:

$$\bar{T} = \frac{4\pi\mu H}{Mk} \left[\tfrac{1}{3} r^3 P(r) \Big|_0^R - \frac{1}{3} \int_0^R r^3 \frac{dP(r)}{dr} dr \right]$$

The first term inside the square brackets vanishes at both limits. If we now use (15.59) we would get

$$\bar{T} = \frac{\mu H G}{3Mk} \int_0^R \frac{1}{r} M(r) \frac{dM(r)}{dr} dr \tag{15.71}$$

We now use (15.64) to obtain

$$\bar{T} = \frac{\mu H G}{3Mk} \left(\frac{4\pi}{3} \right)^{1/3} \int_0^R [\bar{\rho}(r)]^{1/3} [M(r)]^{2/3} \frac{dM(r)}{dr} dr \tag{15.72}$$

We also define the mean pressure from the following equation (cf. (15.67)):

$$\bar{P} = \frac{1}{M} \int P(r) dM(r) = \frac{1}{M} \int_0^R P(r) \frac{dM(r)}{dr} dr \tag{15.73}$$

We integrate by parts to obtain

$$\bar{P} = \frac{1}{M} \left[P(r) M(r) \Big|_0^R - \int_0^R \frac{dP(r)}{dr} M(r) dr \right]$$

$$= \frac{1}{M} \int_0^R \frac{G}{4\pi r^4} M(r) \frac{dM(r)}{dr} M(r) dr$$

$$= \frac{G}{4\pi M} \left(\frac{4\pi}{3} \right)^{4/3} \int_0^R [\bar{\rho}(r)]^{4/3} M^{2/3}(r) \frac{dM(r)}{dr} dr$$

where in the last two steps we have used (15.59) and (15.63). We rewrite the expressions for $P_c, \bar{P}$ and $\bar{T}$:

$$P_c = \frac{G}{4\pi} \left(\frac{4\pi}{3} \right)^{4/3} \int_0^R [\bar{\rho}(r)]^{4/3} M^{-1/3}(r) \frac{dM(r)}{dr} dr \tag{15.74}$$

$$\bar{P} = \frac{G}{4\pi M} \left(\frac{4\pi}{3}\right)^{4/3} \int_0^R [\bar{\rho}(r)]^{4/3} M^{2/3}(r) \frac{dM(r)}{dr} dr \qquad (15.75)$$

and

$$\bar{T} = \frac{\mu H G}{3Mk} \left(\frac{4\pi}{3}\right)^{1/3} \int_0^R [\bar{\rho}(r)]^{1/3} M^{2/3}(r) \frac{dM(r)}{dr} dr \qquad (15.76)$$

where the expression for $\bar{T}$ assumes the perfect gas law.

15.4 Numerical estimates of P_c, $\bar{P}$ and $\bar{T}$ by assuming uniform density inside the star

In order to obtain an order of magnitude estimate of the pressure and temperature inside the star we assume that $\bar{\rho}(r)$ is independent of r and is given by

$$\bar{\rho}(r) = \bar{\rho} = \frac{M}{(4\pi/3)R^3} \qquad (15.77)$$

In each of the three equations (15.74)–(15.76), $\bar{\rho}(r)$ can therefore be taken outside the integral and in each case we are left with an integral of the form

$$\int_0^R [M(r)]^p \frac{dM(r)}{dr} dr$$

which can easily be evaluated:

$$\int_0^R [M(r)]^p \frac{dM(r)}{dr} dr = \frac{1}{(p+1)} [M(r)]^{p+1} \Big|_0^R = \frac{M^{p+1}}{(p+1)} \qquad (15.78)$$

Equations (15.72)–(15.74) thus become

$$P_c \approx \frac{3}{8\pi} \frac{GM^2}{R^4} \approx 1.33 \times 10^9 \left(\frac{M}{M_\odot}\right)^2 \left(\frac{R_\odot}{R}\right)^4 \text{ atmospheres} \qquad (15.79)$$

$$\bar{P} \approx \frac{3}{20\pi} \frac{GM^2}{R^4} \approx 5.30 \times 10^8 \left(\frac{M}{M_\odot}\right)^2 \left(\frac{R_\odot}{R}\right)^4 \text{ atmospheres} \qquad (15.80)$$

and

$$\bar{T} = \frac{1}{5} \frac{\mu H}{k} \frac{GM}{R} \approx 4.61 \times 10^6 \mu \left(\frac{M}{M_\odot}\right) \left(\frac{R_\odot}{R}\right) K \qquad (15.81)$$

where $R_\odot (\approx 6.960 \times 10^{10} \text{ cm})$ and $M_\odot (\approx 1.989 \times 10^{33} \text{ g})$ represent the radius and mass of the Sun. Thus, inside the Sun, the average temperature is about 10^7K. It is because of such high temperatures that nuclear reactions can be sustained inside the Sun. We may mention here that although (15.81) does give the right order for $\bar{T}$, the value of P_c as given by (15.79) is much smaller than that obtained from more sophisticated models. This is because of the neglect of density variation in deriving (15.79) (see Section 15.10 for a better model describing the Sun).

15.5   Some useful theorems

If we do not assume uniform density inside the star (i.e., 15.75) it is possible to derive some important theorems which can be used to obtain upper and lower bounds for pressure and temperature inside the star.

We refer to (15.62) and note that in the domain of integration $r < R$ and therefore we get the following lower bound for P_c:

$$P_c > \frac{G}{4\pi R^4} \int_0^R M(r)\frac{dM(r)}{dr}dr$$

or

$$P_c > \frac{G}{8\pi}\frac{M^2}{R^4} \tag{15.82}$$

which may be rewritten in the form

$$P_c > 0.442 \times 10^9 \left(\frac{M}{M_\odot}\right)^2\left(\frac{R_\odot}{R}\right)^4 \text{ atmospheres} \tag{15.83}$$

We can get a more stringent inequality if we make the realistic assumption that the density does not increase outside, i.e.,

$$\bar{\rho} \leqslant \bar{\rho}(r) \leqslant \rho_c \tag{15.84}$$

where ρ_c represents the density at the center of the star, $\bar{\rho}(r)$ represents the mean density within r (see 15.63) and $\bar{\rho}$ represents the mean density of the star (see 15.77). Using (15.77) and (15.84) we can write

$$\bar{\rho} \geqslant 1.41\left(\frac{M}{M_\odot}\right)\left(\frac{R_\odot}{R}\right)^3 \text{ g cm}^{-3} \tag{15.85}$$

Thus for Sun $\rho_c \geqslant 1.41 \text{ g cm}^{-3}$ and for Sirius B $\rho_c \geqslant 1.49 \text{ g cm}^{-3}$.

We use (15.84) in (15.74) to obtain

$$\frac{G}{4\pi}\left(\frac{4\pi}{3}\right)^{4/3}\bar{\rho}^{4/3}\tfrac{3}{2}M^{2/3} \leqslant P_c \leqslant \frac{G}{4\pi}\left(\frac{4\pi}{3}\right)^{4/3}\rho_c^{4/3}\tfrac{3}{2}M^{2/3}$$

or

$$\frac{3}{8\pi}\frac{GM^2}{R^4} \leqslant P_c \leqslant \tfrac{1}{2}G\left(\frac{4\pi}{3}\right)^{1/3}\rho_c^{4/3}M^{2/3} \tag{15.86}$$

where in the last step we have used (15.77). Thus we get a more stringent inequality (cf. 15.83)

$$P_c \geqslant \frac{3}{8\pi}\frac{GM^2}{R^4} \approx 1.326 \times 10^9 \left(\frac{M}{M_\odot}\right)^2\left(\frac{R_\odot}{R}\right)^4 \text{ atmospheres} \tag{15.87}$$

the equality sign holding for a homogeneous star. Similarly, we have the following inequalities for the average pressure and average temperature:

$$\bar{P} \geqslant \frac{3}{20\pi} \frac{GM^2}{R^4} \approx 5.4 \times 10^8 \left(\frac{M}{M_\odot}\right)^2 \left(\frac{R_\odot}{R}\right)^4 \text{ atmospheres} \quad (15.88)$$

$$\bar{T} \geqslant \frac{1}{5} \frac{\mu H}{k} \frac{GM}{R} \approx 4.61 \times 10^6 \mu \left(\frac{M}{M_\odot}\right) \left(\frac{R_\odot}{R}\right) K \quad (15.89)$$

Once again the equality sign holds for a homogeneous star (see (15.79)–(15.81)).

15.6 The gravitational potential energy and the virial theorem
In this section we will first calculate the gravitational potential energy associated with a star and will also derive the virial theorem and show how it can be used to understand the evolution of a star.

15.6.1 The gravitational potential energy
We will assume the zero of the potential energy of the star to correspond to the state when the entire mass of the star is diffused at infinity. We consider the gradual building up of the star and assume that we have already built up to radius r. The additional work done to increase the radius by dr would be given by

$$- GM(r)dM(r) \int_r^\infty \frac{d\xi}{\xi^2} = - \frac{GM(r)dM(r)}{r}$$

Thus the total gravitational potential energy of the star is given by

$$\Omega = - \int_0^R \frac{GM(r)dM(r)}{r} \quad (15.90)$$

or

$$\Omega = - 4\pi G \int_0^R rM(r)\rho(r)dr \quad (15.91)$$

where we have used (15.50). If we now use (15.57) we would obtain

$$\Omega = 4\pi \int_0^R \frac{dP(r)}{dr} r^3 dr \quad (15.92)$$

We integrate by parts to get

$$\Omega = 4\pi \left[P(r)r^3 \Big|_0^R - 3 \int_0^r r^2 P(r)dr \right]$$

$$= - 3 \int P dV \quad (15.93)$$

where we have used (15.61) and $dV = 4\pi r^2 dr$ represents the volume element. Now, if the pressure is entirely due to thermal motions then kinetic theory tells us that

$$P = \tfrac{2}{3}\varepsilon_{\text{kin}} \tag{15.94}$$

where ε_{kin} represents the translational kinetic energy per unit volume. Substituting in (15.93) we get

$$\Omega = -2K \tag{15.95}$$

where K represents the total kinetic energy of the star. We will also derive the above equation from the virial theorem and then discuss its consequences; however, before doing so we may mention that for a star of uniform density

$$\rho(r) = \bar\rho = \frac{M}{(4\pi/3)R^3} \tag{15.96}$$

and (15.91) becomes

$$\Omega = -4\pi G \int_0^R r\left[\bar\rho\frac{4\pi}{3}r^3\right]\bar\rho dr$$

$$= -\frac{3}{5}\frac{GM^2}{R} \tag{15.97}$$

Although the above equation is strictly valid only for a star of uniform density, it does give the correct order for Ω for an actual star (cf. (15.202)).

15.6.2 The virial theorem[†]

We consider a system of N particles. Let (x_i, y_i, z_i) represent the (instantaneous) Cartesian coordinates of the ith particle ($i = 1, 2, 3, \ldots, N$) and let X_i, Y_i and Z_i represent the Cartesian components of the force ($\mathbf{F}_i$) acting on it. Thus the equations of motion for the ith particle are

$$m_i\frac{d^2x_i}{dt^2} = X_i \tag{15.98}$$

$$m_i\frac{d^2y_i}{dt^2} = Y_i \tag{15.99}$$

and

$$m_i\frac{d^2z_i}{dt^2} = Z_i \tag{15.100}$$

[†] The theorem was first given by Poincaré (see, e.g., Jeans 1928).

Now

$$\frac{1}{2}\frac{d^2}{dt^2}(m_i x_i^{\ 2}) = \frac{d}{dt}\left(m_i x_i \frac{dx_i}{dt}\right)$$

$$= m_i\left(\frac{dx_i}{dt}\right)^2 + x_i X_i \tag{15.101}$$

where we have used (15.98). Thus

$$\frac{1}{2}\frac{d^2}{dt^2}\sum_i\left[m_i(x_i^{\ 2} + y_i^{\ 2} + z_i^{\ 2})\right]$$

$$= 2\sum_i \tfrac{1}{2} m_i\left[\left(\frac{dx_i}{dt}\right)^2 + \left(\frac{dy_i}{dt}\right)^2 + \left(\frac{dz_i}{dt}\right)^2\right]$$

$$+ \sum_i\left[x_i X_i + y_i Y_i + z_i Z_i\right]$$

or

$$\frac{1}{2}\frac{d^2 I}{dt^2} = 2K + \sum_i \mathbf{F}_i \cdot \mathbf{r}_i \tag{15.102}$$

where

$$K = \sum_i \tfrac{1}{2} m_i\left[\left(\frac{dx_i}{dt}\right)^2 + \left(\frac{dy_i}{dt}\right)^2 + \left(\frac{dz_i}{dt}\right)^2\right] \tag{15.103}$$

represents the total kinetic energy of the system and

$$I = \sum_i m_i(x_i^{\ 2} + y_i^{\ 2} + z_i^{\ 2}) = \sum_i m_i r_i^{\ 2} \tag{15.104}$$

represents the moment of inertia of the system about the origin of the coordinate system. For a system in a steady state[†], we may set $d^2 I/dt^2$ equal to zero so that (15.102) takes the form

$$2K + W = 0 \tag{15.105}$$

where

$$W = \sum_i \mathbf{F}_i \cdot \mathbf{r}_i \tag{15.106}$$

is known as the *virial* of the system. Equation (15.105) is usually referred to as the *virial theorem*.[‡] If we assume that only gravitational forces are acting

---

[†] For quasi steady state conditions (e.g., a star with turbulent motions) we may take a time average of (15.102) and set the left hand side equal to zero. Thus (15.105) will still be valid provided the quantities are taken as appropriate time averages (see, Goldstein 1950, Section 3.4).

[‡] The relativistic generalization of the virial theorem has been discussed by Cox and Giuli (1968a) Section 17.2.

on particles then

$$\mathbf{F}_i = -\sum_{\substack{j \\ (j \neq i)}} G \frac{m_i m_j}{|\mathbf{r}_i - \mathbf{r}_j|^3} (\mathbf{r}_i - \mathbf{r}_j) \qquad (15.107)$$

where the summation is over all possible values of j excepting $j = i$. Thus

$$W = \sum_i \mathbf{F}_i \cdot \mathbf{r}_i = -G \sum_i \sum_{\substack{j \\ j \neq i}} \frac{m_i m_j}{|\mathbf{r}_i - \mathbf{r}_j|^3} \mathbf{r}_i \cdot (\mathbf{r}_i - \mathbf{r}_j) \qquad (15.108)$$

In the above summation if we interchange the indices i and j we would get

$$W = -G \sum_j \sum_{\substack{i \\ i \neq j}} \frac{m_i m_j}{|\mathbf{r}_i - \mathbf{r}_j|^3} \mathbf{r}_j \cdot (\mathbf{r}_j - \mathbf{r}_i) \qquad (15.109)$$

Thus we may write

$$W = -\frac{G}{2} \sum_i \sum_{\substack{j \\ j \neq i}} \frac{m_i m_j}{|\mathbf{r}_i - \mathbf{r}_j|^3} (\mathbf{r}_i - \mathbf{r}_j) \cdot (\mathbf{r}_i - \mathbf{r}_j)$$

$$= -\frac{G}{2} \sum_i \sum_{\substack{j \\ j \neq i}} \frac{m_i m_j}{|\mathbf{r}_i - \mathbf{r}_j|} \qquad (15.110)$$

The above expression is simply Ω, where Ω represents the total gravitational potential energy of the system; this follows from the fact that the potential energy of the particle of mass m_i due to m_j is

$$\frac{G m_i m_j}{|\mathbf{r}_i - \mathbf{r}_j|}$$

and the factor $(\frac{1}{2})$ in (15.110) takes care of the fact that in carrying out the double summation, each pair has been counted twice. Thus the virial is given by[†]

$$W = \Omega \qquad (15.111)$$

and (105) becomes[‡]

[†] The gravitational force obeys the inverse square law. For forces varying as r^v, $W = -(v + 1)\Omega$ (see, e.g., Goldstein 1950, Section 3.4); thus, for $v = -2$ we immediately get (15.111).

[‡] We must point out that (15.112) is consistent with (9.103). This is due to the fact that the pressure at the boundary of the star is zero. Indeed, if we consider a perfect gas consisting of non-interacting particles (enclosed in a spherical chamber) then the force is due to the surface only and it can easily be shown that $W = -3PV$ (see Section 9.4). If we substitute this in (15.105) and use the fact that for a perfect gas $K = \frac{3}{2} NkT$, we would immediately get the perfect gas law $PV = NkT$.

$$2K + \Omega = 0 \tag{15.112}$$

Now, if we consider a small mass dm of a perfect gas (consisting of dN particles) then the kinetic energy associated with it would be given by

$$dK = \tfrac{3}{2}kTdN = \tfrac{3}{2}RT\frac{dm}{\mu} \tag{15.113}$$

where μ represents the mean molecular weight (see 15.9). Since $c_p - c_v = R$, we may write

$$dK = \tfrac{3}{2}(\gamma - 1)\frac{c_v T}{\mu}\,dm \tag{15.114}$$

where $\gamma = c_p/c_v$ represents the ratio of specific heats. Now, $c_v T/\mu$ represents the internal energy per unit mass; thus

$$dK = \tfrac{3}{2}(\gamma - 1)dU \tag{15.115}$$

For the whole system we would have

$$K = \tfrac{3}{2}(\gamma - 1)U \tag{15.116}$$

It may be noted that for an ideal monatomic gas, $\gamma = \tfrac{5}{3}$, and $K = U$ (obviously!).

Substituting (15.116) in (15.112) we would have

$$U = -\frac{1}{3(\gamma - 1)}\Omega \tag{15.117}$$

The total energy of the system would therefore be given by

$$E = U + \Omega = \frac{\gamma - \tfrac{4}{3}}{\gamma - 1}\Omega \tag{15.118}$$

The above equation shows that for $\gamma > \tfrac{4}{3}$, E is negative and therefore the star is stable. We must remember that Ω is a negative quantity (see (15.97)).

It is interesting to note that since $\Omega \approx -GM^2/R$ (see (15.97)), as the star contracts Ω becomes more negative and the change in total energy would be

$$\Delta E = \frac{\gamma - \tfrac{4}{3}}{\gamma - 1}\Delta\Omega \tag{15.119}$$

Since $\Delta\Omega$ is negative, ΔE would also be negative and the amount $-\Delta E$ is radiated away. Further, (15.117) tells us that

$$\Delta U = -\frac{1}{3(\gamma - 1)}\Delta\Omega \tag{15.120}$$

is positive, which is responsible for the increase in temperature of the star. Thus, as the star contracts, its internal energy increases and it gets hotter and at the same time it radiates energy! Physically, this is due to the fact that the loss in the potential energy is *greater* than the increase in its internal energy and the difference is radiated away. For example, for $\gamma = \frac{5}{3}$

$$\Delta U = -\tfrac{1}{2}\Delta\Omega \tag{15.121}$$

i.e., only half of the decrease in the potential energy is used up in increasing its internal energy.

15.7 Qualitative understanding of the evolution of a star

Overall energy conservation and hydrostatic equilibrium determine the evolution of a star. We start with the relation which gives the dependence of the average temperature on the mass and radius of the star (see 15.81)

$$\bar{T} \approx \frac{1}{5}\frac{\mu H}{k}\frac{GM}{R} \approx 4.61 \times 10^{6}\,\mu\left(\frac{M}{M_{\odot}}\right)\left(\frac{R_{\odot}}{R}\right) K \tag{15.122}$$

Although the above equation is strictly valid only for a star of uniform density, it does give the correct order for $\bar{T}$ for an actual star. Of course, in deriving (15.122) we have neglected the radiation pressure, which is justified for the Sun and for stars whose mass is less then 3–$4\,M_{\odot}$ (see Section 15.8).

Now, during the formation of the star, R is large and therefore $\bar{T}$ is small. Because of gravitational attraction, the star contracts, R becomes smaller and (according to 15.122) the average temperature increases. The more it contracts, the hotter it gets. Eventually, the temperatures become high enough for thermonuclear reactions to occur, and in the steady state the gas (and the radiation) pressure support the star, the energy radiated out is supplied from thermonuclear reactions and no further gravitational contraction occurs. This steady state is of course temporary because when the nuclear fuel (capable of causing thermonuclear reactions) becomes depleted the star starts to contract again and the gravitational potential energy appears in the form of thermal energy. Thus the star will keep on contracting and thereby increasing its central density and temperature. This process of contraction cannot continue indefinitely because when the density of the star becomes high enough, the electrons supply the necessary pressure to support the star.

We may mention here that one of the important thermonuclear reactions is the transformation of four protons to two helium nuclei:

$$\begin{bmatrix} {}_1H^1 + {}_1H^1 \rightarrow {}_1H^2 + e^+ + \nu + 1.18\,\text{MeV} \\ {}_1H^2 + {}_1H^2 \rightarrow {}_2He^3 + \gamma + 5.49\,\text{MeV} \end{bmatrix} \times 2 \tag{15.123}$$

$$ {}_2He^3 + {}_2He^3 \rightarrow {}_2He^4 + {}_1H^1 + {}_1H^1 + 12.86\,\text{MeV}$$

The reaction thus releases about 26.2 MeV of energy. If we consider a mass of pure hydrogen at $T = 5 \times 10^6$ K having a density of 100 g/cm^{-3} then the rate of energy release can be shown to be about 0.2 erg g^{-1}s^{-1}. At $T = 15 \times 10^6$ K and 30×10^6 K the rate of energy release will be about 50 and 500 erg g^{-1}s^{-1} respectively (For further details, the reader may refer to Aller (1954), Reeves (1965) etc.; the last reference also gives details of many other types of thermonuclear reactions which occur inside the star).

Although the main source of energy in most stars is thermonuclear reaction, it is possible that the temperature inside the stars may not become high enough ($\gtrsim 10^7$ K) for thermonuclear reactions to occur. This is due to the fact that as the star undergoes contraction, the electron density may become high enough to form a degenerate gas. We will show below that when this happens the temperature attains a maximum value ($T_{\max}$) and then starts decreasing again. Only if $T_{\max}$ is high enough, will thermonuclear reactions occur, otherwise the star will (eventually) 'cool-off' without thermonuclear reactions.

We have seen in Chapter 6 that for the electron gas to be degenerate we must have

$$\frac{\varepsilon_{F_0}}{kT} \gtrsim 1 \tag{15.124}$$

or

$$(3\pi^2)^{2/3} \frac{\hbar^2}{2m} \frac{n^{2/3}}{kT} \gtrsim 1 \tag{15.125}$$

Now, the pressure due to the electron gas (assumed to be completely degenerate) is given by (see (6.29))

$$P \approx (3\pi^2)^{2/3} \frac{\hbar^2}{5m} n^{5/3} \tag{15.126}$$

which should also be taken into account in calculating the pressure (see (15.68)). For the sake of simplicity we assume only hydrogen to be present in the star so that the total pressure is given by[†]

$$P(r) \approx n_p(r)kT + (3\pi^2)^{2/3} \frac{\hbar^2}{5m} n_e^{5/3}(r) \tag{15.127}$$

[†] The first term on the right hand side of (15.127) is consistent with the right hand side of (15.68) if we consider only the proton gas.

where $n_p(r)$ and $n_e(r)$ represent the proton density and electron density respectively – we are assuming the protons form a non-degenerate gas and the electrons form a degenerate gas. Thus (cf. (15.73))

$$k\bar{T} \approx \frac{1}{M}\left[\int_0^R \frac{P(r)}{n_p(r)}\frac{dM(r)}{dr}\,dr - (3\pi^2)^{2/3}\frac{\hbar^2}{5m}\int_0^R \frac{n_e^{5/3}(r)}{n_p(r)}\frac{dM(r)}{dr}\,dr\right]$$

(15.128)

Since we are interested in an order of magnitude calculation, we assume the density to be constant (i.e., n_e and n_p to be independent of r) and if we now use the method used in Section 15.3 we would get

$$k\bar{T} \approx \frac{1}{5}\frac{H}{k}\frac{GM}{R} - (3\pi^2)^{2/3}\frac{\hbar^2}{5m}\frac{n_e^{5/3}}{n_p}$$

(15.129)

Let N represent the total number of protons present in the star which would be equal to the total number of electrons present. Thus

$$n_e \approx n_p \approx \frac{N}{(4\pi/3)R^3} \approx \frac{1}{l^3}$$

(15.130)

where

$$l \approx \left[\frac{(4\pi/3)R^3}{N}\right]^{1/3}$$

(15.131)

represents the average distance between protons (or electrons), alternatively, we may associate a volume l^3 with each proton (or electron). Thus

$$k\bar{T} \approx \frac{1}{5}\frac{HG(NH)}{(3N/4\pi)^{1/3}l} - \frac{(3\pi^2)^{2/3}}{5}\frac{\hbar^2}{ml^2}$$

(15.132)

where we have used the relation

$$M = NH$$

(15.133)

Equation (15.132) may be written in the form

$$k\bar{T} \approx 0.32\left(\frac{N}{N_0}\right)^{2/3}\frac{\hbar c}{l} - 1.91\frac{\hbar^2}{ml^2}$$

(15.134)

where

$$N_0 = \left[\frac{\hbar c}{GH^2}\right]^{3/2}$$

$$\approx \left[\frac{1.05 \times 10^{-27} \times 3 \times 10^{10}}{6.67 \times 10^{-8} \times (1.67 \times 10^{-24})^2}\right]^{3/2} \approx 10^{57}$$

(15.135)

A qualitative plot of $k\bar{T}$ with l is shown in Fig. 15.3. It is seen that as the star contracts (i.e., as l decreases from ∞), $k\bar{T}$ first increases, attains a maximum value and then decreases again. The maximum value of $k\bar{T}$ can easily be calculated by elementary differentiation of (15.134):

$$\frac{d(k\bar{T})}{dl} \approx -0.32\left(\frac{N}{N_0}\right)^{2/3}\frac{hc}{l^2} + \frac{3.82\,\hbar^2}{ml^3} \tag{15.136}$$

Setting the above expression to zero would give

$$l \approx l_0 \approx 12\frac{\hbar}{mc}\left(\frac{N}{N_0}\right)^{-2/3} \tag{15.137}$$

Substituting $l = l_0$ in (15.134) we get the maximum value of $k\bar{T}$:

$$(k\bar{T})_{max} \approx 0.013\,mc^2\left(\frac{N}{N_0}\right)^{4/3}$$

$$\approx 6.5 \times 10^3 \left(\frac{N}{N_0}\right)^{4/3} \text{(eV)} \tag{15.138}$$

where we have used the fact that $mc^2 \approx 0.5$ MeV. Now for thermonuclear reactions to occur $T \gtrsim 10^7$ K implying $k\bar{T} \gtrsim 0.8 \times 10^3$ eV. Thus for thermonuclear reactions to occur we must have

$$\frac{N}{N_0} \gtrsim \frac{1}{10} \tag{15.139}$$

It is interesting to note that the number of protons in the Sun, $N_\odot$, is approximately given by

$$N_\odot \approx \frac{M_\odot}{m_p} \approx \frac{2 \times 10^{33}\,\text{g}}{1.67 \times 10^{-24}\,\text{g}} \approx 10^{57} \tag{15.140}$$

Fig. 15.3. A qualitative plot of $k\bar{T}$ as a function of l which represents the average distance between two protons.

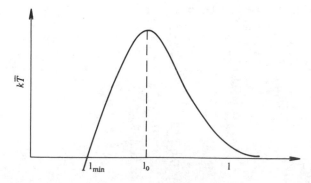

Thus $N_0 \approx N_\odot$ and therefore, according to (15.139) we have the remarkable result that there cannot be any star whose mass is less than (about) 1/10th of that of the Sun – which is indeed consistent with observations. Figure 15.3 also shows that the star will eventually contract to a minimum radius for which $T = 0$ and $l = l_{min} \simeq \frac{1}{2} l_0$. Since $l < l_{min}$ corresponds to negative temperatures, smaller radii are physically impossible. While the star is cooling from $T = T_{max}$ to $T = 0$ it radiates a lot in a relatively shorter interval of time. During this stage the stars are called white dwarfs (see Section 15.11).

15.8   The contribution due to radiation pressure

Until now we have neglected radiation pressure. In this section we will show that for most stars the radiation pressure makes a negligible contribution to the total pressure.

The radiation pressure is given by (5.19):

$$P_r = \tfrac{1}{3} a T^4 \tag{15.141}$$

where

$$a = \frac{\pi^2}{15} \frac{k^4}{(\hbar c)^3}$$

$$\approx 7.57 \times 10^{-15} \, \text{dyne cm}^{-2} \, \text{K}^{-4}$$

is known as the radiation pressure constant. As an example, we consider the centre of the Sun where $T \approx 2 \times 10^7 \, \text{K}$ so that

$$P_{rc} \approx 4 \times 10^{14} \, \text{dyne cm}^{-2}$$

$$\approx 4 \times 10^8 \, \text{atmospheres} \tag{15.142}$$

where the additional subscript c implies that we are referring to the *center* of the star. Now, the gas kinetic pressure is given by

$$P_g \approx \frac{k}{\mu H} \rho T \tag{15.143}$$

Thus

$$P_{gc} \approx \frac{k}{\mu H} \rho_c T_c$$

$$\approx \frac{1.38 \times 10^{-16}}{0.5 \times 1.67 \times 10^{-24}} \times 76 \times 2 \times 10^7 \approx 2.5 \times 10^{17} \, \text{dyne cm}^{-2}$$

$$= 2.5 \times 10^{11} \, \text{atmospheres} \tag{15.144}$$

where ρ_c represents the density at the center of the Sun which we have assumed to be $76 \, \text{g cm}^{-2}$ (value taken from Appendix III of Chandrasekhar (1939): see also Section 15.10). If we compare (15.142) and (15.144) we find that at the center of the Sun the radiation pressure is negligible in comparison to the gas kinetic pressure.

In general, it is customary to introduce a parameter β defined as the fraction of the total pressure which is due to the gas

$$\beta = \frac{P_g}{P} = \frac{P_g}{P_g + P_r} \tag{15.145}$$

where $P(= P_g + P_r)$ represents the total pressure. Thus

$$1 - \beta = \frac{P_r}{P_g + P_r} = \frac{P_r}{P} \tag{15.146}$$

represents the fraction of the total pressure that is due to radiation. Dividing (15.145) by (15.146) we obtain

$$P_g = \frac{\beta}{1 - \beta} P_r \tag{15.147}$$

In the above equation we substitute for P_g and P_r from (15.143) and (15.141) to obtain

$$\frac{k}{\mu H} \rho T = \frac{\beta}{1 - \beta} \tfrac{1}{3} a T^4 \tag{15.148}$$

or

$$T = \left[\frac{3k}{a\mu H} \frac{1 - \beta}{\beta} \right]^{1/3} \rho^{1/3} \tag{15.149}$$

Now

$$P = \frac{1}{\beta} P_g = \frac{1}{\beta} \frac{k}{\mu H} \rho T \tag{15.150}$$

or

$$P = \frac{3}{a} \left[\left(\frac{k}{\mu H} \right)^4 \frac{(1 - \beta)}{\beta^4} \right]^{1/3} \rho^{4/3} \tag{15.151}$$

Thus if P_c represents the total pressure at the center of the star then

$$P_c = \left[\frac{3}{a} \left(\frac{k}{\mu H} \right)^4 \frac{(1 - \beta_c)}{\beta_c{}^4} \right]^{1/3} \rho_c{}^{4/3} \tag{15.152}$$

Now, according to (15.86)

$$P_c \leqslant \frac{1}{2} \left(\frac{4\pi}{3} \right)^{1/3} G \rho_c{}^{4/3} M^{2/3} \tag{15.153}$$

Thus

$$\frac{1 - \beta_c}{\beta_c{}^4} \leqslant \frac{1}{8} \times \frac{4\pi}{3} \frac{a}{3} \left(\frac{\mu H}{k} \right)^4 G^3 M^2$$

$$\leqslant \frac{\pi}{18} \times 7.57 \times 10^{-15} \times \left[\frac{1.67 \times 10^{-24}}{1.38 \times 10^{-16}} \right]^4$$

$$\times (6.668 \times 10^{-8})^3 (1.99 \times 10^{33})^2 \times \mu^4 \left(\frac{M}{M_\odot} \right)^2$$

$$\leqslant 0.0332 \left[\mu^2 \frac{M}{M_\odot} \right]^2 \tag{15.154}$$

We define β^* such that

$$\frac{1 - \beta^*}{\beta^{*4}} \equiv 0.0332 \left[\mu^2 \frac{M}{M_\odot} \right]^2 \tag{15.155}$$

For a given value of $(1 - \beta^*)$ it is easy to calculate $\mu^2 M/M_\odot$ – see Table 15.3. Since the function $(1 - \beta)/\beta^4$ monotonically decreases as β increases from 0 to 1 we have

$$1 - \beta_c \leqslant 1 - \beta^* \tag{15.156}$$

As an example, we consider the Sun. Assuming $\mu \approx 0.5$ we get

$$\mu^2 \frac{M}{M_\odot} \approx 0.25 \tag{15.157}$$

Thus $(1 - \beta^*) \approx 0.002$ implying that only 0.2% of the total pressure is due to radiation which is consistent with (15.142) and (15.144). For Capella,

Table 15.3. *Values of $\mu^2(M/M_\odot)$ for different values of $(1 - \beta^*)$ using (15.155)*

$1 - \beta^*$	$\mu^2 \dfrac{M}{M_\odot}$
0.001	0.174
0.002	0.246
0.003	0.302
0.005	0.392
0.01	0.560
0.02	0.808
0.03	1.010
0.05	1.360
0.1	2.143
0.2	3.835
0.5	15.52
1.0	∞

$M \approx 4.2 M_\odot$ and assuming $\mu \approx 0.5$ we get

$$\mu^2 \frac{M}{M_\odot} \approx 1 \tag{15.158}$$

and $(1 - \beta^*) \approx 0.03$ implying that only 3% of the total pressure is due to radiation.[†]

Table 15.3 indicates that the radiation pressure becomes more and more important as the mass of the star increases. Indeed for $M/M_\odot \gtrsim 50$ the radiation pressure is so large that the star becomes unstable resulting in expansion and scattering itself in space. This also follows from the virial theorem because as the radiation pressure starts dominating, the effective ratio of specific heats, γ, tends to $\frac{4}{3}$ (see Section 5.3) and therefore the total energy (15.118) tends to zero.

15.9   The polytropic model

In this section we will make a study of the equilibrium configuration of a star with the assumption that the pressure–density relationship is given by

$$P(r) = K[\rho(r)]^{1 + 1/n} \tag{15.159}$$

The above equation describes what is known as the polytropic model, the parameters n (known as the polytropic index) and K are assumed to be constants throughout the star. Although the above relation is somewhat of an idealization, nevertheless the model does provide considerable insight into the structure of actual stars.[‡] For example, for a radially independent entropy, (15.159) does represent the correct equation of state for an ideal gas; $n = \frac{3}{2}$ ($\gamma = \frac{5}{3}$) corresponding to a monatomic gas, $n = \frac{5}{2}$ ($\gamma = \frac{7}{5}$) corresponding to a diatomic gas etc. Another important example is the so-called 'standard model' (introduced by Eddington) in which the quantities μ and $(1 - \beta)$ (see (15.151)) are assumed to be constants throughout the star; thus the pressure–density relationship (see (15.151)) is indeed of the form given by (15.159) with

$$n = 3 \quad \text{and} \quad K = \left[\frac{3}{a} \left(\frac{k}{\mu H} \right)^4 \frac{1 - \beta}{\beta^4} \right]^{1/3} \tag{15.160}$$

[†] We may mention here that the effect of degeneracy is to lower the value of $1 - \beta^*$, thus the radiation pressure is expected to have less effect in degenerate stars (see, for example, Cox and Giuli (1968a), Chapter 11).

[‡] For more sophisticated models of stellar configurations, we refer the reader to such books as Aller and McLaughlin (1965), Chandrasekhar (1939), Cox and Giuli (1968b), Menzel, Bhatnagar and Sen (1963), Schwarzschild (1958), etc.

The 'standard model' has played a very important role in the understanding of stellar structure. We may also mention that the pressure–density relationships corresponding to a completely degenerate electron gas in the non-relativistic and extreme relativistic limit (see (15.15) and (15.42)) are also of the form given by (15.159) with $n = \frac{3}{2}$ and $n = 3$ respectively.

In order to obtain the stellar configuration, we start with the condition of hydrostatic equilibrium (15.58)

$$\frac{1}{r^2}\frac{d}{dr}\left[\frac{r^2}{\rho(r)}\frac{dP(r)}{dr}\right] = -4\pi G\rho(r) \tag{15.161}$$

We introduce a dimensionless variable θ defined by the following equation

$$\rho(r) = \rho_c\theta^n(r) \tag{15.162}$$

where $\rho_c\ (=\rho(r=0))$ represents the density at the centre of the star. Using (15.162), (15.159) becomes

$$P(r) = K[\rho_c^{1+1/n}\theta^{n+1}] \tag{15.163}$$

Substituting (15.162) and (15.163) in (15.161) we get

$$\frac{1}{r^2}\frac{d}{dr}\left[\frac{r^2}{\rho_c\theta^n}K\rho_c^{1+1/n}(n+1)\theta^n\frac{d\theta(r)}{dr}\right] = -4\pi G\rho_c\theta^n(r)$$

or

$$\left[\frac{(n+1)K}{4\pi G}\rho_c^{1/n-1}\right]\frac{1}{r^2}\frac{d}{dr}\left[r^2\frac{d\theta(r)}{dr}\right] = -\theta^n(r) \tag{15.164}$$

In order to simplify the above equation, we introduce the dimensionless variable

$$\xi = r/\alpha \tag{15.165}$$

where we choose

$$\alpha = \left[\frac{(n+1)K}{4\pi G}\rho_c^{1/n-1}\right]^{1/2} \tag{15.166}$$

Thus (15.164) becomes

$$\frac{1}{\xi^2}\frac{d}{d\xi}\left[\xi^2\frac{d\theta(\xi)}{d\xi}\right] = -\theta^n(\xi) \tag{15.167}$$

which is known as the 'Lane–Emden equation of index n'. Using (15.162) and the fact that ρ_c represents the density at the centre of the star ($r = 0$), we get

$$\theta(\xi) = 1 \quad \text{at} \quad \xi = 0 \tag{15.168}$$

Furthermore, if we write (15.167) as

$$\xi\frac{d^2\theta}{d\xi^2} + 2\frac{d\theta}{d\xi} = -\xi\theta^n(\xi) \tag{15.169}$$

we get (for well behaved solutions)

$$\frac{d\theta}{d\xi} = 0 \quad \text{at} \quad \xi = 0 \tag{15.170}$$

Equations (15.168) and (15.170) represent the boundary conditions which should be satisfied by the solution of the Lane–Emden equation. Analytical solutions can be obtained for $n = 0, 3$ and 5. For example, for $n = 0$, (15.167) takes the form

$$\frac{d}{d\xi}\left[\xi^2\frac{d\theta_0(\xi)}{d\xi}\right] = -\xi^2$$

where the subscript zero implies that we are considering the case $n = 0$. Simple integration gives

$$\xi^2\frac{d\theta_0}{d\xi} = -\tfrac{1}{3}\xi^3 + C_1$$

where the constant of integration C_1 would vanish because of (15.170). Division by ξ^2 and subsequent integration gives

$$\theta_0(\xi) = 1 - \frac{\xi^2}{6} \tag{15.171}$$

where the boundary condition given by (15.168) has been incorporated. We may mention here that the case $n = 0$ corresponds to the uniform density model for which (cf. Section 15.4)[†]

$$\left.\begin{array}{l} \rho(r) = \rho_c \quad \text{(constant everywhere)}\\ P(r) = P_c\theta(r) \end{array}\right\} \tag{15.172}$$

For $n = 1$, we have

$$\frac{d}{d\xi}\left[\xi^2\frac{d\theta_1}{d\xi}\right] + \xi^2\theta_1(\xi) = 0 \tag{15.173}$$

The transformation $\theta_1(\xi) = u_1(\xi)/\xi$ readily gives

$$\frac{d^2u_1}{d\xi^2} + u_1(\xi) = 0 \tag{15.174}$$

[†] Notice that for $n = 0$, the right hand side of (15.159) becomes indeterminate.

the general solution of which is given by

$$u_1(\xi) = C_1 \sin \xi + C_2 \cos \xi$$

Thus

$$\theta_1(\xi) = C_1 \frac{\sin \xi}{\xi} + C_2 \frac{\cos \xi}{\xi}$$

The boundary condition given by (15.168) gives $C_2 = 0$ and $C_1 = 1$ giving

$$\theta_1(\xi) = \frac{\sin \xi}{\xi} \tag{15.175}$$

the other boundary condition being automatically satisfied. For $n = 5$ the procedure is rather involved (see, e.g., Chandrasekhar 1939, Chapter IV), the final result is

$$\theta_5(\xi) = [1 + \tfrac{1}{3}\xi^2]^{-1/2} \tag{15.176}$$

which can easily be verified by direct substitution. In general, for an arbitrary value of n, one has to carry out numerical integration of the differential equation. One may also develop a power series solution of the form (Hayashi, Hoshi and Sugimoto 1962)

$$\theta(\xi) = 1 - \tfrac{1}{6}\xi^2 + \frac{n}{120}\xi^4 - \frac{n(8n-5)}{42 \times 360}\xi^6 + \cdots \tag{15.177}$$

For $n = 1.5$ and $n = 3$, the function $\theta_n(\xi)$ (together with its derivatives) have been tabulated by Cox and Giuli (1968b) Appendix A5. For $n = 1$, 1.5, 2, 2.5, 3, 3.5, 4, 4.5 and 5 the function $\theta_n(\xi)$ together with its derivatives are tabulated in Mathematical Tables Vol. II (British Association for the Advancement of Science).

Once the solution of the Lane–Emden equation is known it is possible to determine some of the important properties of the star, these are discussed below.

(a) Radius

Let ξ_1 represent the first zero of $\theta_n(\xi)$. Since $\theta = 0$ implies $\rho = 0$ (see (15.162)), $\xi = \xi_1$ would correspond to the radius of the star. Thus

$$R = \alpha\xi_1 = \left[\frac{(n+1)K}{4\pi G}\right]^{1/2} \rho_c^{1-n/2n}\xi_1 \tag{15.178}$$

It can easily be seen from (15.172), (15.175) and (15.176) that for $n = 0$, 1 and 5, $\xi_1 = \sqrt{6}$, π and ∞ respectively. Values of ξ_1 for other values of n are given in Table 15.4. It may be noted that for $n = 1$, $\xi_1 = \pi$ and (15.178) becomes

$$R(n=1) = \left[\frac{K}{4\pi G} \right]^{1/2} \pi \qquad (15.179)$$

independent of ρ_c. The case $n = 5$ (for which $\xi_1 = \infty$) corresponds to an infinite radius of the star (although the total mass is finite – see (15.184)). For $n > 5$, $\theta(\xi)$ never attains zero value, thus they do not correspond to any stars. We should mention that the case $n = 0$ cannot be discussed as the right hand side of (15.178) becomes indeterminate.

(b) Mass

The mass of the star, enclosed inside a sphere of radius r, is given by (15.52)

$$M(r) = \int_0^r 4\pi r^2 \rho(r) \, dr$$

Thus

$$M(\xi) = 4\pi \alpha^3 \int_0^\xi \xi^2 \rho(\xi) \, d\xi$$

$$= -4\pi \alpha^3 \rho_c \int_0^\xi \xi^2 \theta^n(\xi) \, d\xi \quad \text{(using (15.162))}$$

$$= -4\pi \alpha^3 \rho_c \int_0^\xi \frac{d}{d\xi} \left[\xi^2 \frac{d\theta(\xi)}{d\xi} \right] d\xi \quad \text{(using (15.167))}$$

Table 15.4. *Values of ξ_1 and $(d\theta_n/d\xi)_{\xi = \xi_1}$ for typical values of the polytropic index n*

n	ξ_1	$\left(\dfrac{d\theta_n}{d\xi} \right)_{\xi=\xi_1}$ [†]
0	2.4494	$-0.816\,53$
0.5	2.7528	$-0.499\,76$
1.0	3.141\,59	$-0.318\,31$
1.5	3.653\,75	$-0.203\,30$
2.0	4.352\,87	$-0.127\,25$
2.5	5.355\,28	$-7.626\,48 \times 10^{-2}$
3.0	6.896\,85	$-4.242\,98 \times 10^{-2}$
3.5	9.535\,81	$-2.079\,10 \times 10^{-2}$
4.0	14.971\,55	$-7.752\,94 \times 10^{-3}$
4.5	31.836\,46	$-6.050\,82 \times 10^{-5}$
5.0	∞	0

Table adapted from Chandrasekhar (1939).

[†] The quantity $\xi_1{}^2 (d\theta_n/d\xi)_{\xi = \xi_1}$ tends to the finite limit 1.732\,05.

or

$$M(\xi) = 4\pi\alpha^3\rho \left[-\xi^2 \frac{d\theta(\xi)}{d\xi} \right] \qquad (15.180)$$

Substituting for α from (15.166) we get

$$M(\xi) = 4\pi \left[\frac{(n+1)K}{4\pi G} \right]^{3/2} \rho_c^{3/2n-1/2} \left[-\xi^2 \frac{d\theta(\xi)}{d\xi} \right] \qquad (15.181)$$

The total mass of the star is therefore given by

$$M = 4\pi \left[\frac{(n+1)K}{4\pi G} \right]^{3/2} \rho_c^{3/2n-1/2} \left[-\xi^2 \frac{d\theta(\xi)}{d\xi} \right]_{\xi=\xi_1} \qquad (15.182)$$

It may be noted that for $n = 3$, M is independent of ρ_c and using the values given in Table 15.4 we get

$$M(n=3) \approx 25.36 \left[\frac{K}{\pi G} \right]^{3/2} \qquad (15.183)$$

Furthermore, for $n = 5$ we have $\xi_1 = \infty$ and

$$\lim_{\xi \to \infty} \left[-\xi^2 \frac{d\theta_5(\xi)}{d\xi} \right] = \lim_{\xi \to \infty} \left[\xi^2 \frac{d}{d\xi}(1 + \tfrac{1}{3}\xi^2)^{-1/2} \right]$$

$$= \lim_{\xi \to \infty} \left[\tfrac{1}{3}\xi^3(1 + \tfrac{1}{3}\xi^2)^{-3/2} \right]$$

$$= \sqrt{3}$$

Thus

$$M(n=5) = 4\sqrt{3}\pi \left[\frac{3K}{2\pi G} \right]^{3/2} \rho_c^{-1/5} \qquad (15.184)$$

showing a finite mass for the star in spite of the fact that its radius is infinite.

(c) The mass–radius relationship

If we eliminate ρ_c from (15.178) and (15.182) we would get

$$M^{(n-1)/n} R^{(3-n)/n} = \frac{(n+1)K}{G(4\pi)^{1/n}} \left[-\xi_1^{1+1/n} \left(\frac{d\theta}{d\xi} \right)_{\xi_1}^{1-1/n} \right] \qquad (15.185)$$

For a given value of n, the quantity inside the square brackets has a well defined value. Thus assuming a value of n and knowing M and R (15.185) can be used to determine K.

(d) The central density

In terms of the variable ξ, the mean density within r (see (15.63)) is given by

$$\bar{\rho}(\xi) = \frac{M(\xi)}{(4\pi/3)\alpha^3\xi^3} \tag{15.186}$$

Using (15.180) we get

$$\frac{\rho_c}{\bar{\rho}(\xi)} = \left[-\frac{3}{\xi}\left(\frac{d\theta}{d\xi}\right) \right]^{-1} \tag{15.187}$$

or

$$\frac{\rho_c}{\bar{\rho}} = \left[-\frac{3}{\xi_1}\left(\frac{d\theta}{d\xi}\right)_{\xi_1} \right]^{-1} \tag{15.188}$$

where

$$\bar{\rho} = \frac{M}{(4\pi/3)R^3} = \frac{M}{(4\pi/3)\alpha^3\xi_1^3} \tag{15.189}$$

represents the mean density of the star. Using the values given in Table 15.4 we can easily calculate $\rho_c/\bar{\rho}$; for example, for $n = 3$ (which is approximately applicable to the Sun – see Section 15.10) we have $\rho_c/\bar{\rho} \approx 54.18$. It may be noted that for $n = 0$, (15.187) becomes

$$\frac{\rho_c}{\bar{\rho}(\xi)} = \left[-\frac{3}{\xi}\frac{d\theta_0}{d\xi} \right]^{-1} = 1 \tag{15.190}$$

where we have used (15.171). Thus the case $n = 0$ corresponds to the uniform density model.

(e) The central pressure
Using (15.159) we get

$$P_c = K\rho_c^{1+1/n} \tag{15.191}$$

If we now substitute for K and ρ_c from (15.185) and (15.188) respectively, we would get

$$P_c = W_n \frac{M^2 G}{R^4} \tag{15.192}$$

where

$$W_n = \left[4\pi(n+1)\left(\frac{d\theta}{d\xi}\right)^2 \xi_1 \right]^{-1}$$

and use has been made of (15.189). For a given value of n, the value of P_c can easily be calculated by using Table 15.4. For example, for $n = 3$, $M = M_\odot$ and $R = R_\odot$ we get

$$P_c \approx 1.24 \times 10^{17} \, \text{dyne cm}^{-2}$$

We should mention that for $n = 0$, if we use (15.171) and the fact that $\xi_1 = \sqrt{6}$, we would obtain $W_0 = \frac{3}{8}\pi$ making (15.192) consistent with (15.79).

(f) The gravitational potential energy
The total gravitational potential energy is given by (see (15.90))

$$
\begin{aligned}
\Omega &= -\int_0^R \frac{GM(r)\mathrm{d}M(r)}{r} \\
&= -G\int_0^R \frac{1}{r}\left[M(r)\frac{\mathrm{d}M(r)}{\mathrm{d}r}\right]\mathrm{d}r \\
&= -G\left[\frac{1}{r}\frac{1}{2}M^2(r)\Big|_0^R + \frac{1}{2}\int_0^R \frac{M^2(r)}{r^2}\mathrm{d}r\right] \\
&= -\frac{GM^2}{2R} - \frac{1}{2}\int_0^R \frac{\mathrm{d}V(r)}{\mathrm{d}r}M(r)\mathrm{d}r
\end{aligned}
\tag{15.193}
$$

where $V(r)$ represents the gravitational potential and is given by

$$
\frac{\mathrm{d}V(r)}{\mathrm{d}r} = \frac{GM(r)}{r^2}
\tag{15.194}
$$

For $r \geqslant R$, $M(r) = M = \text{constant}$, thus integrating the above equation (from r to ∞) we get

$$
V(r) = -\frac{GM}{r} \quad (r \geqslant R)
\tag{15.195}
$$

where we have assumed that $V(\infty) = 0$. We next write down the condition for hydrostatic equilibrium (15.57)

$$
\frac{\mathrm{d}P(r)}{\mathrm{d}r} = -\frac{GM(r)}{r^2}\rho(r)
\tag{15.196}
$$

or

$$
\frac{1}{\rho(r)}\frac{\mathrm{d}P(r)}{\mathrm{d}r} = -\frac{\mathrm{d}V(r)}{\mathrm{d}r}
\tag{15.197}
$$

Now,

$$
\frac{\mathrm{d}}{\mathrm{d}r}\left[\frac{P(r)}{\rho(r)}\right] = \frac{1}{\rho(r)}\frac{\mathrm{d}P(r)}{\mathrm{d}r} - \frac{P(r)}{\rho^2(r)}\frac{\mathrm{d}\rho(r)}{\mathrm{d}r}
\tag{15.198}
$$

But

$$
\frac{\mathrm{d}}{\mathrm{d}r}\frac{P(r)}{\rho(r)} = K\frac{\mathrm{d}}{\mathrm{d}r}[\rho(r)]^{1/n} \quad \text{(using (15.159))}
$$

$$
= \frac{K}{n}[\rho(r)]^{1/n-1}\frac{\mathrm{d}\rho(r)}{\mathrm{d}r}
$$

or

$$n\frac{d}{dr}\left[\frac{P(r)}{\rho(r)}\right] = \frac{P(r)}{\rho^2(r)}\frac{d\rho(r)}{dr} \tag{15.199}$$

Substituting in (15.198) we get

$$\frac{1}{\rho(r)}\frac{dP(r)}{dr} = (n+1)\frac{d}{dr}\left[\frac{P(r)}{\rho(r)}\right] = -\frac{dV(r)}{dr} \tag{15.200}$$

where in the last step we have made use of (15.197). Thus (15.193) becomes

$$\begin{aligned}
\Omega &= -\frac{GM^2}{2R} + \frac{(n+1)}{2}\int_0^R \frac{d}{dr}\left[\frac{P(r)}{\rho(r)}\right]M(r)\,dr \\
&= -\frac{GM^2}{2R} + \frac{(n+1)}{2}\left[\left.\frac{P(r)}{\rho(r)}M(r)\right|_0^R - \int_0^R \frac{P(r)}{\rho(r)}\frac{dM(r)}{dr}\,dr\right] \\
&= -\frac{GM^2}{2R} - \frac{(n+1)}{2}\int_0^R P(r)4\pi r^2\,dr \tag{15.201}
\end{aligned}$$

where we have used (15.51). If we now use (15.93) we would get

$$\Omega = -\frac{GM^2}{2R} + \frac{(n+1)\Omega}{6}$$

or

$$\Omega = -\frac{3}{5-n}\frac{GM^2}{R} \tag{15.202}$$

(g) The internal energy
If we neglect radiation pressure then we can use (15.117) to calculate the internal energy of the star:

$$U = -\frac{1}{3(\gamma-1)}\Omega = \frac{1}{(\gamma-1)(5-n)}\frac{GM^2}{2R} \tag{15.203}$$

Further, if we assume convective adiabatic equilibrium (see Chandrasekhar (1939) for details) then the exponent of ρ in (15.159) will be the ratio of specific heats γ so that

$$1 + \frac{1}{n} = \gamma \tag{15.204}$$

15.10   The standard model

The 'standard model' was introduced by Eddington and according to the model, the pressure–density relationship is assumed to be given by (see (15.151))

$$P = K\rho^{4/3} \tag{15.205}$$

where

$$K = \left[\frac{3}{a}\left(\frac{k}{\mu H}\right)^4 \frac{1-\beta}{\beta^4}\right]^{1/3} \tag{15.206}$$

The parameters $(1 - \beta)$ and μ (and therefore K) are assumed to be constants throughout the star. We may mention here that the standard model has several shortcomings, nevertheless it has played a very important role in the early understanding of stellar structure and it does approximate to most stellar models in temperature and density variations – for a detailed discussion on the standard model see, e.g., Chandrasekhar (1939), Cox and Giuli (1968b) and other similar books.

If we compare (15.205) with (15.159) we find that the star configuration will correspond to the Lane–Emden polytrope of index $n = 3$. Thus the formulae derived in the previous section can be readily applied. Because of the importance of the standard model we have tabulated $\theta_3(\xi)$ and $\theta_3'(\xi)$ in Table 15.5; a more detailed table can be found in Cox and Giuli (1968b). We first note that

Table 15.5. *Solution of the Lane–Emden equation of index 3*

ξ	$\theta_3(\xi)$	$\theta_3'(\xi)$
0.0	1.000 00	− 0.000 00
0.5	0.959 83	− 0.154 84
1.0	0.855 05	− 0.252 13
1.5	0.719 50	− 0.279 91
2.0	0.582 85	− 0.261 49
2.5	0.461 13	− 0.223 97
3.0	0.359 23	− 0.184 05
3.5	0.276 26	− 0.148 85
4.0	0.209 28	− 0.120 17
4.5	0.155 07	− 0.097 62
5.0	0.110 82	− 0.080 13
5.5	0.074 28	− 0.066 58
6.0	0.043 74	− 0.056 04
6.5	0.017 87	− 0.047 77
6.896 848 6	0.000 00	− 0.042 43

Note: Table adapted from Cox and Giuli (1968b), originally obtained from the British Association for the Advancement of Science.

$$\text{and} \qquad \left.\begin{array}{l} \xi_1(n=3) \approx 6.897 \\[2mm] \dfrac{d\theta_3}{d\xi}\bigg|_{\xi=\xi_1} \approx -4.243 \times 10^{-2} \end{array}\right\} \qquad (15.207)$$

Thus the expression for the mass of the star (15.182) for $n = 3$ and K given by (15.206) becomes

$$M = 4\pi \left[\frac{1}{\pi G}\left\{\left(\frac{k}{\mu H}\right)^4 \frac{3}{a}\frac{1-\beta}{\beta^4}\right\}^{1/3}\right]^{3/2}\left[-\xi_1^2 \frac{d\theta_3}{d\xi}\bigg|_{\xi=\xi_1}\right] \qquad (15.208)$$

or

$$\left(\frac{M}{M_\odot}\right)\mu^2 \approx \frac{4}{\sqrt{\pi}}\frac{1}{[6.672 \times 10^{-8}]^{3/2}} \times \left(\frac{1.38 \times 10^{-16}}{1.67 \times 10^{-24}}\right)^2$$

$$\times \left(\frac{3}{7.564 \times 10^{-15}}\right)^{1/2} \times \left(\frac{1-\beta}{\beta^4}\right)^{1/2}$$

$$\times [(6.897)^2 \times 4.243 \times 10^{-2}] \times \frac{1}{1.989 \times 10^{33}}$$

$$\approx 18.08\left[\frac{1-\beta}{\beta^4}\right]^{1/2} \qquad (15.209)$$

Thus for a given value of β we can very easily calculate $(M/M_\odot)\mu^2$ and conversely – see Table 15.6.

Using (15.207), the expressions for the central density (15.188) and (15.189) and the central pressure (15.191) and (15.192) can be written in the form

$$\rho_c \approx 54.18\bar{\rho} = 54.18\frac{M}{(4\pi/3)R^3}$$

Table 15.6. *Calculated values of $(M/M_\odot)\mu^2$ for different values of $(1-\beta)$ using (15.209)*

$1-\beta$	$(M/M_\odot)\mu^2$
0.003	0.996
0.016	2.362
0.025	3.007
0.045	4.205
0.1	7.059
0.5	51.14
1.0	∞

$$\approx 76.3 \times \left(\frac{M}{M_\odot}\right) \times \left(\frac{R_\odot}{R}\right)^3 \text{g cm}^{-3} \tag{15.210}$$

and

$$P_c \approx 11.05 \frac{M^2 G}{R^4}$$

$$\approx 1.243 \times 10^{17} \left(\frac{M}{M_\odot}\right)^2 \left(\frac{R_\odot}{R}\right)^4 \text{dyne cm}^{-2} \tag{15.211}$$

We should mention here that once ρ_c and P_c are known the spatial variations of ρ and P can easily be calculated from the following relations (see (15.162) and (15.159))

$$\rho(\xi) = \rho_c \theta_3^3(\xi) \tag{15.212}$$

and

$$P(\xi) = K\rho_c^{1+1/n} \left(\frac{\rho}{\rho_c}\right)^{1+1/n} = P_c \theta_3^4(\xi) \tag{15.213}$$

where we have used (15.190) and the fact that $n = 3$. If we now use the values given in Table 15.5 we can calculate $\rho(\xi)$ and $P(\xi)$. The actual r variation can easily be calculated by noting that

$$r = \frac{\xi}{\alpha}$$

where

$$\alpha = \left[\frac{(n+1)K}{4\pi G} \rho_c^{1/n-1}\right]^{1/2}$$

$$= \frac{1}{(\pi G)^{1/2}} \left[\frac{3}{a}\left(\frac{k}{\mu H}\right)^4 \frac{1-\beta}{\beta^4}\right]^{1/6} \rho_c^{-1/3} \tag{15.214}$$

Thus the parameter α (and hence the actual spatial variation) is completely determined if we know the values of μ and β. The corresponding temperature variation can be determined from (15.143) and (15.145):

$$P = \frac{1}{\beta}P_g = \frac{1}{\beta}\frac{k}{\mu H}\rho T \tag{15.215}$$

or

$$T(\xi) = (\beta\mu)\frac{H}{k}\frac{P_c}{\rho_c}\theta_3(\xi) \tag{15.216}$$

or

$$T(\xi) \approx 19.70 \times 10^6 (\beta\mu)\left(\frac{M}{M_\odot}\right)\left(\frac{R_\odot}{R}\right)\theta_3(\xi) \text{K} \tag{15.217}$$

Table 15.7. *Values of* ρ_c, P_c, β, T_c, *and* $\bar{T}$ *for the Sun, Sirius A and Capella A using the standard moded and* $\mu = 1$

	$M/M_\odot$	$R/R_\odot$	$\rho_c (\mathrm{g\,cm^{-3}})$	$P_c (\mathrm{dyne\,cm^{-2}})$	β	$T_c(\mathrm{K})$	$\bar{T}(\mathrm{K})$
Sun	1.00	1.00	76	1.2×10^{17}	0.997	20×10^6	12×10^6
Sirius A	2.34	1.78	32	6.8×10^{16}	0.984	25×10^6	15×10^6
Capella A	4.18	15.9	7.9×10^{-2}	3.4×10^{13}	0.955	4.9×10^6	2.9×10^6

Note: The table has been in part adapted from Chandrasekhar (1939).

Thus the temperature at the centre of the star is given by

$$T_c \approx 19.70 \times 10^6 \beta\mu \left(\frac{M}{M_\odot}\right)\left(\frac{R_\odot}{R}\right) K \tag{15.218}$$

The average temperature can also be calculated:

$$\bar{T} = \frac{1}{M}\int T(r)\,dM(r) \tag{15.67}$$

$$= \frac{1}{M}\frac{\beta\mu H}{k}\int \frac{P(r)}{\rho(r)}\frac{dM(r)}{dr}\,dr \quad \text{(using (15.215))}$$

$$= \frac{1}{M}\frac{\beta\mu H}{k}\int P(r)4\pi r^2\,dr \quad \text{(using (15.51))}$$

$$= \frac{1}{M}\frac{\beta\mu H}{k}[-\tfrac{1}{3}\Omega] \quad \text{(using (15.93))}$$

$$= \frac{1}{2}\frac{\beta\mu H}{k}\frac{GM}{R} \quad \text{(using (15.202))} \tag{15.219}$$

or

$$\bar{T} \approx 11.54 \times 10^6 \beta\mu \left(\frac{M}{M_\odot}\right)\left(\frac{R_\odot}{R}\right) K \tag{15.220}$$

We next apply the formulae derived in this section to the Sun, Sirius A and Capella A. We assume $\mu \approx 1.0$. The value of β is first calculated by using (15.209). The values of ρ_c and P_c (which do not require the values of β and μ) are calculated by using (15.210) and (15.211) and the values of T_c and $\bar{T}$ are calculated by using (15.218) and 15.220). The results are given in Table 15.7. The numerical values given in the table are indeed representative of the values found from more sophisticated models.

15.11 The white dwarf stars

As discussed earlier, a star, after exhausting its nuclear fuel, will start to contract again. This contraction usually leads to very high densities so that the electrons may form a degenerate gas. This is indeed the case for white dwarf stars[†] which represent the final stage is stellar evolution.

The pressure due to electrons forming a completely degenerate gas is

---

[†] The white dwarf stars are not only characterized by high densities, but they also have small radii and low luminosities – for a detailed account we refer the reader to Chandrasekhar (1939), Cox and Giuli (1968b), Mestel (1965), Schatzman (1958), Schwarzschild (1958) etc. The book by Schatzman (1958) deals entirely with white dwarf stars.

given by (see (6.63), (6.69), 6.70) and (6.71)):

$$P = A f(x) \tag{15.221}$$

where

$$A = \frac{m^4 c^5}{24\pi^2 \hbar^3} \approx 6.02 \times 10^{22} \text{ c.g.s. units} \tag{15.222}$$

$$f(x) = (2x^3 - 3x)(1 + x^2)^{1/2} + 3 \sinh^{-1} x \tag{15.223}$$

$$x = x(r) = \frac{p_{FO}}{mc} = \frac{\hbar}{mc}[3\pi^2 n_e(r)]^{1/3} \approx 1.195 \times 10^{-10}[n_e(r)]^{1/3} \tag{15.224}$$

where n_e (measured in cm^{-3}) represents the electron density. Using (15.12) and (15.224) we get the density

$$\rho(r) = Bx^3(r) \tag{15.225}$$

where

$$B = \frac{1}{3\pi^2}\left(\frac{mc}{\hbar}\right)^3 \mu' H \approx 9.81 \times 10^5 \mu' \tag{15.226}$$

and

$$\mu' \approx \frac{1}{1 + x} \quad (15.13) \tag{15.227}$$

We may mention here that in a white dwarf star the densities are such that, whereas the electron gas is in a degenerate state the gas consisting of protons and other nuclei is in a non-degenerate state, so that the perfect gas law is applicable to these nuclei. However, the pressure due to the gas formed by the nuclei is usually negligible in comparison with the pressure due to the degenerate electron gas and therefore we will consider only the pressure due to the degenerate electron gas.

Equations (15.221) and (15.225) define the equation of state and using these we can solve the equation of hydrostatic equilibrium (15.58) to determine the pressure and density variation inside the star. We will discuss this solution in Section 15.11.2, however, before we do so we will obtain the star configuration in the extreme relativistic limit (i.e., for $x \gg 1$).

15.11.1 *Solution of the equation of hydrostatic equilibrium for a completely degenerate gas in the extreme relativistic limit*

In the extreme relativistic limit (i.e., for $x \gg 1$) we have

$$f(x) \to 2x^4 \tag{15.228}$$

(see (6.80)). Thus, we have

$$P \approx 2x^4 A \tag{15.229}$$

If we now substitute the value of x from (15.225) we would get

$$P \approx K_3 \rho^{4/3} \tag{15.230}$$

where

$$K_3 = \frac{2A}{B^{4/3}} \approx 1.235 \times 10^{15} (\mu')^{-4/3} \tag{15.231}$$

Comparing with (15.159) we find that the star configuration will correspond to the Lane–Emden polytrope of index $n = 3^{\dagger}$. Thus, using (15.182) and (15.207) we would get

$$M = \frac{4}{\sqrt{\pi}} \left[\frac{1.235 \times 10^{15}}{6.672 \times 10^{-8}} (\mu')^{-4/3} \right]^{3/2} [(6.897)^2 \times 4.243 \times 10^{-2}]$$

or

$$\frac{M}{M_\odot} \approx \frac{5.75}{\mu'^2} \tag{15.232}$$

The above equation represents the limiting mass for white dwarf stars and experimental observations confirm this (see Section 15.11.2). This limiting mass for a white dwarf star is usually referred to as the *Chandrasekhar limit*. It may be noted that as $x \to \infty$, the density ρ would also tend to ∞ and therefore from (15.178) it readily follows that the radius of the star will tend to zero! Indeed, we will show in the following section that the greater the mass of a white dwarf star, the smaller will be its radius!

15.11.2 The general solution corresponding to a completely degenerate gas

Using (15.221) and (15.225), the equation describing hydrostatic equilibrium (15.58)

$$\frac{1}{r^2} \frac{d}{dr} \left[\frac{r^2}{\rho(r)} \frac{dP}{dr} \right] = -4\pi G \rho(r) \tag{15.233}$$

becomes

$$\frac{A}{B} \frac{1}{r^2} \frac{d}{dr} \left[\frac{r^2}{x^3} \frac{df(x)}{dr} \right] = -4\pi G B [x(r)]^3 \tag{15.234}$$

---

† We may mention here that in the non-relativistic limit, the pressure is proportional to $\rho^{5/3}$ (see (6.29)) and thus the star configuration will correspond to the Lane–Emden polytrope of index 1.5.

The above equation can be easily transformed to the following form (see Appendix 8):

$$\frac{1}{r^2}\frac{d}{dr}\left[r^2\frac{dy}{dr}\right] = -\frac{\pi GB^2}{2A}[y^2 - 1]^{3/2} \tag{15.235}$$

where

$$y^2(r) = x^2(r) + 1 = \left[\frac{\rho(r)}{B}\right]^{2/3} + 1 \tag{15.236}$$

Introducing the variable

$$\zeta = r\mu' \tag{15.237}$$

(15.235) and (15.236) become

$$\frac{1}{\zeta^2}\frac{d}{d\zeta}\left[\zeta^2\frac{dy(\zeta)}{d\zeta}\right] = -\frac{\pi G(B/\mu')^2}{2A}[y^2(\zeta) - 1]^{3/2}$$

$$\approx -1.68 \times 10^{-18}[y^2(\zeta) - 1]^{3/2} \tag{15.238}$$

and

$$y^2(\zeta) \approx \left[\frac{1}{9.81 \times 10^5}\frac{\rho(\zeta)}{\mu'}\right]^{2/3} + 1 \tag{15.239}$$

where all quantities are now measured in c.g.s. units.

Assuming μ' to be constant throughout the star, the numerical method for solving (15.238) is simple. We first note that

$$y^2(\zeta = 0) \approx 1.013 \times 10^{-4}\left[\frac{\rho_c}{\mu'}\right]^{2/3} + 1 \tag{15.240}$$

Table 15.8. *The mass and radius for a completely degenerate configuration corresponding to different values of the central density*

$\dfrac{1}{y^2(\zeta=0)}$	$\dfrac{\rho_c}{\mu}(\text{g cm}^{-3})$	$\mu'\dfrac{R}{R_\odot}$	$\mu'^2\dfrac{M}{M_\odot}$
0	∞	0	5.75
0.01	9.67×10^8	5.93×10^{-3}	5.51
0.02	3.37×10^8	7.82×10^{-3}	5.32
0.05	8.13×10^7	1.10×10^{-2}	4.87
0.1	2.65×10^7	1.43×10^{-2}	4.33
0.2	7.85×10^6	1.85×10^{-2}	3.54
0.4	1.80×10^6	2.47×10^{-2}	2.45
0.6	5.34×10^5	3.09×10^{-2}	1.62
0.8	1.23×10^5	4.01×10^{-2}	0.88
1.0	0	∞	0

Note: Table adapted from Chandrasekhar (1939).

and

$$\left.\frac{dy}{d\zeta}\right|_{\zeta=0} = 0 \qquad (15.241)$$

the last condition follows directly from (15.238). We start with a given value of ρ_c/μ' (and hence of $y^2(\zeta=0)$) and using (15.241) we can integrate (15.238) step by step to obtain[†] $y(\zeta)$ from which we can calculate $x(\zeta)$, $\rho(\zeta)/\mu'$ and $P(\zeta)$ using (15.236), (15.239) and (15.221) respectively. The family of curves corresponding to different initial values of ρ_c/μ' will represent a universal set.

Let ζ_1 represent the value of ζ where $P(\zeta)$ becomes zero; this value of

Fig. 15.4. The mass–radius relationship for white dwarf stars using Table 15.8. The upper and lower curves correspond to $\mu' = 1$ and $\mu' = 2$ respectively. The approximate position of seven white dwarf stars is also given showing the high preponderance of heavy elements. (Figure adapted from Schatzman (1958).)

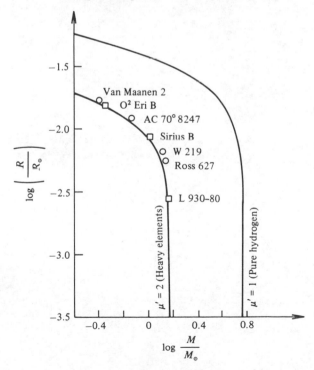

[†] For the detailed results of such numerical calculations, we refer the reader to Chandrasekhar (1935), A. Riez (1949) and other similar papers.

ζ will correspond to the radius of the star, thus

$$R = \frac{\zeta_1}{\mu'} \tag{15.242}$$

The mass of the star can be calculated by evaluating the integral

$$M = \int_0^R \rho(r) 4\pi r^2 \, dr = \frac{1}{\mu'^3} \left[\int_0^{\zeta_1} \rho(\zeta) 4\pi \zeta^2 \, d\zeta \right]$$

or

$$\mu'^2 M = \int_0^{\zeta_1} \left[\frac{\rho(\zeta)}{\mu'} \right] 4\pi \zeta^2 \, d\zeta \tag{15.243}$$

For different values of $y(\xi = 0)$ (and hence of ρ_c/μ' – see (15.240)) the values of $\mu' R$ and $\mu'^2 M$ are tabulated in Table 15.8. The mass–radius relationship for $\mu' = 1$ and $\mu' = 2$ is plotted in Fig. 15.4 and we find that most of the white dwarf stars lie approximately on the $\mu' = 2$ curve showing the presence of heavy elements inside the star.

16

Equations of state in elementary particle physics

16.1 Overview

Spontaneous symmetry breaking is playing a major role in the theoretical physics of elementary particles (see, e.g., Salam (1969) and Weinberg (1967) in unifying weak and electromagnetic interactions). Moreover, it has been realized that there exist formal analogies between spontaneously broken symmetries and the phenomenon of superconductivity (Nambu and Jona-Lasinio (1961) and Goldstone (1961)). However, there is a basic difference between the broken symmetries in high energy physics and in many-body physics such as superconductivity. In elementary particle physics there is no *direct* evidence that the various symmetry breakings are indeed of a spontaneous nature, while in a 'many-body state' the symmetry can be changed by heating the system and thus proving the spontaneous character of the breaking. The main difficulty with high energy physics of elementary particles is to define the temperature of the system for a very small number of particles. The main question is how to use many-body physics for two colliding 'elementary' particles. Therefore, most of the literature on this subject is described by field theory at zero temperature ($T = 0$!).

In the seventies having phase transitions in gauge models was suggested for electromagnetic and weak interactions (Kirzhnits and Linde (1972); Weinberg (1974) and Dolan and Jackiw (1974)) starting with the analogy between the Landau–Ginsburg equation for superconductors and the Lagrangian of the Salam–Weinberg model. However, these authors considered the change of symmetry for an elementary particle system in a heat bath of macroscopic dimensions, so these field theories at finite temperature can be used only in astrophysical considerations.

The idea of describing high energy physics by field theory at finite temperature rather than at zero temperature was suggested, (Eliezer 1975;

Eliezer and Weiner 1976, 1978) implying that the phase transitions predicted within unified gauge theories should be looked at not only in the astrophysical domain but also in elementary-particle interactions at very high energies. In a unified theory of weak and electromagnetic interactions one might expect, for example, that in reactions induced by very energetic neutrinos, the parity, the strangeness, CP, etc. might become conserved quantities. For hadron physics of strong interactions the $SU(3)$ symmetry might be restored at high energies and at large transverse momentum, which correspond to high temperature (Eliezer and Weiner 1974). However, it must be emphasized that all the ideas of using the concept of temperature in elementary particle physics are of a very speculative nature and so far no direct experimental evidence has been observed to show the necessity of defining a temperature as an internal parameter of elementary particle physics.

In order to show how it is possible to introduce the concepts of equilibrium-statistical mechanics we will describe a model for strong interaction developed mainly by Hagedorn. Even if the application of this model to particle physics is questionable, the formalism developed in the next section can be applied to any macroscopic system, such as the early universe.

We conclude this introduction with a table where units $\hbar = c = k = 1$ are introduced. This system of units is (almost uniquely) used in high energy physics, therefore we shall also work in this way.

16.2   Hagedorn model of strong interactions

16.2.1   Introduction

The idea of describing hadronic matter by statistical thermodynamics had already been suggested by Koppe (1948, 1949) and Fermi (1950) and was later developed into a comprehensive model for strong

Table 16.1. *A conversion table between c.g.s. and* $\hbar = c = k = 1$

$1\,\mathrm{cm} = 5.07 \times 10^{10}\ \mathrm{MeV}^{-1}$
$1\,\mathrm{g} = 5.61 \times 10^{26}\ \mathrm{MeV}$
$1\,\mathrm{s} = 1.52 \times 10^{21}\ \mathrm{MeV}^{-1}$
$1\,\mathrm{K} = 8.62 \times 10^{-11}\ \mathrm{MeV}$
electron mass $m_e = 0.51\ \mathrm{MeV}$
$1\,\mathrm{amu} = 931.50\ \mathrm{MeV}$
$e^2 = \frac{1}{137}$ (e is the electron charge)

interactions by Hagedorn (1971). In order to use statistical mechanics, the system under consideration should have many degrees of freedom. The number of particles created in strong interactions does not imply from a simplistic viewpoint that one has to deal with a large number of degrees of freedom to suit the statistical requirements. However, the strong internal interactions can be described by an infinite number of components representing all possible resonant states. Therefore, even in the process of the creation of only a small number of *real* particles (e.g., ten particles), one has to solve a system with many degrees of freedom due to the particle creation and absorption of *virtual* states (which are allowed by the uncertainty principle[†] $\Delta E \cdot \Delta t \approx 1$). In particular, this picture can be understood using the philosophy that every particle which is able to participate in a strong interaction (i.e. a hadron) is made up of all the other hadrons. (This picture is known in the literature as the 'bootstrap model'.) Therefore, it is conceivable to make the following assumption:

1st postulate: Strong interactions are described by an infinite number of degrees of freedom.

The time-scale of strong interaction is 10^{-23}s, therefore in order to use thermodynamic equilibrium one has to assume that during a period much shorter than 10^{-23}s equilibrium is reached. It is difficult to imagine many collisions during such a short period of time, therefore, Hagedorn assumes an 'instantaneous equilibrium', or equivalently,

2nd postulate: Particles in strong interactions obey a 'pre-established equilibrium'.

The experimental physics of strong interactions is reached by high energy collisions between hadrons. In order to use thermodynamics one should be able to separate the isotropic thermal (i.e. random) motion from the directional motion along the axis of collision. For this purpose Hagedorn suggested the following postulate:

3rd postulate: No turbulence in the thermodynamics of strong interactions.

The meaning of this assumption is that it is possible to find a continuum of local rest frames (Lorentz frames which are moving in the direction of the collision axis) where the observer measures in his vicinity only thermal motion. In order to understand the problem it is useful to imagine two big clouds of gas colliding with very high (center of mass) velocities relative to the thermal motion which is developed during the collision. The third postulate enables us to distinguish between the kinematic variables describing the collision between the hadrons and the thermodynamic variables.

[†] Note: We use the units $\hbar = c = 1$.

Using the three postulates defined above Hagedorn (1971) has developed a thermodynamic model for strong interactions. In Section 16.22 we calculate the partition function and in Section 16.2.3 we use the bootstrap condition in order to be able to calculate the thermodynamic functions in Section 16.2.4. A possible comparison with experiments is made in Section 16.2.5.

16.2.2   The partition function

Although the partition function has already been developed for Bose and Fermi statistics in Chapters 5 and 6, we will repeat briefly the calculations for the relativistic case in order to see how the mass spectrum enters into the calculations. The energy of a particle i with a mass m_i and momentum p_α is given by $\varepsilon_{\alpha i}$[†]

$$\varepsilon_{\alpha i} = (p_\alpha{}^2 + m_i{}^2)^{1/2} \tag{16.1}$$

A quantum state is defined by a $\infty \times \infty$ matrix $\{n\}$ containing a possible complete set of occupation numbers which can take the following values

$$n_{\alpha i} = \begin{cases} n_{\alpha f} = 0, 1 & \text{for fermions} \\ n_{\alpha b} = 0, 1, 2, \ldots \infty & \text{for bosons} \end{cases} \tag{16.2}$$

$n_{\alpha i}$ are the occupation number describing the number of particles of kind i having a momentum p_α. The Pauli exclusion principle is expressed by (16.2).

The energy of the system $E_{\{n\}}$ in a quantum state $\{n\}$ is given by

$$E_{\{n\}} = \sum_{\alpha=1}^{\infty} \sum_{i=1}^{\infty} n_{\alpha i} \varepsilon_{\alpha i} \tag{16.3}$$

The partition function Z is

$$Z(V, T) = \sum_{\{n\}} \exp\left[-\frac{1}{T} \sum_\alpha \sum_i n_{\alpha i} \varepsilon_{\alpha i} \right] \tag{16.4}$$

where the sum $\sum_{\{n\}}$ is over all possible quantum states.

Denoting

$$X_{\alpha i} \equiv \exp(-\varepsilon_{\alpha i}/T) \tag{16.5}$$

one can rewrite (16.4) by[‡]

$$Z(V, T) = \sum_{\{n\}} \prod_\alpha \prod_i X_{\alpha i}{}^{n_{\alpha i}} = \prod_{\alpha, i} \sum_{n_{\alpha i}} X_{\alpha i}{}^{n_{\alpha i}} \tag{16.6}$$

[†] in units $\hbar = c = k = 1$
[‡] See Chapter 5 (5.39) and the footnote.

At this stage we split the subscript i into two: b for bosons and f for fermions and use (16.2) to obtain

$$Z(V, T) = \prod_{\alpha,b} \frac{1}{1 - X_{\alpha b}} \prod_{\alpha,f} (1 + X_{\alpha f}) \tag{16.7}$$

or equivalently

$$\ln Z = - \sum_{\alpha,b} \ln (1 - X_{\alpha b}) + \sum_{\alpha,f} \ln (1 + X_{\alpha f}) \tag{16.8}$$

The momentum summation is changed into appropriate integrals

$$\sum_{\alpha} \to \frac{V}{2\pi^2} \int_0^\infty p^2 dp \tag{16.9}$$

and in a similar way we change the particle summation into appropriate integrals over the mass spectrum

$$\sum_b \to \int_0^\infty \rho_B(m) dm \quad \sum_f \to \int_0^\infty \rho_F(m) dm \tag{16.10}$$

where ρ_B and ρ_F are the mass densities for bosons and fermions respectively. In order to write (16.10) one has to assume a dense spectrum of particles, and this is the case for the Hagedorn model (see 1st postulate). Using (16.9) and (16.10) in (16.8), the partition function becomes

$$\ln Z = \frac{V}{2\pi^2} \int_0^\infty p^2 dp$$
$$\times \left[\int_0^\infty \rho_F(m) \ln (1 + X_{pm}) dm - \int^\infty \rho_B(m) \ln (1 - X_{pm}) dm \right] \tag{16.11}$$

Expanding the logarithms in a Taylor series, one gets

$$\ln Z = \frac{V}{2\pi^2} \sum_{n=1}^\infty \frac{1}{n} \int_0^\infty dm \int_0^\infty dp \left\{ [p^2 (\rho_B(m) - (-1)^n \rho_F(m))] \right.$$
$$\left. \times \exp\left(-\frac{n}{T} (p^2 + m^2)^{1/2} \right) \right\} \tag{16.12}$$

It is also convenient to define the mass density $\rho(m; n)$ for

$n = 1, 2, 3, \ldots,$ by

$$\rho(m, n) \equiv \rho_B(m) - (-1)^n \rho_F(m) \tag{16.13}$$

The momentum integration is done by using the second modified Hankel function K_n, defined by

$$
\int_0^\infty p^{2n} \exp\left(-\frac{(p^2+m^2)^{1/2}}{T}\right) dp
$$

$$
= (2n-1)!!\, m^{2n+1}\left(\frac{T}{m}\right)^n K_{n+1}\left(\frac{m}{T}\right) \tag{16.14}
$$

where $(2n-1)!! = 1.3.5. \ldots (2k-1)$, $(-1)!! = 1$, to obtain from (16.12)

$$
Z(V,T) = \exp\left(\frac{VT}{2\pi^2} \sum_{n=1}^{\infty} \frac{1}{n^2} \int_0^\infty \rho(m,n) m^2 K_2\left(\frac{nm}{T}\right) dm\right) \tag{16.15}
$$

On the other hand, the partition function Z is given by

$$
Z(V,T) = \sum_i e^{-E_i/T} = \int_0^\infty \sigma(E,V) e^{-E/T} dE \tag{16.16}
$$

where $\sigma(E,V)dE$ is the number of energy levels in the energy interval between E and $E + dE$. The physical meaning of $\sigma(E,V)$ is easily understood if one makes the assumption that $\sigma(E,V)e^{-E/T}$ has a sharp maximum at $\bar{E}$ with a width ΔE. In this case the integral of (16.16) is easily estimated

$$
Z(V,T) \approx \sigma(\bar{E},V) e^{-E/T} \Delta E \tag{16.17}
$$

From the thermodynamic relation

$$
\bar{E} = T^2 \frac{\partial}{\partial T}(\ln Z)_V \tag{16.18}
$$

we obtain from (16.17) that $\bar{E}$ is the average energy of the system. Furthermore, the entropy

$$
S = \frac{\partial}{\partial T}(T\ln Z) \tag{16.19}
$$

is estimated from (16.17) to be

$$
S(\bar{E},V) \approx \ln[\sigma(\bar{E},V)\Delta E] \tag{16.20}
$$

Equation (16.20) suggests that the physical (thermodynamic) quantity of interest is $\ln[\sigma(E,V)\Delta E]$, rather than the spectrum $\sigma(E,V)dE$ itself. Therefore in the Hagedorn model it is assumed that two systems are equivalent if they have logarithmically equal energy density levels.

16.2.3   The bootstrap condition[†]

The bootstrap approach was introduced by Chew and Mandelstam (1961) in the context of a dispersion relation to describe the ρ meson. The

---

[†] See, e.g., Chew (1961); Chew and Frautschi (1961).

bootstrap philosophy is formulated by a set of self-consistent equations which describe the interaction between all particles (hadrons) in terms of the exchange of all hadrons between them. The self-consistency is contained in the requirement that the exchanges of hadrons in '*t* channel' and '*u* channel' must describe the existence of all hadrons in the '*s* channel'[†].

In Hagedorn's model every hadron is a 'fireball' of matter which can be described by thermodynamic equilibrium. In this picture a 'fireball' is a statistical equilibrium of all other fireballs (each of them again is a statistical equilibrium of all others, etc., i.e. bootstrap!). In this model the thermodynamics is completely described by the mass spectrum $\rho(m)$ (see (16.15)) or equivalently by the energy density spectrum $\sigma(E, V)$. Hagedorn's bootstrap philosophy is formulated by the requirement that $\rho(m)$ and $\sigma(m, V_0)$ are thermodynamically equivalent, i.e.

$$\lim_{m \to \infty} \left(\frac{\ln \rho(m)}{\ln \sigma(m, V_0)} \right) = 1 \tag{16.21}$$

where V_0 is the volume (i.e. the range) of strong interactions[‡] $\approx (4\pi/3)m_\pi^{-3}$. This equation is the basic formula for Hagedorn's model. Equation (16.21) actually determines the asymptotic behaviour of $\rho(m)$ (or $\sigma(m, V_0)$), but since $\rho(m)$ is supposed to be known experimentally for small m, it can be calculated over the entire domain of its existence. Therefore the thermodynamics of strong interactions are in principle determined from experiments and (16.21).

Equation (16.16) can be rewritten by changing the variable of integration from E to m

$$Z(V, T) = \int_0^\infty \sigma(m, V)e^{-m/T}dm \tag{16.22}$$

in order to compare it with (16.15) and to imply the bootstrap equation (16.21).

1st approximation: $T \gg m_{\max}$, i.e., there is only a finite number of hadrons

$$\int_0^\infty \rho_B(m)dm = \beta \tag{16.23}$$

[†] If '*s* channel' refers to the reaction $1 + 2 \to 3 + 4$ then '*t* channel' describes the reactions $1 + \bar{3} \to \bar{2} + 4$ and $\bar{1} + 3 \to 2 + \bar{4}$ and the '*u* channel' is $1 + \bar{4} \to \bar{2} + 3$ and $\bar{1} + 4 \to 2 + \bar{3}$. The notation $\bar{a}$ denotes the antiparticle of a. Moreover, $\sqrt{s}$, $\sqrt{t}$ and $\sqrt{u}$ represent the centres of mass energies in the s, t and u channels respectively.
[‡] in c.g.s. units: $R \approx \hbar/m_\pi c \approx 1.4 \times 10^{-13}$ cm, is the range of strong interactions.

$$\int_0^\infty \rho_F(m)dm = \phi \tag{16.24}$$

where β and ϕ are finite numbers. In this case the asymptotic value for K_2

$$\lim_{T \to \infty}\left(K_2\left(\frac{nm}{T}\right)\right) \approx \frac{2T^2}{n^2 m^2} \tag{16.25}$$

can be used in (16.15) to obtain the 'zero order' approximation

$$Z^{(0)}(V, T \to \infty) \approx \exp\left(\frac{VT^3}{\pi}\sum_{n=1}^\infty \frac{1}{n^4}\int_0^\infty \rho(m,n)dm\right)$$

$$= \exp\left(\frac{VT^3}{\pi}\left[\beta \sum_{n=1}^\infty \frac{1}{n^4} - \phi \sum_{n=1}^\infty \frac{(-1)^n}{n^4}\right]\right) \tag{16.26}$$

Using the properties of the Riemann ζ-function

$$\sum_{n=1}^\infty \frac{1}{n^4} = \frac{\pi^4}{90}; \quad -\sum_{n=1}^\infty \frac{(-1)^n}{n^4} = \frac{7}{8}\cdot\frac{\pi^4}{90} \tag{16.27}$$

in (16.26), one gets

$$Z^{(0)}(V, T \to \infty) \approx \exp\left(\frac{\gamma \pi^2 VT^3}{90}\right) \tag{16.28}$$

where[†]

$$\gamma \equiv \int_0^\infty \rho_B(m)dm + \frac{7}{8}\int_0^\infty \rho_F(m)dm \equiv \beta + \frac{7}{8}\phi \tag{16.29}$$

The energy of the system is given by

$$E^{(0)} = T^2 \frac{\partial}{\partial T}\{\ln Z^{(0)}\} \approx \frac{\gamma \pi^2 VT^4}{30} \tag{16.30}$$

It is interesting to point out that for a gas of photons $\rho_F = 0$ and $\rho_B = 2\delta(m)$ one gets $\gamma = 2$ and from (16.30)

$$\frac{E^{(0)}}{V} = 4\sigma T^4 \quad \sigma \equiv \frac{\pi^2}{60} \tag{16.31}$$

where σ here is the Stefan–Boltzmann (see (5.25) with $\hbar = c = k = 1$). Using the approximation of (16.17), one gets from (16.28)

$$\sigma^{(0)}(E, V) \approx \frac{1}{\Delta E}Z^{(0)}(V, T)e^{E^{(0)}/T}$$

[†] The factor 7/8 is due to Fermi–Dirac statistics!

$$\approx \frac{1}{\Delta E} \exp\left(\frac{4}{3}\left(\frac{\gamma\pi^2 V}{30}\right)^{1/4} E^{(0)3/4}\right) \tag{16.32}$$

where (16.30) has been used. Changing variables from E to m, we get from the bootstrap equation (16.21) that the mass density should behave like

$$\rho^{(0)}(m) \approx \exp(---m^{3/4}) \tag{16.33}$$

so that our assumption $\int_0^\infty \rho(m)dm < \infty$ (see (16.23) and (16.24)) is wrong. Therefore, this approximation is *not* consistent with the bootstrap condition.

2nd approximation: $m_{\max} \gg T$, i.e., there exist masses much larger than the temperature. It will now be shown, that this condition is consistent with the bootstrap assumption (16.21) and actually an exponential mass density is implied. In this case the asymptotic value for K_2

$$\lim_{m\to\infty}\left(K_2\left[\frac{nm}{T}\right]\right) \approx \left(\frac{\pi T}{2nm}\right)^{1/2} \exp(-nm/T) \tag{16.34}$$

can be used in (16.15) to obtain the asymptotic behaviour

$$\rho(m) \approx \sigma(m, V) \approx \exp(m/T_0) \tag{16.35}$$

Multiplying $\exp(m/T_0)$ by (16.34) the first integral ($n = 1$) in (16.15) will diverge for $T \to T_0$, however, the second term ($n = 2$) in (16.15) will diverge for $T \to 2T_0$, the third ($n = 3$) for $T \to 3T_0$, etc., therefore by assuming (16.35) we have to consider only the divergent first integral ($n = 1$) and to neglect all the others in the sum ($n = 2, 3, \ldots \infty$), which can be shown to add up to a finite constant. In order to satisfy the bootstrap formula (16.21), we have to assume in general the following forms for the mass density spectrum $\rho(m)$ and the energy density of the state spectrum $\sigma(m)$:

$$\lim_{m\to\infty} \rho(m) = f(m)\exp(m/T_0) \tag{16.36}$$

$$\lim_{m\to\infty} \sigma(m) = g(m)\exp(m/T_0) \tag{16.37}$$

Substituting (16.36) into (16.15) and (16.37) into (16.22), one has

$$Z(V_0, T)_{(\rho)} = \exp\left(V_0\left(\frac{T_0}{2\pi}\right)^{3/2} \int_{M_0}^{\infty} f(m)m^{3/2}e^{-m\tau}dm\right.$$

$$\left. + V_0 C(M_0, T_0)\right) \tag{16.38}$$

$$Z(V_0, T)_{(\sigma)} = \int_0^\infty g(m)e^{-m\tau}dm \tag{16.39}$$

where

$$\tau \equiv (T_0 - T)/(T_0 T) \tag{16.40}$$

M_0 is some large mass and $V_0 C(M_0, T_0)$ is a finite term containing all the other terms $(n = 2, 3, \ldots \infty)$ in the sum of (16.15). A possible solution, for $f(m)$ and $g(m)$ to satisfy the bootstrap condition (16.21), is

$$\left.\begin{aligned} f(m) &= \frac{f_0}{m^{5/2}} \\ g(m) &= g_0 m^{a-1} \end{aligned}\right\} \tag{16.41}$$

The integral in (16.38) in the limit $\tau \to 0$ (i.e. $T \to T_0$) is

$$\int_{M_0}^\infty f(m)m^{3/2}e^{-m\tau}dm = f_0 \int_{M_0}^\infty \frac{dm}{m}e^{-m\tau} = f_0 \int_{M_0\tau}^\infty \frac{dx}{x}e^{-x}$$

$$\underset{\tau \to 0}{\to} f_0 \left[\ln \frac{T_0}{T_0 - T} + \ln \frac{T_0}{M_0} \right] \tag{16.42}$$

The integral in (16.39) becomes

$$\int_0^\infty g(m)e^{-m\tau}dm = g_0 \int_0^\infty m^{a-1}e^{-m\tau}dm = \frac{g_0 T_0^{2a}}{(T_0 - T)}\Gamma(a) \tag{16.43}$$

Substituting (16.42) and (16.43) in (16.38) and (16.39) therefore, one gets the partition functions Z

$$Z(V_0, T)_{(\rho)} \underset{\tau \to 0}{\to} \left(\frac{T_0}{T_0 - T} \right)^A \exp \left[V_0 C(M_0, T_0) + A \ln \frac{T_0}{M_0} \right] \tag{16.44}$$

$$Z(V_0, T)_{(\sigma)} \underset{\tau \to 0}{\to} \left(\frac{T_0}{T_0 - T} \right)^a \cdot g_0 T_0^a \Gamma(a) \tag{16.45}$$

where $A \equiv f_0 V_0 (T_0/2\pi)^{3/2}$ is a dimensionless number.

In order to derive the bootstrap equation (16.21) also for $T \to T_0$ we assume

$$a = A = f_0 V_0 \left(\frac{T_0}{2\pi} \right)^{3/2}$$

$$aT_0^a \Gamma(a) = \exp \left[V_0 C(M_0, T_0) + A \ln \frac{T_0}{M_0} \right] \tag{16.46}$$

so that in general now, we have

$$\frac{\ln \rho(m)}{\ln \sigma(m, V_0)} = \frac{m/T_0 + \ln f(m)}{m/T_0 + \ln g(m, V_0)}$$

$$= \frac{m/T_0 + \ln f_0 - \frac{5}{2}\ln m}{m/T_0 + \ln g_0 + (a-1)\ln m}$$

$$\underset{m \to \infty}{\longrightarrow} 1 \, (\text{Q.E.D.}) \tag{16.47}$$

Equations (16.41) which satisfy the bootstrap condition are not unique. Any solution for $f(m)$ of the form m^x with $x \leqslant -\frac{5}{2}$ (or any expression with a leading power m^x) satisfies the bootstrap condition (16.47) even for $T \to T_0$. The solution $x = -\frac{5}{2}$ is the endpoint of a continuum set of solutions $x < -\frac{5}{2}$. Hagedorn favours the solution $x = -\frac{5}{2}$ because in strong interaction $\rho(m)$ should increase as fast as possible and this is achieved by $\rho(m) \approx m^{-5/2} \exp(m/T_0)$. Moreover, using this philosophy T_0 should be as small as possible, and this suggests that T_0 should be equal to the lowest mass of the existing hadrons, i.e., $T_0 \approx m_\pi \approx 150 \, \text{MeV}$.

16.2.4 *The thermodynamic functions: pressure and energy*

(a) $$\rho(m) = f_0 m^{-5/2} \exp(m/T_0) \tag{16.48}$$

We shall calculate the pressure and the energy of the system for temperatures near T_0. The partition function Z is given in (16.44), so that

$$\ln Z(V_0, T \to T_0) = V_0 C(M_0, T_0) + A \ln \frac{T_0}{M_0} + A \ln\left(\frac{T_0}{T_0 - T}\right) \tag{16.49}$$

The pressure P is given by

$$P = T\frac{\partial}{\partial V}(\ln Z) = A_1 + A_2 \ln\left(\frac{T_0}{T_0 - T}\right) \tag{16.50}$$

where (16.46) has been used (i.e. A in (16.49) is proportional to V_0) and A_1 and A_2 are two constants which can easily relate to the other constants defined above. The energy of the system is given by

$$E(V_0, T \to T_0) = T^2\frac{\partial}{\partial T}(\ln Z) = \frac{A T_0^2}{T_0 - T} \tag{16.51}$$

It is interesting to point out that in Hagedorn's model, the partition function Z, the pressure, as well as the energy of the system are only well defined for T smaller than T_0. Therefore in this model, T_0 is the universal *maximum temperature* for matter!

(b) $\rho(m) = f_0 m^{-3} \exp(m/T_0)$ (16.52)

In this case, instead of (16.42) one gets

$$\int_{M_0}^{\infty} f(m) m^{3/2} e^{-m\tau} dm = f_0 \int_{M_0}^{\infty} \frac{dm\, e^{-m\tau}}{m^{3/2}}$$

$$= f_0 \tau^{1/2} \int_{M_0 \tau}^{\infty} dx \frac{e^{-x}}{x^{3/2}} \xrightarrow[\tau \to 0]{} \tau^{1/2} \Gamma(-\tfrac{1}{2})   \quad (16.53)$$

Using this relation in (16.38) one gets[†] (instead of (16.49))

$$\ln Z(V_0, T \to T_0) = -B_1 (T_0 - T)^{1/2} + B_2 \quad (B_1 > 0)   \quad (16.54)$$

so that the pressure and energy of the system are

$$P(T \to T_0) = \frac{B_2}{V_0} - \frac{B_1}{V_0}(T_0 - T)^{1/2}   \quad (16.55)$$

$$E(V_0, T \to T_0) = \tfrac{1}{2} B_1 T_0^2 \cdot \frac{1}{(T_0 - T)^{1/2}}   \quad (16.56)$$

In this case one also has a maximum temperature for hadronic matter.

(c) $\rho(m) = f_0 m^x \exp(m/T_0) \quad x < -\tfrac{7}{2}$ (16.57)

Instead of (16.42) one has

$$\int_{M_0}^{\infty} \frac{dm\, e^{-m\tau}}{m^y} \cong \tau^{y_2 - 1} \Gamma(-y + 1) \quad y_2 > 2   \quad (16.58)$$

Implying, from (16.38)

$$\ln Z(V_0, T \to T_0) = -C_1 (T_0 - T)^{y_1} + C_2, y_1 > 1   \quad (16.59)$$

This partition function gives an energy

$$E(V_0, T \to T_0) = C_1 y_1 T_0^2 (T_0 - T)^{y_1 - 1} + C_3   \quad (16.60)$$

From (16.60) one can see that for $x < -\tfrac{7}{2}$ (also $x = -\tfrac{7}{2}$) the energy is a constant for $T_0 \to T$ and therefore above a certain value of energy it is no longer possible to use this philosophy for thermodynamics. Therefore, Hagedorn's philosophy is applicable only for values of x in the domain

$$-\tfrac{7}{2} < x \leqslant -\tfrac{5}{2}   \quad (16.61)$$

[†] $B_1 > 0$ since $\Gamma(-\tfrac{1}{2}) < 0$.

16.2.5 Transverse momentum distribution

In Section 16.2.1 the difficulty of extracting the 'thermal motion' from the total (e.g. center of mass) energy was mentioned (see the 3rd postulate). The main problem is caused by the directional energy across the axis of interactions. Therefore, in order to make some comparison between experiments and the thermodynamical model, looking for transverse momentum measurements was suggested so that the motion along the axis of collision is excluded. This is, of course, only an approximation, but it might give some insight into the thermodynamics of hadronic matter.

The momentum **p** of a particle is divided into two parts, the longitudinal momentum $p_\parallel$ along the axis of collision, and the transverse momentum p which is perpendicular to the collision axis

$$\mathbf{p} = (p_\parallel, p) \tag{16.62}$$

The volume element, in momentum space using this notation, is

$$d^3 p = 2\pi p \, dp \, dp_\parallel \tag{16.63}$$

assuming azymuthal (i.e., cylindrical) symmetry. To a first approximation, the probability of measuring a transverse momentum (magnitude) p for a particle with a mass m_i is given by $W_i(p)$

$$W_i(p) = 2\pi p \int_{-\infty}^{\infty} f_{m_i}(p, p_\parallel, T) dp_\parallel \tag{16.64}$$

where

$$f_m(\mathbf{p}, T) d^3 p = G_m \frac{V}{(2\pi)^3} d^3 p \frac{X_{pm}}{1 \pm X_{pm}} \tag{16.65}$$

where X_{pm} was defined in (16.5) and (16.1)

$$X_{pm} = \exp\{-(p_\parallel^2 + p^2 + m^2)^{1/2}/T\} \tag{16.66}$$

The $-$ sign in (16.65) is for bosons while the $+$ sign is for fermions, and G_m is the degeneracy factor given by[†]

$$G_m = \begin{cases} (2J+1)(2I+1) & \text{if spin } (J) \text{ and charge is not observed} \\ (2J+1) & \text{if spin is not observed but charge is} \\ (2I+1) & \text{if spin is observed but charge is not} \\ 1 & \text{if spin and charge are observed} \end{cases} \tag{16.67}$$

[†] I is the isospin quantum number. In (16.67) if for example the spin of pions (π) is measured but the charge is not observed (i.e., we don't know if it is a π^+, π^- or π^0 event), then $G_m = 2 \times 1 + 1 = 3$, etc.

In the following discussion we neglect the factor G_m, although it is understood to be there if necessary. The integral in (16.64) is calculated in the following way,

$$\int_{-\infty}^{\infty} f_m(\mathbf{p}, T)dp_{\parallel}$$

$$= \frac{V}{(2\pi)^3} \int_{-\infty}^{\infty} \frac{X_{pm}dp}{1 \pm X_{pm}} = \frac{V}{(2\pi)^3} \sum_{n=0}^{\infty} (\pm 1)^n \int_{-\infty}^{\infty} X_{pm}^{n+1} dp_{\parallel}$$

$$= \frac{V}{(2\pi)^3} \sum_{n=0}^{\infty} (\mp 1)^n \int_{-\infty}^{\infty} dp_{\parallel} \exp\left[-\frac{(n+1)}{T}(p_{\parallel}^2 + p^2 + m^2)^{1/2} \right]$$

$$= \frac{2V}{(2\pi)^3} \cdot (p^2 + m^2)^{1/2} \sum_{n=0}^{\infty} (\mp 1)^n K_1\left[\frac{(n+1)}{T}(p^2 + m^2)^{1/2} \right]$$

$$(16.68)$$

where K_1 is the Hankel function of order $n = 0$, usually defined by

$$\int_0^{\infty} p^{2n} \exp\left(-\frac{(p^2 + m^2)^{1/2}}{T} \right) dp$$

$$= (2n - 1)!! m^{2n+1} \left(\frac{T}{m} \right)^n K_{n+1}\left(\frac{m}{T} \right) \qquad (16.69)$$

$$\int_0^{\infty} p^{2n+1} \exp\left(-\frac{(p^2 + m^2)^{1/2}}{T} \right) dp$$

$$= 2^n n! \left(\frac{2}{n} \right)^{1/2} m^{2n+2} \left(\frac{T}{m} \right)^{n+1/2} K_{n+3/2}\left(\frac{m}{T} \right) \qquad (16.70)$$

where $(2k - 1)!! = 1 \times 3 \times 5 \ldots (2k - 1)$ and $(-1)!! = 1$. Substituting (16.68) into (16.64) we obtain

$$W(p) = \frac{V_0}{2\pi^2} p(p^2 + m^2)^{1/2} \sum_{n=0}^{\infty} (\mp 1)^n K_1\left[\frac{(n+1)}{T}(p^2 + m^2)^{1/2} \right] \qquad (16.71)$$

where the $+$ sign is for bosons and the $-$ sign for fermions.

We now calculate the mean transverse moments

$$\langle p(m) \rangle = \frac{\displaystyle\int_0^{\infty} pW(p)dp}{\displaystyle\int_0^{\infty} W(p)dp}$$

$$= \frac{\sum\limits_{n=0}^{\infty} (\mp 1)^n \int_0^{\infty} p^2 (p^2 + m^2)^{1/2} K_1 \left[\frac{(n+1)}{T} (p^2 + m^2)^{1/2} \right] dp}{\sum\limits_{n=0}^{\infty} (\mp 1)^n \int_0^{\infty} p(p^2 + m^2)^{1/2} K_1 \left[\frac{(n+1)}{T} (p^2 + m^2)^{1/2} \right] dp}$$

(16.72)

We now use the formula

$$\int_0^{\infty} x^{2k+1} (x^2 + m^2)^{1/2} K_1 \left[\frac{(x^2 + m^2)^{1/2}}{T} \right] dx$$

$$= 2^k \Gamma(k+1) m^{k+2} T^{k+1} K_{k+2} \left(\frac{m}{T} \right)$$

(16.73)

in (16.72) to obtain the exact result, (Imaeda 1967):

$$\langle p(m) \rangle = \left(\frac{mT}{2} \right)^{1/2} \frac{\sum\limits_{n=0}^{\infty} (\mp 1)^n K_{5/2} \left(\frac{(n+1)m}{T} \right)}{\sum\limits_{n=0}^{\infty} (\mp 1)^n K_2 \left(\frac{(n+1)m}{T} \right)}$$

(16.74)

Taking into consideration the asymptotic expansion of $K_n(x)$

$$\lim_{x \to \infty} K_n(x) \approx \left(\frac{\pi}{2x} \right)^{1/2} e^{-x} \left[1 + \frac{(n^2 - \frac{1}{4})}{2x} + \cdots \right]$$

(16.75)

and the definition of $K_{n+1/2}(x)$

$$K_{n+1/2}(x) = \left(\frac{\pi}{2x} \right)^{1/2} e^{-x} \sum_{k=0}^{n} \frac{(n+k)!}{k!(n-k)!} \left(\frac{1}{2x} \right)^k$$

(16.76)

one can neglect all terms except $n = 0$ in the infinite series of (16.74) for $m/T \gg 1$, i.e., large masses relative to the temperature of the system. Therefore, (16.74) can be approximated by

$$\langle p(m, T) \rangle \approx \left(\frac{\pi mT}{2} \right)^{1/2} \frac{K_{5/2} \left(\frac{m}{T} \right)}{K_2 \left(\frac{m}{T} \right)}$$

(16.77)

Again using (16.75) and (16.76) to take only the leading terms for $m/T \gg 1$, we get from (16.77)

$$\langle p \rangle \approx \left(\frac{\pi mT}{2} \right)^{1/2}$$

(16.78)

or equivalently

$$\frac{\langle p \rangle^2}{2m} \approx \frac{\pi}{4} T \qquad (16.79)$$

Equation (16.79) implies that the two degrees of freedom of translational motion give the amount of energy expected from the classical equipartition theorem (see Chapter 4). The result agrees with the non-relativistic kinetic energy, because in our approximations in deriving (16.79) we assumed very large masses, so that the motion in this case is non-relativistic. The factor in (16.79) is $\pi/4$ instead of 1, as expected from the equipartition theory. This came about because we calculated $\langle p \rangle^2$ instead of $\langle p^2 \rangle$ as should be the case for kinetic energy. A similar calculation with the same approximations (i.e. $m \to \infty$) implies exactly

$$\lim_{m \to \infty} \frac{\langle p^2 \rangle}{2m} = T \qquad (16.80)$$

in agreement with the 'classical' equipartition theorem. Equation (16.80) might serve as an 'experimental definition' of temperature in hadronic interactions.

A free particle inside a box and the density of states

We consider a free particle (of mass m) inside a cube of volume L^3. Since inside the box, the potential energy is zero, the Schrödinger equation will be given by

$$\nabla^2 \psi + \frac{2m}{\hbar^2} \varepsilon \psi = 0 \quad \left. \begin{array}{l} 0 < x < L \\ 0 < y < L \\ 0 < z < L \end{array} \right\} \tag{A1.1}$$

The particle cannot go outside the box and therefore we must solve (A1.1) subject to the boundary condition that ψ should vanish everywhere on the surface of the cube. We use the method of separation of variables to solve (A1.1) and write

$$\psi(x, y, z) = X(x)Y(y)Z(z) \tag{A1.2}$$

Substituting in (A1.1) and dividing by ψ we obtain

$$\frac{1}{X} \frac{d^2 X}{dx^2} + \frac{1}{Y} \frac{d^2 Y}{dy^2} + \frac{1}{Z} \frac{d^2 Z}{dz^2} + k^2 = 0 \tag{A1.3}$$

where

$$k^2 = \frac{2m\varepsilon}{\hbar^2} \tag{A1.4}$$

The variables have indeed separated out and we may write

$$\frac{1}{X} \frac{d^2 X}{dx^2} = -k_x^2 \tag{A1.5}$$

$$\frac{1}{Y} \frac{d^2 Y}{dy^2} = -k_y^2 \tag{A1.6}$$

and

$$\frac{1}{Z}\frac{d^2Z}{dz^2} = -k_z^2 \tag{A1.7}$$

where k_x^2, k_y^2 and k_z^2 are constants subject to the condition that

$$k_x^2 + k_y^2 + k_z^2 = k^2 \tag{A1.8}$$

In (A1.5)–(A1.7), we have set each term equal to a negative constant; otherwise the boundary conditions cannot be satisfied. The solution of (A1.5) is

$$X(x) = A \sin k_x x + B \cos k_x x \tag{A1.9}$$

and since ψ has to vanish at *all* points on the surfaces $x = 0$ and $x = L$ we must have $B = 0$ and

$$k_x = n_x \pi/L \quad n_x = 1, 2, 3, \ldots \tag{A1.10}$$

Similarly, the allowed values of k_y and k_z are

$$k_y = n_y \pi/L \quad n_y = 1, 2, 3, \ldots \tag{A1.11}$$

$$k_z = n_z \pi/L \quad n_z = 1, 2, 3, \ldots \tag{A1.12}$$

It may be noted that negative values of n_x, n_y and n_z do not lead to an independent solution. The wave function remains the same except for a trivial change of sign. Using (A1.4) the permitted values of ε are given by

$$\varepsilon = \frac{\pi^2 \hbar^2}{2mL^2}(n_x^2 + n_y^2 + n_z^2) \tag{A1.13}$$

Now the number of states whose x component of $\mathbf{k}$ lies between k_x and $k_x + dk_x$ would simply be the number of integers lying between $k_x L/\pi$ and $(k_x + dk_x)L/\pi$. This number would be approximately equal to $L dk_x/\pi$. Similarly, the number of states whose y and z components of $\mathbf{k}$ lie between k_y and $k_y + dk_y$ and $k_z + dk_z$ would respectively, be

$$\frac{L}{\pi}dk_y \quad \text{and} \quad \frac{L}{\pi}dk_z$$

Thus there will be

$$\left(\frac{L}{\pi}dk_x\right)\left(\frac{L}{\pi}dk_y\right)\left(\frac{L}{\pi}dk_z\right) = \frac{V}{\pi^3}dk_x dk_y dk_z \tag{A1.14}$$

states in the range $dk_x dk_y dk_z$ of $\mathbf{k}$; here $V(= L^3)$ represents the volume of the box. Since $\mathbf{p} = \hbar \mathbf{k}$, the number of states per unit volume in the $\mathbf{p}$ space (momentum space) would be

$$\frac{V}{\pi^3 \hbar^3} \tag{A1.15}$$

If $P(p)dp$ represents the number of states whose $|\mathbf{p}|$ lies between p and $p + dp$ then

$$P(p)dp = \frac{1}{8} \times \frac{V}{\pi^3 \hbar^3} 4\pi p^2 dp = \frac{V}{h^3} 4\pi p^2 dp \tag{A1.16}$$

where the factor $4\pi p^2 dp$ represents the volume element (in the $\mathbf{p}$ space) for $|\mathbf{p}|$ to lie between p and $p + dp$ and the factor $\frac{1}{8}$ is due to the fact that k_x, k_y and k_z can take only positive values (see (A1.10)–(A1.12)) so that while counting the states in the $\mathbf{k}$ (or $\mathbf{p}$) space we must consider only the positive octant. Now, if $g(\varepsilon)d\varepsilon$ represents the number of states with energy $(= p^2/2m)$ lying between ε and $\varepsilon + d\varepsilon$ then

$$g(\varepsilon)d\varepsilon = P(p)dp = \frac{V}{8\pi^3 \hbar^3} 4\pi p^2 dp \tag{A1.17}$$

Using the relation $p^2 = 2m\varepsilon$ we get

$$g(\varepsilon)d\varepsilon = \frac{(2m)^{3/2} V}{4\pi^2 \hbar^3} \varepsilon^{1/2} d\varepsilon \tag{A1.18}$$

for the density of states. If there are degeneracies associated with each energy state then the density of states will be given by

$$g(\varepsilon)d\varepsilon = G \frac{(2m)^{3/2} V}{4\pi^2 \hbar^3} \varepsilon^{1/2} d\varepsilon \tag{A1.19}$$

where G is known as the degeneracy parameter. For example, for an electron gas, each electron has two additional degrees of freedom due to its spin and therefore there will be twice as many states as given by (A1.18) and G will be equal to 2.

It is of interest to mention that instead of using the boundary condition that ψ vanishes at the surface of the cube, it is often more convenient to use periodic boundary conditions according to which

$$\left.\begin{array}{l} \psi(x + L, y, z) = \psi(x, y, z) \\ \psi(x, y + L, z) = \psi(x, y, z) \\ \Psi(x, y, z + L) = \psi(x, y, z) \end{array}\right\} \tag{A1.20}$$

Then instead of choosing the sine and cosine functions as solutions (see (A1.9)) it is more convenient to choose

$$e^{ik_x x}, e^{ik_y y} \quad \text{and} \quad e^{ik_z z} \tag{A1.21}$$

as solutions. Using (A1.20) we get

$$e^{ik_xL} = 1 = e^{ik_yL} = e^{ik_zL} \tag{A1.22}$$

Thus the allowed values of k_x are

$$k_x = \frac{2n_x\pi}{L} \quad n_x = 0, \pm 1, \pm 2,\ldots$$

Similarly

$$k_y = \frac{2n_y\pi}{L} \quad n_y = 0, \pm 1, \pm 2,\ldots \tag{A1.23}$$

and

$$k_z = \frac{2n_z\pi}{L} \quad n_z = 0, \pm 1, \pm 2,\ldots$$

Notice that a reversal of sign of k_x changes the wave function, thus $n_x = 2$ and $n_x = -2$ correspond to two different states although they both correspond to the same energy eigenvalue which is now given by

$$\varepsilon = \frac{2\pi\hbar^2}{mL^2}(n_x{}^2 + n_y{}^2 + n_z{}^2) \tag{A1.24}$$

Proceeding as before we will now have (cf. (A1.14))

$$\left(\frac{L}{2\pi}dk_x\right)\left(\frac{L}{2\pi}dk_y\right)\left(\frac{L}{2\pi}dk_z\right) = \frac{V}{8\pi^3}dk_xdk_ydk_z \tag{A1.25}$$

states in the range $dk_xdk_ydk_z$. Further, the number of states per unit volume in the **p** space would be (cf. (A1.15))

$$\frac{V}{8\pi^3\hbar^3}\left(=\frac{V}{h^3}\right) \tag{A1.26}$$

implying that a volume h^3 in the phase space corresponds to one state. It may be noted that the number of states whose $|\mathbf{p}|$ lies between p and $p + dp$ would be

$$P(p)dp = \frac{V}{8\pi^3\hbar^3}4\pi p^2dp \tag{A1.27}$$

which is identical to (A1.16) and thus we would get the same expression for $g(\varepsilon)$ (see (A1.18)). We may note that in (A1.27) we do not have an extra $\frac{1}{8}$ factor because p_x, p_y and p_z can now also take negative values. Thus we see that the two boundary conditions lead to the same expression for the density of states.

The Stirling formula

In this appendix we will derive an approximate formula for $n!$ and $\ln n!$ when n is a large number. We note that

$$\ln n! = \ln 1 + \ln 2 + \cdots + \ln n \qquad (A2.1)$$

when n is large we have

$$\ln n! \approx \int_1^n \ln x\, dx = n \ln n - n + 1 \qquad (A2.2)$$

or, since n is assumed to be a large number

$$\ln n! \approx n \ln n - n \qquad (A2.3)$$

which is the Stirling formula. A more accurate formula is given by

$$n! \approx \left(\frac{n}{e}\right)^n (2\pi n)^{1/2} \left[1 + \frac{1}{12n} + \frac{1}{228n^2} + \cdots \right] \qquad (A2.4)$$

Table of Fermi–Dirac functions[†]

In this appendix we have tabulated the values of the functions,

$$\tfrac{2}{3}F_{3/2}(\eta), F_{1/2}(\eta) = F(\eta) = F, wF', w^2F'' \quad \text{and} \quad w^3F'''$$

where,

$$F_k(\eta) \equiv \int_0^\infty \frac{x^k \mathrm{d}x}{e^{x-\eta}+1} \quad F'(\eta) = \frac{\mathrm{d}}{\mathrm{d}\eta}F(\eta)$$

and $w = 0.1$, which represents the interval in the argument. We may note the following relation,

$$F_k'(\eta)\left[= \frac{\mathrm{d}}{\mathrm{d}\eta}F_k(\eta) \right] = kF_{k-1}(\eta) \quad k > 0$$

Thus

$$F(\eta)[= F_{1/2}(\eta)] = \frac{\mathrm{d}}{\mathrm{d}\eta}[\tfrac{2}{3}F_{3/2}(\eta)].$$

In the table the dot symbol after the last printed digit indicates that the next digit lies between 3 and 7. The functions are listed to the sixth decimal place for $-4.0 \leqslant \eta \leqslant +4.0$, and to the fifth decimal place for $4.0 \leqslant \eta \leqslant 20.0$. As an example, for $\eta = 2.0$, we have, $wF' = 0.129\,770$, $w^2F'' = 0.003\,706$ and $w^3F''' = -0.000\,047$, giving $F' = 1.29770$, $F'' = 0.3706$ and $F''' = -0.047$. One can check these values by using (9.218). Similarly, for $\eta = 20$, we have $wF' = 0.44675$ and $w^2F'' = 0.00112$, giving,

$$F' = 4.4675 \quad \text{and} \quad F'' = 0.112.$$

The w^3F''' values are not given for $\eta > 8.0$ as they are smaller than $0.000\,01$.

[†] The following table is reproduced from McDougall and Stoner (1938). Beer, Chase and Choquard (1955) have extended the McDougall–Stoner table to include the values of $F_{5/2}$, $F_{7/2}$, $F_{9/2}$ and $F_{11/2}$.

η	$\frac{2}{3}F_{3/2}$	F	wF'	w^2F''	w^3F'''
− 4.0	0.016 179˙	0.016 128	1 602˙	158	15˙
− 3.9	0.017 875	0.017 812	1 768˙	174˙	17
− 3.8	0.019 748	0.019 670˙	1 952	192	18˙
− 3.7	0.021 816	0.021 721˙˙	2 153˙	211˙	20˙
− 3.6	0.024 099	0.023 984˙˙	2 376	233	22˙
− 3.5	0.026 620˙	0.026 480˙	2 620˙	256˙	24˙
− 3.4	0.029 404	0.029 233˙	2 889˙	310˙	27
− 3.3	0.032 476˙	0.032 269	3 186	310˙	29˙
− 3.2	0.035 868	0.035 615	3 511˙	341˙	32˙
− 3.1	0.039 611	0.039 303	3 870	375	35˙
− 3.0	0.043 741	0.043 366˙	4 263	412	38˙
− 2.9	0.048 298	0.047 842	4 695	452	42
− 2.8	0.053 324˙	0.052 770	5 168	496	45˙
− 2.7	0.058 868˙	0.058 194	5 687˙	543	49˙
− 2.6	0.064 981˙	0.064 161˙	6 256	595	54
− 2.5	0.071 720˙	0.070 724˙	6 879	651	58
− 2.4	0.079 148	0.077 938˙	7 559˙	711	63
− 2.3	0.087 332	0.085 864	8 303	776˙	68
− 2.2	0.096 347	0.094 566˙	9 114	846˙	73
− 2.1	0.016 273˙	0.104 116	9 997˙	922	78
− 2.0	0.177 200˙	0.114 588	10 959˙	1 003	83˙
− 1.9	0.129 224˙	0.126 063	12 005	1 089	89
− 1.8	0.142 449˙	0.138 627˙	13 139	1 180˙	94˙
− 1.7	0.156 989˙	0.152 373	14 368	1 278	100
− 1.6	0.172 967	0.167 397	15 697	1 381	105˙
− 1.5	0.190 515	0.183 802	17 131	1 489	110˙
− 1.4	0.209 777	0.201 696	18 676	1 602	116
− 1.3	0.230 907˙	0.221 193	20 337	1 720˙	120˙
− 1.2	0.254 073	0.242 410˙	22 118	1 843	124˙
− 1.1	0.279 451	0.265 471	24 024	1 969	128
− 1.0	0.307 232˙	0.290 501	26 057˙	2 098˙	131
− 0.9	0.337 621	0.317 630	28 222	2 231	133
− 0.8	0.370 833	0.346 989˙	30 520	2 364˙	134
− 0.7	0.407 098	0.378 714	32 951˙	2 499	134˙
− 0.6	0.446 659	0.412 937	35 517˙	2 633	133˙
− 0.5	0.489 773	0.449 793	38 217	2 766	131˙
− 0.4	0.536 710	0.489 414˙	41 048	2 896	128˙
− 0.3	0.587 752˙	0.531 931˙	44 007˙	3 022	124
− 0.2	0.643 197	0.577 470˙	47 091	3 144	119
− 0.1	0.703 351	0.626 152˙	50 293	3 259˙	112˙
0.0	0.768 536	0.678 094	53 608	3 368˙	105
0.1	0.839 082	0.733 403	57 027˙	3 470	97
0.2	0.915 332	0.792 181˙	60 544	3 562	88
0.3	0.997 637	0.854 521	64 149	3 646	78˙
0.4	1.086 358	0.920 505˙˙	67 832˙	3 719˙	69

(Contd.)

η	$\frac{2}{3}F_{3/2}$	F	wF'	w^2F''	w^3F'''
0.5	1.181 862˙	0.990 209	71 584˙	3 783	58˙
0.6	1.284 526	1.063 694˙	75 395	3 836˙	48
0.7	1.394 729	1.141 015˙	79 254	3 880	38
0.8	1.512 858	1.222 215˙	83 151˙	3 913	28
0.9	1.639 302˙	1.307 327˙	87 076˙	3 936	18˙
1.0	1.774 455	1.396 375	91 020˙	3 950	9˙
1.1	1.918 709˙	1.489 372	94 974	3 955	1
1.2	2.072 461	1.586 323˙	98 928	3 952	− 7
1.3	2.236 106	1.687 226	103 875	3 941	− 14˙
1.4	2.410 037˙	1.792 068˙	106 807˙	3 923	− 21˙
1.5	2.594 650	1.900 833˙	110 718˙	3 898˙	− 27˙
1.6	2.790 334˙	2.013 496˙	114 602	3 868˙	− 32˙
1.7	2.997 478˙	2.130 027	118 453˙	3 833˙	− 37
1.8	3.216 467˙	2.250 391	122 267˙	3 794˙	− 41
1.9	3.447 683	2.374 548˙	126 041	3 751˙	− 44˙
2.0	3.691 502	2.502 458	129 770	3 706	− 47
2.1	3.948 298	2.634 072˙	133 451˙	3 657˙	− 49
2.2	4.218 438˙	2.769 344˙	137 084	3 607˙	− 50˙
2.3	4.502 287	2.908 224	140 666	3 556	− 52
2.4	4.800 202	3.050 659˙	144 196	3 504	− 53
2.5	5.112 536	3.196 598˙	147 673	3 450˙	− 53˙
2.6	5.439 637	3.345 988	151 097	3 397	− 53˙
2.7	5.781 847	3.498 775	154 467˙	3 343˙	− 53˙
2.8	6.139 503	3.654 905˙	157 784˙	3 290˙	− 53
2.9	6.512 937˙	3.814 326˙	161 049	3 238	− 52
3.0	6.902 476˙	3.976 985˙	164 261	3 186	− 51˙
3.1	7.308 441	4.142 831	167 421	3 135	− 50˙
3.2	7.731 147	4.311 811	170 531	3 085	− 50
3.3	8.170 906	4.483 876˙	173 591	3 035˙	− 48˙
3.4	8.628 023˙	4.658 977˙	176 602˙	2 987˙	− 47˙
3.5	9.102 801	4.837 066	179 566˙	2 941	− 46
3.6	9.595 535	5.018 095	183 484˙	2 895	− 45
3.7	10.106 516˙	5.202 020	185 357	2 850˙	− 44
3.8	10.636 034	5.388 795	188 186	2 807˙	− 42˙
3.9	11.184 369	5.578 378	190 972˙	2 765˙	− 41
4.0	11.751 801˙	5.770 726˙	193 717˙	2 725	− 40
4.1	12.338 60˙	5.965 80	196 42	2 68˙	− 4
4.2	12.945 05	6.163 56	199 09	2 65	− 4
4.3	13.571 40˙	6.363 96	201 72	2 61	− 3˙
4.4	14.217 93	6.566 98	204 31	2 57˙	− 3˙
4.5	14.884 89	6.772 57˙	206 87	2 54	− 3˙
4.6	15.572 53	6.980 70˙	209 39	2 50˙	− 3˙
4.7	16.281 11	7.191 34˙	211 88	2 47˙	− 3
4.8	17.010 88	7.404 45˙	214 34	2 44	− 3
4.9	17.762 08˙	7.620 01	216 76˙	2 41	− 3

η	$\frac{2}{3}F_{3/2}$	F	wF'	w^2F''	w^3F'''
5.0	18.534 96˙	7.837 97˙	219 16˙	2 38	− 3
5.1	19.329 76	8.058 32˙	221 53	2 35˙	− 3
5.2	20.146 71	8.281 03	223 87	2 32˙	− 3
5.3	20.986 04	8.506 06	226 18˙	2 30	− 2˙
5.4	21.847 99˙	8.733 39	228 47	2 27˙	− 2˙
5.5	22.732 79˙	8.962 99˙	230 73˙	2 25	− 2˙
5.6	23.640 67	9.194 85	232 97	2 22˙	− 2˙
5.7	24.571 84	9.428 93	235 18˙	2 20	− 2˙
5.8	25.526 53	9.665 21	237 37˙	2 18	− 2
5.9	26.504 95˙	9.903 67	239 54	2 16	− 2
6.0	27.506 33˙	10.144 28˙	241 69	2 13˙	− 2
6.1	28.533 88˙	10.387 03˙	243 81	2 11˙	− 2
6.2	29.584 81˙	10.631 90	245 91˙	2 09˙	− 2
6.3	30.660 33˙	10.878 86	248 00	2 07˙	− 2
6.4	31.760 65˙	11.127 89˙	250 06˙	2 05˙	− 2
6.5	32.885 98	11.378 98˙	252 11	2 03˙	− 2
6.6	34.036 52	11.632 11˙	254 14	2 02	− 2
6.7	35.212 47	11.887 26	256 15	2 00	− 2
6.8	36.414 04	12.144 40˙	258 14	1 98˙	− 2˙
6.9	37.641 42	12.403 54	260 12	1 97	− 1˙
7.0	38.894 81	12.664 64	262 08	1 95	− 1˙
7.1	40.174 41	12.927 69	264 02	1 93˙	− 1˙
7.2	41.480 41˙	13.192 67˙	265 95	1 92	− 1˙
7.3	42.813 01	13.459 58	267 86	1 90˙	− 1˙
7.4	44.172 39˙	13.728 39˙	269 76	1 89	− 1˙
7.5	45.558 75	13.999 10	271 64˙	1 87˙	− 1˙
7.6	44.972 27˙	14.271 68	273 51˙	1 86	− 1˙
7.7	48.413 15	14.546 12	275 37	1 85	− 1˙
7.8	49.881 56	14.822 41	277 21	1 83˙	− 1˙
7.9	51.377 69˙	15.100 53˙	279 04	1 82	− 1
8.0	52.901 73	15.380 48˙	280 85˙	1 81	
8.1	54.453 85	15.662 24	282 66	1 80	
8.2	56.034 24	15.945 80	284 45	1 78˙	
8.3	57.643 07	16.231 14˙	286 23	1 77˙	
8.4	59.280 52˙	16.518 26	288 00	1 76	
8.5	60.946 78	16.807 14	289 75˙	1 75	
8.6	62.642 01	17.097 76˙	291 50	1 74	
8.7	64.366 39	17.390 13	293 23˙	1 73	
8.8	66.120 09˙	17.684 23	294 95˙	1 72	
8.9	67.903 29˙	17.980 04	296 67	1 71	
9.0	69.716 16	18.277 56	298 37	1 69˙	
9.1	71.558 86˙	18.576 77˙	300 06	1 68˙	
9.2	73.431 57	18.877 68	301 74˙	1 67˙	
9.3	75.334 45˙	19.180 26	303 41˙	1 67	
9.4	77.267 68	19.484 51	305 08	1 66	
9.5	79.231 41	19.790 41	306 73	1 65	

(*Contd.*)

η	$\frac{2}{3}F_{3/2}$	F	wF'	w^2F''
9.6	81.225 82	20.097 96˙	308 37	1 64
9.7	83.251 06	20.407 15˙	310 01	1 63
9.8	85.307 30	20.717 97˙	311 63	1 62
9.9	87.394 71	21.030 42	313 25	1 61
10.0	89.513 44	21.344 47	314 86	1 60˙
10.1	91.663 65˙	21.660 13	316 45˙	1 59˙
10.2	93.845 52	21.977 38	318 04˙	1 58˙
10.3	96.059 18˙	22.296 22	319 63	1 58
10.4	98.304 81˙	22.616 64	321 20˙	1 57
10.5	100.582 56˙	22.938 62˙	322 77	1 56
10.6	102.892 59	23.262 17˙	324 33	1 55˙
10.7	105.235 05	23.587 28	325 88	1 54˙
10.8	107.610 10	23.913 93	327 42˙	1 54
10.9	110.017 89	24.242 12˙	328 96	1 53
11.0	112.458 57˙	24.571 84˙	330 48˙	1 52˙
11.1	114.932 31	24.903 09˙	332 01	1 52
11.2	117.439 24˙	25.235 86	333 52	1 51
11.3	119.979 53	25.570 13˙	335 03	1 50˙
11.4	122.553 32	25.905 91˙	336 53	1 49˙
11.5	125.160 76˙	26.243 19	338 02	1 49
11.6	127.802 01	26.581 95˙	339 51	1 48
11.7	130.477 20˙	26.922 20˙	340 98˙	1 47˙
11.8	133.186 50	27.263 93	342 46	1 47
11.9	135.930 04	27.607 12	343 92˙	1 46˙
12.0	138.707 97˙	27.951 78	345 38˙	1 45˙
12.1	141.520 44˙	28.297 89	346 84	1 45
12.2	144.367 60	28.645 45˙	348 29	1 44˙
12.3	147.249 58˙	28.994 46˙	349 73	1 44
12.4	150.166 54	29.344 91	351 16˙	1 43
12.5	153.118 61˙	29.696 79	352 59˙	1 42˙
12.6	156.105 94˙	30.050 09˙	354 01˙	1 42
12.7	159.128 68	30.404 82	355 43˙	1 41˙
12.8	162.186 96	30.760 96	356 84˙	1 41
12.9	165.280 92	31.118 51	358 25	1 40˙
13.0	168.410 71	31.477 46˙	359 65	1 40
13.1	171.576 46	31.837 81˙	361 05	1 39
13.2	174.778 31˙	32.199 56	362 43˙	1 38˙
13.3	780.016 42	32.562 68˙	363 82	1 38
13.4	181.290 90	32.927 20	365 20	1 37˙˙
13.5	184.601 90	33.293 08˙	366 57	1 37
13.6	187.949 56	33.660 34	367 94	1 36˙
13.7	191.334 01˙	34.028 96	369 30˙	1 36
13.8	194.755 40	34.398 94˙	370 66	1 35˙
13.9	198.213 85	34.770 28	372 01˙	1 35
14.0	201.709 50	35.142 97	373 36	1 34˙

η	$\frac{2}{3}F_{3/2}$	F	wF'	w^2F''	w^3F'''
14.1	205.242 49	35.517 00·	374 70·	1 34	
14.2	208.812 95	35.892 38	376 04	1 33·	
14.3	212.421 01	36.269 08·	377 37·	1 33	
14.4	216.066 81	36.647 12·	378 70·	1 32·	
14.5	219.750 48	37.026 49	380 03	1 32	
14.6	223.472 15	37.407 18	381 34·	1 31·	
14.7	227.231 96	37.789 18	382 66	1 31	
14.8	231.030 03	38.172 50	383 97	1 31	
14.9	234.866 50	38.557 12	385 27·	1 30·	
15.0	238.741 50	38.943 04·	386 57·	1 30	
15.1	242.655 15·	39.330 27	387 87	1 29·	
15.2	246.607 59·	39.718 79	389 16·	1 29	
15.3	250.598 95·	40.108 59·	390 45	1 28·	
15.4	254.629 36	40.499 69	391 73·	1 28	
15.5	258.698 93·	40.892 06·	393 01·	1 27·	
15.6	262.807 81·	41.285 71·	394 29	1 27	
15.7	266.956 12	41.680 64	395 56	1 27	
15.8	271.143 98·	42.076 83	396 82·	1 26·	
15.9	275.371 53	42.474 29	398 09	1 26	
16.0	279.638 88·	42.873 00·	399 34	1 25·	
16.1	283.946 17	43.272 98	400 60	1 25	
16.2	288.293 52	43.674 20·	401 85	1 25	
16.3	292.681 05·	44.076 68	403 09	1 24·	
16.4	297.108 90	44.480 39·	404 34	1 24	
16.5	301.577 17·	44.885 35·	405 58	1 23·	
16.6	306.086 01	45.291 55	406 81	1 23·	
16.7	310.635 53	45.698 98	408 04·	1 23	
16.8	315.225 85	46.107 63·	409 27	1 22·	
16.9	319.857 09·	46.517 52	410 49·	1 22	
17.0	324.529 39	46.928 62·	411 71·	1 22	
17.1	329.242 86	47.340 95	412 93	1 21·	
17.2	333.997 62	47.754 48·	414 14·	1 21	
17.3	338.793 80	48.169 23·	415 35	1 21	
17.4	343.631 51	48.585 19	416 56	1 20·	
17.5	348.510 87·	49.002 35	417 76	1 20	
17.6	353.432 02	49.420 71	418 96	1 19·	
17.7	358.395 06	49.840 26·	420 15·	1 19·	
17.8	363.400 11	50.261 01	421 34·	1 19	
17.9	368.447 30	50.682 95	422 53·	1 18·	
18.0	373.536 74	51.106 08	423 72	1 18·	
18.1	378.668 55·	51.530 39	424 90	1 18	
18.2	383.842 86	51.955 87·	426 02	1 17·	
18.3	389.059 77	52.382 54	427 25	1 17·	
18.4	394.319 40·	52.810 38	428 42·	1 17	
18.5	399.621 88·	53.239 39	429 59	1 17	

(Contd.)

η	$\frac{2}{3}F_{3/2}$	F	wF'	w^2F''
18.6	404.967 32	53.669 56˙	430 76	1 16˙
18.7	410.355 83˙	54.100 90	431 92	1 16
18.8	415.787 54	54.533 40	433 08	1 16
18.9	421.262 55˙	54.967 06	434 23˙	1 15˙
19.0	426.780 99	55.401 87	435 38˙	1 15
19.1	432.342 97	55.837 83	436 53˙	1 15
19.2	437.948 59˙	56.274 94	437 68˙	1 14˙
19.3	443.497 99˙	56.713 20	438 82˙	1 14
19.4	449.291 27˙	57.152 59˙	439 97	1 14
19.5	455.028 55	57.593 13	441 10˙	1 13˙
19.6	460.809 94	58.034 80˙	442 24	1 13˙
19.7	466.635 55	58.477 61	443 37	1 13
19.8	472.505 50	58.921 54˙	444 50	1 13
19.9	478.419 89˙	59.366 61	445 62˙	1 12˙
20.0	484.378 85˙	59.812 79˙	446 75	1 12

Derivation of the virial theorem result

We first prove that

$$I_k'(\eta) = kI_{k-1}(\eta) \tag{A4.1}$$

The proof is simple. Starting from the definition of the Fermi–Dirac function (see (9.158)) we have

$$I_k'(\eta) = \frac{\mathrm{d}}{\mathrm{d}\eta} I_k(\eta) = \frac{\mathrm{d}}{\mathrm{d}\eta} \left[\int_0^\infty \frac{y^k \mathrm{d}y}{e^{y-\eta}+1} \right]$$

$$= -\int_0^\infty y^k \frac{\mathrm{d}}{\mathrm{d}y} \left[\frac{1}{e^{y-\eta}+1} \right] \mathrm{d}y$$

$$= -y^k \left[\frac{1}{e^{y-\eta}+1} \right] \Big|_0^\infty + k \int_0^\infty \frac{y^{k-1} \mathrm{d}y}{e^{y-\eta}+1} \tag{A4.2}$$

where in the last step we have integrated by parts. The first term on the right hand side of (A4.2) vanishes at both limits so we readily obtain (A4.1). Now, the kinetic energy term is given by (see (9.166)).

$$E_{\mathrm{kin}} = \frac{ZkTa}{\Psi(0)} J_1 \tag{A4.3}$$

where

$$J_1 \equiv \int_0^1 \mathrm{d}\xi \, \xi^2 I_{3/2} \left[\frac{\Psi(\xi)}{\xi} \right] \tag{A4.4}$$

Integrating by parts, we get

$$J_1 = \left[\tfrac{1}{3}\xi^3 I_{3/2}\left(\frac{\Psi}{\xi}\right) \right]_0^1 - \frac{1}{3} \int_0^1 \xi^3 \frac{\mathrm{d}}{\mathrm{d}\xi} \left[I_{3/2}\left(\frac{\Psi}{\xi}\right) \right] \mathrm{d}\xi$$

$$= \tfrac{1}{3} I_{3/2}[\Psi(1)] - \frac{1}{3} \int_0^1 \xi^3 I_{3/2}'\left(\frac{\Psi(\xi)}{\xi}\right) \frac{\mathrm{d}}{\mathrm{d}\xi} \left[\frac{\Psi(\xi)}{\xi} \right] \tag{A4.5}$$

where prime denotes differentiation with respect to the argument and in writing the first term we have used the following relation

$$\left\{ \xi^3 I_{3/2}\left(\frac{\Psi(\xi)}{\xi}\right) \right\}_{\xi=0} \to 0 \tag{A4.6}$$

This follows from the fact that

$$I_{1/2}(\eta) \approx \eta^{3/2} \quad \text{for} \quad \eta \to \infty \tag{A4.7}$$

$$I_{3/2}(\eta) \approx \eta^{5/2} \quad \text{for} \quad \eta \to \infty \tag{A4.8}$$

Thus

$$\eta^{-3} I_{3/2}(\eta) \approx \eta^{-1/2} \quad \text{for} \quad \eta \to \infty \tag{A4.9}$$

Returning to (A4.5) we have

$$J_1 = \tfrac{1}{3} I_{3/2}[\Psi(1)] - \frac{1}{2}\int_0^1 \xi[\xi\Psi'(\xi) - \Psi(\xi)] I_{1/2}\left[\frac{\Psi}{\xi}\right] d\xi \tag{A4.10}$$

where we have used (A4.1). We write the above equation in the form

$$J_1 = \tfrac{1}{3} I_{3/2}[\Psi(1)] + \tfrac{1}{2}(J_2 - J_3) \tag{A4.11}$$

where

$$J_2 = \int_0^1 \xi\Psi(\xi) I_{1/2}\left[\frac{\Psi(\xi)}{\xi}\right] d\xi \tag{A4.12}$$

and

$$J_3 = \int_0^1 \xi^2 \Psi'(\xi) I_{1/2}\left[\frac{\Psi(\xi)}{\xi}\right] d\xi \tag{A4.13}$$

Thus

$$2E_{\text{kin}} = \frac{ZkTa}{\Psi(0)}[\tfrac{2}{3} I_{3/2}(\Psi(1)) + J_2 - J_3]$$

$$= 3PV + \frac{ZkTa}{\Psi(0)}(J_2 - J_3) \tag{A4.14}$$

where we have used (9.163). In order to evaluate J_2 and J_3 we use the Thomas–Fermi equation (see (9.155))

$$\Psi'' = a\xi I_{1/2}\left[\frac{\Psi(\xi)}{\xi}\right] \tag{A4.15}$$

Thus

$$J_2 = \frac{1}{a}\int_0^1 \Psi''(\xi)\Psi(\xi) d\xi$$

$$= \frac{1}{a}\left[\Psi'(\xi)\Psi(\xi)\Big|_0^1 - \int_0^1 \Psi'^2(\xi) d\xi \right]$$

or

$$aJ_2 = [\Psi(1)]^2 - \Psi'(0)\Psi(0) - \int_0^1 \Psi'^2(\xi)d\xi \qquad (A4.16)$$

where we have used the boundary condition (see (9.160)): Similarly, $\Psi'(1) = \Psi(1)$

$$aJ_3 = \int_0^1 \xi\Psi'\Psi''(\xi)d\xi = \frac{1}{2}\int_0^1 \xi\frac{d}{d\xi}[\Psi'(\xi)]^2 d\xi$$

$$= \frac{1}{2}\left[\xi\Psi'^2(\xi)\Big|_0^1 - \int_0^1 \Psi'^2(\xi)d\xi \right]$$

$$= \frac{1}{2}\left[\Psi^2(1) - \int_0^1 \Psi'^2(\xi)d\xi \right] \qquad (A4.18)$$

Thus

$$2E_{kin} = 3PV + \frac{ZkT}{2\Psi(0)}\left[\Psi^2(1) - 2\Psi'(0)\Psi(0) \right.$$

$$\left. - \int_0^1 \Psi'^2(\xi)d\xi \right] \qquad (A4.19)$$

Now, (9.172) gives us

$$E_{pot} = -\frac{ZkTa}{2\Psi(0)}J_4 \qquad (A4.20)$$

where

$$J_4 = \int_0^1 d\xi I_{1/2}\left[\frac{\Psi(\xi)}{\xi} \right]\xi[\Psi(\xi) - \xi\Psi(1) + \Psi(0)]$$

or

$$aJ_4 = \int_0^1 d\xi\Psi''\Psi - \Psi(1)\int_0^1 \xi\Psi''d\xi + \Psi(0)\int_0^1 \Psi''(\xi)d\xi \qquad (A4.21)$$

where we have used (A4.15). We evaluate each term

$$\int_0^1 d\xi\Psi''\Psi = aJ_2 = [\Psi(1)]^2 - \Psi'(0)\Psi(0) - \int_0^1 \Psi'^2(\xi)d\xi \qquad (A4.22)$$

(see (A4.16)). Further

$$\int_0^1 \xi\Psi''(\xi)d\xi = \xi\Psi'(\xi)\Big|_0^1 - \int_0^1 \Psi'(\xi)d\xi$$

$$= \Psi'(1) - \Psi(\xi)|_0^1$$

$$= \Psi(0) \qquad (A4.23)$$

where we have used (A4.17). And

$$\int_0^1 \Psi''(\xi)d\xi = \Psi'(\xi)\Big|_0^1 = \Psi'(1) - \Psi'(0) \tag{A4.24}$$

Thus

$$aJ_4 = [\Psi(1)]^2 - \Psi'(0)\Psi(0) - \int_0^1 \Psi'^2(\xi)d\xi$$

$$- \Psi(1)\Psi(0) + \Psi(0)[\Psi(1) - \Psi'(0)]$$

$$= [\Psi(1)]^2 - 2\Psi'(0)\Psi(0) - \int_0^1 \Psi'^2(\xi)d\xi \tag{A4.25}$$

Substituting in (A4.20) we get

$$E_{\text{pot}} = -\frac{ZkT}{2\Psi(0)}\left[[\Psi(1)]^2 - 2\Psi'(0)\Psi(0) - \int_0^1 \Psi'^2(\xi)d\xi \right] \tag{A4.26}$$

Substituting the above expression in (A4.19) we immediately get

$$2E_{\text{kin}} + E_{\text{pot}} = 3PV \tag{A4.27}$$

which shows that the virial theorem remains valid in spite of the simplifying assumptions of the Thomas–Fermi model.

Tables of Thomas–Fermi corrected equation of state

In this appendix we have tabulated TF pressures and energy with their corrections for scaled temperatures from 10^{-2} to 10^{-5} eV. The first column represents ZV which is the scaled volume. The second and fourth columns ($P/Z^{10/3}$ and $E/Z^{7/3}$) represent the TF pressure and energy. The third and the fifth columns ($DP/Z^{8/3}$ and $DE/Z^{5/3}$) represent the corrections to the pressure and energy and contain both quantum and exchange corrections. The units of thermodynamic functions are,

$ZV: Å^3/$Atom $(1Å^3 = 10^{-24} \, cm^3)$,

$P/Z^{10/3}$ and $DP/Z^{8/3}$: Mbars $(1 \, Mbar = 10^{12} \, dynes \, cm^{-2})$,

$E/Z^{7/3}$ and $DE/Z^{5/3}$: Mbars cm^3 g^{-1} (1 Mbar cm^3/g

$= 10^{12} \, ergs \, g^{-1})$.

Table A5.1. $kT/Z^{4/3} = 1.0\mathrm{E} - 02$

ZV	$P/Z^{10/3}$	$-DP/Z^{8/3}$	$E/Z^{7/3}$	$-DE/Z^{5/3}$
3.060E + 09	2.983E − 13	9.324E − 13	4.833E − 03	− 5.483E − 01
1.290E + 09	6.697E − 13	2.107E − 12	4.137E − 03	− 3.593E − 01
5.440E + 08	1.495E − 12	4.721E − 12	3.632E − 03	− 2.312E − 01
2.294E + 08	3.315E − 12	1.046E − 11	3.230E − 03	− 1.485E − 01
9.675E + 07	7.301E − 12	2.306E − 11	2.890E − 03	− 9.658E − 02
4.080E + 07	1.598E − 11	5.050E − 11	2.591E − 03	− 6.410E − 02
1.720E + 07	5.475E − 11	1.107E − 10	2.323E − 03	− 4.368E − 02
7.254E + 06	7.530E − 11	2.449E − 10	2.081E − 03	− 3.037E − 02
1.290E + 06	3.540E − 10	1.296E − 09	1.666E − 03	− 1.534E − 02
5.440E + 05	7.763E − 10	3.194E − 09	1.488E − 03	− 1.081E − 02
2.294E + 05	1.735E − 09	8.461E − 09	1.327E − 03	− 7.613E − 03
9.675E + 04	4.000E − 09	2.464E − 08	1.181E − 03	− 5.074E − 03
4.080E + 04	9.734E − 09	8.123E − 08	1.049E − 03	− 2.901E − 03
2.649E + 04	1.566E − 08	1.558E − 07	9.879E − 04	− 1.781E − 03
1.720E + 04	2.600E − 08	3.120E − 07	9.314E − 04	− 9.556E − 04
1.117E + 04	4.497E − 08	6.521E − 07	8.814E − 04	6.768E − 04
7.254E + 03	8.233E − 08	1.417E − 06	8.430E − 04	2.926E − 03
3.060E + 03	3.559E − 07	7.139E − 06	8.658E − 04	9.345E − 03
2.612E + 03	4.872E − 07	9.611E − 06	9.038E − 04	1.136E − 02
2.213E + 03	6.883E − 07	1.315E − 05	9.657E − 04	1.346E − 02
1.987E + 03	8.690E − 07	1.607E − 05	1.022E − 03	1.496E − 02
1.792E + 03	1.093E − 06	1.952E − 05	1.090E − 03	1.695E − 02
1.601E + 03	1.411E − 06	2.414E − 05	1.183E − 03	1.865E − 02
1.437E + 03	1.812E − 06	2.961E − 05	1.295E − 03	2.115E − 02
1.290E + 03	2.337E − 06	3.635E − 05	1.434E − 03	2.344E − 02
1.158E + 03	3.024E − 06	4.464E − 05	1.605E − 03	2.695E − 02
1.082E + 03	3.567E − 06	5.087E − 05	1.732E − 03	2.840E − 02
9.335E + 02	5.107E − 06	6.751E − 05	2.065E − 03	3.372E − 02
8.378E + 02	6.658E − 06	8.313E − 05	2.370E − 03	3.715E − 02
6.752E + 02	1.134E − 05	1.261E − 04	3.173E − 03	4.691E − 02
5.440E + 02	1.936E − 05	1.913E − 04	4.316E − 03	5.909E − 02
3.532E + 02	5.604E − 05	4.384E − 04	8.150E − 03	9.259E − 02
2.294E + 02	1.596E − 04	9.904E − 04	1.543E − 02	1.426E − 01
1.491E + 02	4.443E − 04	2.196E − 03	2.881E − 02	2.158E − 01

Source: The tables have been adapted with permission from S. L. McCarthy's report 'The Kirzhnits Corrections to the Thomas–Fermi Equation of State' University of California, Lawrence Radiation Laboratory Report UCRL 14364, (1965).

Table A5.2. $kT/Z^{4/3} = 5.0E - 02$

ZV	$P/Z^{10/3}$	$-DP/Z^{8/3}$	$E/Z^{7/3}$	$-DE/Z^{5/3}$
9.150E + 08	1.391E − 11	1.906E − 11	5.419E − 02	− 2.866E − 01
3.858E + 08	3.145E − 11	4.350E − 11	4.968E − 02	− 1.790E − 01
1.627E + 08	7.060E − 11	9.891E − 11	4.559E − 02	− 1.110E − 01
6.860E + 07	1.577E − 10	2.224E − 10	4.131E − 02	− 6.899E − 02
2.894E + 07	3.497E − 10	4.955E − 10	3.743E − 02	− 4.315E − 02
1.220E + 07	7.703E − 10	1.095E − 09	3.376E − 02	− 2.709E − 02
5.145E + 08	1.655E − 09	2.412E − 09	3.032E − 02	− 1.535E − 02
2.169E + 06	3.668E − 09	5.335E − 09	2.710E − 02	− 1.050E − 02
9.150E + 05	7.955E − 09	1.196E − 08	2.414E − 02	− 6.045E − 03
3.858E + 05	1.726E − 08	2.757E − 08	2.142E − 02	− 2.930E − 03
1.627E + 05	3.765E − 08	6.642E − 08	1.895E − 02	− 5.425E − 04
6.860E + 04	8.321E − 08	1.704E − 07	1.672E − 02	1.410E − 03
2.894E + 04	1.882E − 07	4.755E − 07	1.470E − 02	3.646E − 03
1.220E + 04	4.457E − 07	1.480E − 06	1.289E − 02	6.686E − 03
7.924E + 03	6.972E − 07	2.738E − 06	1.207E − 02	8.846E − 03
5.145E + 03	1.121E − 06	5.257E − 06	1.130E − 02	1.182E − 02
3.402E + 03	1.817E − 06	1.016E − 05	1.064E − 02	1.607E − 02
3.341E + 03	1.858E − 06	1.047E − 05	1.061E − 02	1.649E − 02
2.169E + 03	3.206E − 06	2.159E − 05	1.004E − 02	2.266E − 02
9.150E + 02	1.146E − 05	9.814E − 05	9.666E − 03	4.868E − 02
7.815E + 02	1.499E − 05	1.298E − 04	9.828E − 03	5.571E − 02
6.619E + 02	2.012E − 05	1.737E − 04	1.014E − 02	6.403E − 02
5.942E + 02	2.455E − 05	2.099E − 04	1.044E − 02	7.043E − 02
5.359E + 02	2.986E − 05	2.513E − 04	1.081E − 02	7.687E − 02
4.788E + 02	3.719E − 05	3.057E − 04	1.133E − 02	8.408E − 02
4.298E + 02	4.616E − 05	3.686E − 04	1.197E − 02	9.196E − 02
3.858E + 02	5.759E − 05	4.444E − 04	1.277E − 02	1.014E − 01
3.531E + 02	6.931E − 05	5.182E − 04	1.356E − 02	1.087E − 01
3.464E + 02	7.215E − 05	5.356E − 04	1.375E − 02	1.108E − 01
3.235E + 02	8.341E − 05	6.028E − 04	1.448E − 02	1.167E − 01
2.792E + 02	1.146E − 04	7.782E − 04	1.639E − 02	1.328E − 01
2.506E + 02	1.453E − 04	9.388E − 04	1.814E − 02	1.449E − 01
2.019E + 02	2.348E − 04	1.367E − 03	2.269E − 02	1.744E − 01
1.056E + 02	1.016E − 03	4.238E − 03	4.970E − 02	3.049E − 01
6.860E + 01	2.682E − 03	8.901E − 03	8.674E − 02	4.408E − 01
4.458E + 01	6.956E − 03	1.858E − 02	1.508E − 01	6.297E − 01
2.894E + 01	1.775E − 02	3.809E − 02	2.587E − 01	8.851E − 01

Table A5.3. $kT/Z^{4/3} = 2.0E - 01$

ZV	$P/Z^{10/3}$	$- DP/Z^{8/3}$	$E/Z^{7/3}$	$- DE/Z^{5/3}$
3.253E + 08	3.467E − 10	2.161E − 10	4.420E − 01	− 1.715E − 01
1.364E + 08	7.893E − 10	5.034E − 10	4.089E − 01	− 1.089E − 01
5.752E + 07	1.790E − 09	1.164E − 09	3.763E − 01	− 6.849E − 02
2.425E + 07	4.033E − 09	2.672E − 09	3.446E − 01	− 4.332E − 02
1.025E + 07	9.039E − 09	6.071E − 09	3.138E − 01	− 2.750E − 02
4.314E + 06	2.012E − 08	1.367E − 08	2.840E − 01	− 1.870E − 02
1.819E + 06	4.445E − 08	3.035E − 08	2.555E − 01	− 1.421E − 02
7.670E + 05	9.749E − 08	6.815E − 08	2.284E − 01	− 1.006E − 02
3.253E + 05	2.126E − 07	1.528E − 07	2.030E − 01	− 7.951E − 03
1.564E + 05	4.625E − 07	3.490E − 07	1.793E − 01	− 7.102E − 03
5.706E + 04	1.014E − 06	8.311E − 07	1.573E − 01	− 5.471E − 03
2.425E + 04	2.208E − 06	2.053E − 06	1.377E − 01	− 3.076E − 03
1.693E + 04	3.071E − 06	3.060E − 06	1.300E − 01	− 1.575E − 03
1.025E + 04	4.916E − 06	5.501E − 06	1.198E − 01	1.596E − 03
6.664E + 03	7.406E − 06	9.321E − 06	1.116E − 01	5.486E − 03
4.314E + 03	1.126E − 05	1.623E − 05	1.039E − 01	1.065E − 02
3.500E + 03	1.582E − 05	2.135E − 05	1.004E − 01	1.027E − 02
2.802E + 03	1.735E − 03	2.907E − 05	9.666E − 02	1.815E − 02
1.819E + 03	2.711E − 05	5.568E − 05	8.997E − 02	2.854E − 02
1.681E + 03	2.948E − 05	6.028E − 05	8.880E − 02	3.045E − 02
1.203E + 03	4.244E − 05	9.958E − 05	8.414E − 02	4.306E − 02
1.181E + 03	4.331E − 05	1.024E − 04	8.390E − 02	4.385E − 02
7.670E + 02	7.117E − 05	2.015E − 04	7.863E − 02	6.611E − 02
5.653E + 02	1.054E − 04	3.295E − 04	7.553E − 02	1.554E − 01
3.255E + 02	2.178E − 04	8.309E − 04	7.257E − 02	1.448E − 01
2.762E + 02	2.739E − 04	1.081E − 03	7.250E − 02	1.663E − 01
2.540E + 02	3.518E − 04	1.425E − 03	7.285E − 02	1.915E − 01
2.101E + 02	4.163E − 04	1.703E − 03	7.357E − 02	2.095E − 01
1.895E + 02	4.914E − 04	2.019E − 03	7.464E − 02	2.278E − 01
1.693E + 02	5.917E − 04	2.427E − 03	7.626E − 02	2.496E − 01
1.520E + 02	7.105E − 04	2.893E − 03	7.834E − 02	2.717E − 01
1.364E + 02	8.574E − 04	3.445E − 03	8.103E − 02	2.955E − 01
1.248E + 02	1.004E − 03	3.973E − 03	8.375E − 02	3.163E − 01
1.225E + 02	1.059E − 03	4.097E − 03	8.441E − 02	3.207E − 01
1.144E + 02	1.176E − 03	4.570E − 03	8.697E − 02	3.374E − 01
9.070E + 01	1.547E − 03	5.777E − 03	9.365E − 02	3.767E − 01
8.839E + 01	1.900E − 03	6.836E − 03	9.998E − 02	4.077E − 01
7.140E + 01	2.043E − 03	9.656E − 03	1.165E − 01	4.768E − 01
5.755E + 01	1.083E − 02	2.678E − 02	2.124E − 01	7.624E − 01
2.423E + 01	2.660E − 02	5.292E − 02	3.398E − 01	1.040E + 00
1.576E + 01	6.482E − 02	1.052E − 01	5.512E − 01	1.407E + 00
1.025E + 01	1.564E − 01	2.027E − 01	8.921E − 01	1.894E + 00
4.314E + 00	8.683E − 01	7.441E − 01	2.249E + 00	3.282E + 00

Table A5.4. $kT/Z^{4/3} = 1.0E + 00$

ZV	$P/Z^{10/3}$	$-DP/Z^{8/3}$	$E/Z^{7/3}$	$-DE/Z^{5/3}$
9.675E + 07	1.162E − 08	2.085E − 09	3.762E + 00	− 8.801E − 01
4.060E + 07	2.690E − 08	5.082E − 09	3.554E + 00	− 7.968E − 01
1.720E + 07	6.206E − 08	1.254E − 08	3.343E + 00	− 7.054E − 01
7.254E + 06	1.427E − 07	2.978E − 08	3.127E + 00	− 6.250E − 01
3.060E + 06	3.262E − 07	7.137E − 08	2.907E + 00	− 5.520E − 01
1.290E + 06	7.416E − 07	1.696E − 07	2.685E + 00	− 4.899E − 01
5.440E + 05	1.675E − 06	3.991E − 07	2.460E + 00	− 4.352E − 01
2.294E + 05	3.755E − 06	9.301E − 07	2.234E + 00	− 3.903E − 01
9.675E + 04	8.340E − 06	2.155E − 06	2.010E + 00	− 3.529E − 01
6.283E + 04	1.240E − 05	3.282E − 06	1.900E + 00	− 3.360E − 01
4.080E + 04	1.841E − 05	5.008E − 06	1.791E + 00	− 3.213E − 04
2.649E + 04	2.729E − 05	7.672E − 06	1.685E + 00	− 3.049E − 01
1.205E + 04	5.580E − 05	1.687E − 05	1.498E + 00	− 2.848E − 01
1.117E + 04	5.985E − 05	1.839E − 05	1.480E + 00	− 2.744E − 01
7.254E + 03	8.864E − 05	2.892E − 05	1.382E + 00	− 2.588E − 01
5.068E + 03	1.229E − 04	4.257E − 05	1.304E + 00	− 2.428E − 01
3.060E + 03	1.952E − 04	7.480E − 05	1.198E + 00	− 2.189E − 01
1.933E + 03	2.914E − 04	1.241E − 04	1.112E + 00	− 1.934E − 01
1.290E + 03	4.356E − 04	2.096E − 04	1.032E + 00	− 1.628E − 01
1.049E + 03	5.296E − 04	2.718E − 04	9.950E − 01	− 1.455E − 01
8.378E + 02	6.566E − 04	3.634E − 04	9.560E − 01	− 1.244E − 01
5.440E + 02	9.987E − 04	6.467E − 04	8.855E − 01	− 7.551E − 02
5.027E + 02	1.080E − 03	7.205E − 04	8.732E − 01	− 6.495E − 02
3.597E + 02	1.509E − 03	1.150E − 03	8.233E − 01	− 1.492E − 02
3.532E + 02	1.537E − 03	1.181E − 03	8.209E − 01	− 1.176E − 02
2.294E + 02	2.405E − 03	2.212E − 03	7.634E − 01	7.287E − 02
1.689E + 02	3.344E − 03	3.501E − 03	7.279E − 01	1.538E − 01
9.675E + 01	6.319E − 03	8.253E − 03	6.776E − 01	3.384E − 01
8.261E + 01	7.648E − 03	1.056E − 02	6.680E − 01	4.075E − 01
6.998E + 01	9.400E − 03	1.368E − 02	6.608E − 01	4.872E − 01
6.283E + 01	1.079E − 02	1.619E − 02	6.580E − 01	5.430E − 01
5.666E + 01	1.234E − 02	1.902E − 02	6.568E − 01	6.005E − 01
5.063E + 01	1.434E − 02	2.263E − 02	6.574E − 01	6.667E − 01
4.545E + 01	1.661E − 02	2.676E − 02	6.602E − 01	7.340E − 01
4.080E + 01	1.931E − 02	3.159E − 02	6.654E − 01	8.050E − 01
3.733E + 01	2.191E − 02	3.618E − 02	6.718E − 01	8.661E − 01
3.663E + 01	2.253E − 02	3.724E − 02	6.734E − 01	8.797E − 01
3.421E + 01	2.488E − 02	4.128E − 02	6.799E − 01	9.244E − 01
2.952E + 01	3.099E − 02	5.153E − 02	6.989E − 01	1.040E + 00
2.650E + 01	3.656E − 02	6.051E − 02	7.175E − 01	1.126E + 00
2.133E + 01	5.145E − 02	8.300E − 02	7.687E − 01	1.509E + 00
1.117E + 01	1.556E − 01	2.085E − 01	1.088E + 00	1.965E + 00
7.254E + 00	3.442E − 01	3.816E − 01	1.518E + 00	2.524E + 00
4.714E + 00	7.666E − 01	6.992E − 01	2.220E + 00	3.227E + 00
3.060E + 00	1.731E + 00	1.286E + 00	3.324E + 00	4.110E + 00
1.290E + 00	8.696E + 00	4.932E + 00	7.472E + 00	6.543E + 00
5.440E − 01	4.224E + 01	1.438E + 01	1.627E + 01	1.007E + 01

Table A5.5. $kT/Z^{4/3} = 5.0\mathrm{E}+00$

ZV	$P/Z^{10/3}$	$-DP/Z^{8/3}$	$E/Z^{7/3}$	$-DE/Z^{5/3}$
2.894E + 07	2.621E − 07	4.791E − 09	1.687E + 01	− 5.260E + 00
1.220E + 07	6.180E − 07	1.264E − 08	1.655E + 01	− 4.246E + 00
5.145E + 06	1.455E − 06	3.302E − 08	1.616E + 01	− 4.110E + 00
2.169E + 06	3.421E − 06	8.652E − 08	1.575E + 01	− 3.951E + 00
9.150E + 05	8.026E − 06	2.275E − 07	1.530E + 01	− 3.753E + 00
3.858E + 05	1.878E − 05	5.968E − 07	1.479E + 01	− 3.577E + 00
1.627E + 05	4.380E − 05	1.568E − 06	1.422E + 01	− 3.368E + 00
6.800E + 04	1.017E − 04	4.114E − 06	1.358E + 01	− 3.146E + 00
2.894E + 04	2.348E − 04	1.076E − 05	1.286E + 01	− 2.916E + 00
1.879E + 04	3.559E − 04	1.739E − 05	1.247E + 01	− 2.798E + 00
1.220E + 04	5.386E − 04	2.809E − 05	1.205E + 01	− 2.679E + 00
7.924E + 03	8.131E − 04	4.538E − 05	1.161E + 01	− 2.559E + 00
5.145E + 03	1.225E − 03	7.345E − 05	1.115E + 01	− 2.438E + 00
3.341E + 03	1.845E − 03	1.193E − 04	1.067E + 01	− 2.316E + 00
2.169E + 03	2.766E − 03	1.947E − 04	1.017E + 01	− 2.191E + 00
1.516E + 03	3.870E − 03	2.942E − 04	9.744E + 00	− 2.084E + 00
9.150E + 02	6.200E − 03	5.330E − 04	9.136E + 00	− 1.928E + 00
5.931E + 02	9.287E − 03	9.019E − 04	8.612E + 00	− 1.783E + 00
3.858E + 02	1.387E − 02	1.546E − 03	8.095E + 00	− 1.627E + 00
3.139E + 02	1.682E − 02	2.016E − 03	7.849E + 00	− 1.545E + 00
2.506E + 02	2.076E − 02	2.707E − 03	7.586E + 00	− 1.451E + 00
1.627E + 02	3.117E − 02	4.839E − 03	7.093E + 00	− 1.247E + 00
1.504E + 02	3.359E − 02	5.393E − 03	7.006E + 00	− 1.207E + 00
1.076E + 02	4.619E − 02	8.607E − 03	6.644E + 00	− 1.019E + 00
1.056E + 02	4.700E − 02	8.832E − 03	6.625E + 00	− 1.008E + 00
6.860E + 01	7.132E − 02	1.643E − 02	6.190E + 00	− 7.196E − 01
5.052E + 01	9.671E − 02	2.644E − 02	5.933E + 00	− 2.818E − 01
2.894E + 01	1.695E − 01	5.969E − 02	5.462E + 00	8.190E − 02
2.471E + 01	2.000E − 01	7.593E − 02	5.336E + 00	2.681E − 01
2.093E + 01	2.385E − 01	9.787E − 02	5.257E + 00	4.842E − 01
1.879E + 01	2.679E − 01	1.154E − 01	5.200E + 00	6.344E − 01
1.695E + 01	2.999E − 01	1.352E − 01	5.152E + 00	7.861E − 01
1.514E + 01	3.397E − 01	1.607E − 01	5.107E + 00	9.612E − 01
1.359E + 01	3.835E − 01	1.895E − 01	5.071E + 00	1.138E + 00
1.220E + 01	4.344E − 01	2.254E − 01	5.050E + 00	1.411E + 00
1.116E + 01	4.807E − 01	2.558E − 01	5.028E + 00	1.486E + 00
1.095E + 01	4.916E − 01	2.633E − 01	5.025E + 00	1.521E + 00
1.023E + 01	5.327E − 01	2.920E − 01	5.019E + 00	1.651E + 00
8.828E + 00	6.353E − 01	3.645E − 01	5.021E + 00	1.941E + 00
7.924E + 00	7.248E − 01	4.283E − 01	5.037E + 00	2.165E + 00
6.386E + 00	9.504E − 01	5.888E − 01	5.111E + 00	2.639E + 00
3.341E + 00	2.294E + 00	1.470E + 00	5.811E + 00	4.242E + 00
2.169E + 00	4.380E + 00	2.613E + 00	6.885E + 00	5.431E + 00
1.410E + 00	8.726E + 00	4.555E + 00	8.722E + 00	6.734E + 00
9.150E − 01	1.801E + 01	7.919E + 00	1.167E + 01	8.226E + 00
3.838E − 01	8.075E + 01	2.433E + 01	2.277E + 01	1.210E + 01
1.627E − 01	3.675E + 02	7.665E + 01	4.570E + 01	1.761E + 01
6.860E − 02	1.656E + 03	2.432E + 02	9.051E + 01	2.523E + 01

Table A5.6. $kT/Z^{4/3} = 1.0E + 01$

ZV	$P/Z^{10/3}$	$-DP/Z^{8/3}$	$E/Z^{7/3}$	$-DE/Z^{5/3}$
1.721E + 07	9.121E − 07	4.433E − 09	2.755E + 01	− 7.330E + 00
7.255E + 06	2.158E − 06	1.165E − 08	2.725E + 01	− 6.745E + 00
3.053E + 06	5.100E − 06	3.172E − 08	2.695E + 01	− 5.393E + 00
1.290E + 06	1.205E − 05	8.451E − 08	2.658E + 01	− 5.469E + 00
5.441E + 05	2.845E − 05	2.285E − 07	2.622E + 01	− 5.251E + 00
2.294E + 05	6.700E − 05	6.098E − 07	2.572E + 01	− 5.185E + 00
9.674E + 04	1.576E − 04	1.655E − 06	2.518E + 01	− 5.026E + 00
4.079E + 04	3.700E − 04	4.521E − 06	2.456E + 01	− 4.846E + 00
1.721E + 04	8.654E − 04	1.245E − 05	2.382E + 01	− 4.642E + 00
1.117E + 04	1.322E − 03	2.067E − 05	2.539E + 01	− 4.531E + 00
7.258E + 03	2.016E − 03	3.446E − 05	2.292E + 01	− 4.412E + 00
4.712E + 03	3.071E − 03	5.760E − 05	2.242E + 01	− 4.285E + 00
3.059E + 03	4.671E − 03	9.667E − 05	2.186E + 01	− 4.151E + 00
1.986E + 03	7.093E − 03	1.630E − 04	2.125E + 01	− 4.007E + 00
1.290E + 03	1.075E − 02	2.767E − 04	2.060E + 01	− 3.854E + 00
9.015E + 02	1.516E − 02	4.321E − 04	2.001E + 01	− 3.717E + 00
5.441E + 02	2.435E − 02	8.200E − 04	1.915E + 01	− 3.508E + 00
3.527E + 02	3.700E − 02	1.445E − 03	1.833E + 01	− 3.309E + 00
2.294E + 02	5.575E − 02	2.589E − 03	1.750E + 01	− 3.087E + 00
1.833E + 02	6.779E − 02	3.410E − 03	1.709E + 01	− 2.970E + 00
1.490E + 02	8.390E − 02	4.667E − 03	1.664E + 01	− 2.834E + 00
9.674E + 01	1.265E − 01	8.639E − 03	1.577E + 01	− 2.539E + 00
8.940E + 01	1.361E − 01	9.687E − 03	1.561E + 01	− 2.480E + 00
6.397E + 01	1.872E − 01	1.585E − 02	1.495E + 01	− 2.206E + 00
6.281E + 01	1.905E − 01	1.628E − 02	1.491E + 01	− 2.190E + 00
4.079E + 01	2.889E − 01	3.117E − 02	1.408E + 01	− 1.771E + 00
3.004E + 01	3.871E − 01	4.950E − 02	1.331E + 01	− 1.448E + 00
1.721E + 01	6.708E − 01	1.185E − 01	1.260E + 01	− 6.254E − 01
1.469E + 01	7.863E − 01	1.515E − 01	1.237E + 01	− 3.594E − 01
1.245E + 01	9.309E − 01	1.966E − 01	1.214E + 01	− 5.212E − 02
1.117E + 01	1.040E + 00	2.329E − 01	1.201E + 01	1.619E − 01
1.008E + 01	1.158E + 00	2.759E − 01	1.189E + 01	3.764E − 01
9.004E + 00	1.303E + 00	3.269E − 01	1.177E + 01	6.244E − 01
8.082E + 00	1.460E + 00	3.872E − 01	1.167E + 01	8.757E − 01
7.255E + 00	1.642E + 00	4.621E − 01	1.160E + 01	1.218E + 00
6.639E + 00	1.805E + 00	5.265E − 01	1.152E + 01	1.365E + 00
6.515E + 00	1.843E + 00	5.421E − 01	1.151E + 01	1.415E + 00
6.083E + 00	1.936E + 00	6.027E − 01	1.147E + 01	1.597E + 00
5.249E + 00	2.339E + 00	7.570E − 01	1.142E + 01	2.010E + 00
4.712E + 00	2.643E + 00	8.936E − 01	1.140E + 01	2.329E + 00
3.797E + 00	3.391E + 00	1.240E + 00	1.142E + 01	3.008E + 00
1.986E + 00	7.560E + 00	3.194E + 00	1.221E + 01	5.339E + 00
1.298E + 00	1.559E + 01	5.769E + 00	1.362E + 01	7.069E + 00
8.385E − 01	2.552E + 01	1.011E + 01	1.616E + 01	8.086E + 00
5.441E − 01	5.006E + 01	1.741E + 01	2.034E + 01	1.084E + 01
2.294E − 01	2.096E + 02	5.152E + 01	3.648E + 01	1.556E + 01
9.674E − 02	9.227E + 02	1.565E + 02	7.018E + 01	2.205E + 01
4.079E − 02	4.085E + 03	4.930E + 02	1.538E + 02	3.120E + 01

Table A5.7. $kT/Z^{4/3} = 5.0\text{E} + 01$

ZV	$P/Z^{10/3}$	$-DP/Z^{8/3}$	$E/Z^{7/3}$	$-DE/Z^{5/3}$
5.146E + 06	1.553E − 05	3.087E − 09	9.215E + 01	− 7.055E + 00
2.170E + 06	5.682E − 05	7.262E − 09	9.199E + 01	− 6.737E + 00
9.149E + 05	8.727E − 05	1.892E − 08	9.185E + 01	− 7.581E + 00
3.858E + 05	2.060E − 04	4.738E − 08	9.162E + 01	− 7.539E + 00
1.627E + 05	4.900E − 04	1.570E − 07	9.146E + 01	− 7.125E + 00
6.861E + 04	1.161E − 03	3.902E − 07	9.108E + 01	− 6.857E + 00
2.895E + 04	2.748E − 03	1.076E − 06	9.075E + 01	− 6.866E + 00
1.220E + 04	6.506E − 03	5.026E − 06	9.028E + 01	− 6.829E + 00
5.146E + 03	1.558E − 02	8.655E − 06	8.975E + 01	− 6.780E + 00
5.541E + 03	2.565E − 02	1.474E − 05	8.939E + 01	− 6.753E + 00
2.170E + 03	3.636E − 02	2.558E − 05	8.902E + 01	− 6.722E + 00
1.409E + 03	5.586E − 02	4.415E − 05	8.859E + 01	− 6.687E + 00
9.149E + 02	8.583E − 02	7.785E − 05	8.811E + 01	− 6.647E + 00
5.941E + 02	1.518E − 01	1.393E − 04	8.755E + 01	− 6.601E + 00
3.858E + 02	2.025E − 01	2.538E − 04	8.692E + 01	− 6.547E + 00
2.695E + 02	2.886E − 01	4.243E − 04	8.633E + 01	− 6.494E + 00
1.627E + 02	4.755E − 01	8.935E − 04	8.537E + 01	− 6.404E + 00
1.055E + 02	7.297E − 01	1.758E − 03	8.441E + 01	− 6.305E + 00
6.861E + 01	1.115E + 00	3.510E − 03	8.334E + 01	− 6.180E + 00
5.581E + 01	1.367E + 00	4.933E − 03	8.277E + 01	− 6.108E + 00
4.456E + 01	1.705E + 00	7.197E − 03	8.211E + 01	− 6.018E + 00
2.893E + 01	2.607E + 00	1.510E − 02	8.072E + 01	− 5.801E + 00
2.674E + 01	2.816E + 00	1.732E − 02	8.045E + 01	− 5.754E + 00
1.915E + 01	3.911E + 00	3.123E − 02	7.925E + 01	− 5.522E + 00
1.879E + 01	3.981E + 00	3.226E − 02	7.918E + 01	− 5.507E + 00
1.220E + 01	6.077E + 00	6.990E − 02	7.750E + 01	− 5.106E + 00
8.984E + 00	8.201E + 00	1.216E − 01	7.624E + 01	− 4.730E + 00
5.445E + 00	1.417E + 01	5.546E − 01	7.383E + 01	− 3.790E + 00
2.695E + 00	2.689E + 01	1.075E + 00	7.112E + 01	− 2.105E + 00
2.170E + 00	3.335E + 01	1.569E + 00	7.024E + 01	− 1.407E + 00
3.858E − 01	2.090E + 02	2.895E + 01	7.050E + 01	9.583E + 00
2.507E − 01	3.501E + 02	5.512E + 01	7.489E + 01	1.370E + 01
1.627E − 01	6.050E + 02	1.012E + 02	8.246E + 01	1.805E + 01
6.861E − 02	2.043E + 03	3.051E + 03	1.156E + 02	2.701E + 01
2.893E − 02	7.915E + 03	0.737E + 03	1.908E + 02	3.707E + 01
1.220E − 02	3.300E + 04	2.588E + 03	3.427E + 02	5.041E + 01
5.146E − 03	1.409E + 05	7.941E + 03	6.301E + 02	6.870E + 01
2.170E − 03	6.043E + 05	2.484E + 04	1.159E + 03	9.530E + 01

Table A5.8. $kT/Z^{4/3} = 1.0E + 02$

ZV	$P/Z^{10/3}$	$-DP/Z^{8/3}$	$E/Z^{7/3}$	$-DE/Z^{5/3}$
3.060E + 06	5.231E − 05	2.549E − 09	1.682E + 02	− 6.975E + 00
1.290E + 06	1.240E − 04	6.958E − 09	1.681E + 02	− 6.931E + 00
5.440E + 05	2.941E − 04	1.696E − 08	1.608E + 02	− 6.780E + 00
2.294E + 05	6.971E − 04	3.985E − 08	1.678E + 02	− 7.498E + 00
9.675E + 04	1.652E − 03	1.254E − 07	1.675E + 02	− 7.144E + 00
4.080E + 04	3.916E − 03	3.243E − 07	1.673E + 02	− 7.012E + 00
1.720E + 04	9.279E − 03	9.290E − 07	1.670E + 02	− 6.989E + 00
7.254E + 03	2.198E − 02	2.383E − 06	1.666E + 02	− 7.060E + 00
3.060E + 03	5.205E − 02	6.867E − 06	1.660E + 02	− 7.060E + 00
1.987E + 03	8.008E − 02	1.178E − 05	1.657E + 02	− 7.050E + 00
1.290E + 03	1.252E − 01	2.052E − 05	1.654E + 02	− 7.037E + 00
8.578E + 02	1.895E − 01	3.627E − 05	1.650E + 02	− 7.021E + 00
5.440E + 02	2.915E − 01	6.529E − 05	1.645E + 02	− 7.002E + 00
3.532E + 02	4.482E − 01	1.200E − 04	1.659E + 02	− 6.980E + 00
2.294E + 02	6.890E − 01	2.258E − 04	1.633E + 02	− 6.953E + 00
1.603E + 02	9.844E − 01	3.894E − 04	1.627E + 02	− 6.925E + 00
9.675E + 01	1.626E + 00	8.653E − 04	1.618E + 02	− 6.875E + 00
6.272E + 01	2.502E + 00	1.770E − 03	1.609E + 02	− 6.818E + 00
4.080E + 01	3.834E + 00	3.693E − 03	1.598E + 02	− 6.742E + 00
3.319E + 01	4.706E + 00	5.302E − 03	1.592E + 02	− 6.697E + 00
2.649E + 01	5.883E + 00	7.910E − 03	1.585E + 02	− 6.639E + 00
1.720E + 01	9.024E + 00	1.729E − 02	1.571E + 02	− 6.494E + 00
1.590E + 01	9.758E + 00	1.999E − 02	1.568E + 02	− 6.461E + 00
1.158E + 01	1.359E + 01	3.715E − 02	1.556E + 02	− 6.298E + 00
7.254E + 00	2.121E + 01	8.646E − 02	1.537E + 02	− 5.992E + 00
5.342E + 00	2.870E + 01	1.555E − 01	1.525E + 02	− 5.690E + 00
3.060E + 00	4.979E + 01	4.461E − 01	1.496E + 02	− 4.942E + 00
1.290E + 00	1.174E + 02	2.289E + 00	1.453E + 02	− 2.754E + 00
2.294E − 01	6 986E + 02	5.066E + 01	1.440E + 02	9.114E + 00
1.491E − 01	1.135E + 03	1.017E + 02	1.493E + 02	1.432E + 01
9.675E − 02	1.881E + 03	1.975E + 02	1.581E + 02	2.017E + 01
4.080E − 02	5.765E + 03	6.460E + 02	2.003E + 02	3.292E + 01
1.720E − 02	2.053E + 04	1.870E + 03	3.011E + 02	4.608E + 01
7.254E − 03	8.176E + 04	5.383E + 03	5.127E + 02	6.192E + 01
3.060E − 03	3.423E + 05	1.601E + 04	9.205E + 02	8.315E + 01
1.290E − 03	1.455E + 06	5.004E + 04	1.674E + 03	1.127E + 02

Table A5.9. $kT/Z^{4/3} = 5.0E + 02$

ZV	$P/Z^{10/3}$	$-DP/Z^{8/3}$	$E/Z^{7/3}$	$-DE/Z^{5/3}$
9.150E + 05	8.752E − 04	2.786E − 09	7.696E + 02	− 6.632E + 00
3.853E + 05	2.076E − 03	3.872E − 09	7.695E + 02	− 6.873E + 00
1.627E + 05	4.922E − 03	9.329E − 09	7.693E + 02	− 7.034E + 00
6.869E + 04	1.167E − 02	2.407E − 08	7.692E + 02	− 7.136E + 00
2.894E + 04	2.767E − 02	7.064E − 08	7.694E + 02	− 6.946E + 00
1.220E + 04	6.561E − 02	2.006E − 07	7.687E + 02	− 7.012E + 00
5.145E + 03	1.536E − 01	4.941E − 07	7.684E + 02	− 7.316E + 00
2.169E + 03	3.688E − 01	2.025E − 06	7.679E + 02	− 7.095E + 00
9.150E + 02	8.740E − 01	4.426E − 06	7.675E + 02	− 7.166E + 00
5.942E + 02	1.346E + 00	7.758E − 06	7.670E + 02	− 7.267E + 00
3.858E + 02	2.072E + 00	1.457E − 05	7.665E + 02	− 7.286E + 00
2.566E + 02	3.189E + 00	2.809E − 05	7.660E + 02	− 7.240E + 00
1.627E + 02	4.910E + 00	5.555E − 05	7.655E + 02	− 7.259E + 00
1.056E + 02	7.559E + 00	1.124E − 04	7.648E + 02	− 7.249E + 00
7.860E + 01	1.163E + 01	2.355E − 04	7.641E + 02	− 7.243E + 00
2.894E + 01	2.755E + 01	1.111E − 03	7.625E + 02	− 7.225E + 00
1.220E + 01	6.523E + 01	5.646E − 03	7.599E + 02	− 7.174E + 00
5.145E + 00	1.544E + 02	3.000E − 02	7.568E + 02	− 7.061E + 00
2.169E + 00	3.655E + 02	1.634E − 01	7.529E + 02	− 6.800E + 00
1.598E + 00	4.959E + 02	2.986E − 01	7.513E + 02	− 6.642E + 00
9.150E − 01	8.645E + 02	8.995E − 01	7.481E + 02	− 6.198E + 00
3.858E − 01	2.048E + 03	4.963E + 00	7.429E + 02	− 4.804E + 00
6.860E − 02	1.169E + 04	1.418E + 02	7.406E + 02	5.262E + 00
2.894E − 02	2.882E + 04	6.922E + 02	7.609E + 02	− 1.888E + 01
1.290E − 02	7.529E + 04	2.949E + 03	8.278E + 02	4.132E + 01
5.145E − 03	2.183E + 05	1.041E + 04	1.003E + 03	6.985E + 01
2.169E − 03	7.289E + 05	3.111E + 04	1.412E + 03	9.919E + 01
9.150E − 04	2.764E + 06	8.835E + 04	2.273E + 03	1.315E + 02
3.858E − 04	1.127E + 07	2.608E + 05	3.942E + 03	1.742E + 02
1.627E − 04	4.725E + 07	7.977E + 05	7.020E + 03	2.323E + 02
6.860E − 05	1.997E + 08	2.490E + 06	1.258E + 04	3.109E + 02

Table A5.10. $kT/Z^{4/3} = 1.0E + 03$

ZV	$P/Z^{10/3}$	$-DP/Z^{8/3}$	$E/Z^{7/3}$	$-DE/Z^{5/3}$
5.441E + 05	2.944E − 03	3.063E − 09	1.520E + 03	− 6.686E + 00
2.204E + 05	6.982E − 03	9.153E − 09	1.520E + 03	− 6.617E + 00
9.674E + 04	1.656E − 02	1.207E − 08	1.520E + 03	− 6.785E + 00
4.079E + 04	3.927E − 02	2.253E − 08	1.519E + 03	− 6.934E + 00
1.721E + 04	9.308E − 02	5.494E − 08	1.419E + 03	− 7.054E + 00
7.255E + 03	2.207E − 01	1.148E − 07	1.519E + 03	− 7.140E + 00
3.059E + 03	5.254E − 01	3.804E − 07	1.519E + 03	− 7.101E + 00
1.290E + 03	1.241E + 00	1.649E − 06	1.518E + 03	− 7.116E + 00
5.441E + 02	2.942E + 00	2.962E − 06	1.517E + 03	− 7.342E + 00
2.294E + 02	6.975E + 00	1.471E − 05	1.516E + 03	− 7.223E + 00
9.674E + 01	1.653E + 01	6.172E − 05	1.515E + 03	− 7.217E + 00
4.079E + 01	3.919E + 01	2.806E − 04	1.514E + 03	− 7.259E + 00
1.721E + 01	9.286E + 01	1.415E − 03	1.512E + 03	− 7.252E + 00
7.255E + 00	2.200E + 02	7.502E − 03	1.509E + 03	− 7.220E + 00
3.059E + 00	5.212E + 02	4.087E − 02	1.505E + 03	− 7.135E + 00
1.290E + 00	1.235E + 03	2.256E − 01	1.501E + 03	− 6.924E + 00
9.500E − 01	1.676E + 03	4.144E − 01	1.499E + 03	− 6.795E + 00
5.441E − 01	2.925E + 03	1.259E + 00	1.496E + 03	− 6.430E + 00
2.294E − 01	6.929E + 03	6.981E + 00	1.490E + 03	− 5.275E + 00
4.079E − 02	3.932E + 04	2.077E + 02	1.489E + 03	3.454E + 00
1.721E − 02	9.544E + 04	1.071E + 03	1.513E + 03	1.652E + 01
7.255E − 03	2.403E + 05	5.017E + 03	1.593E + 03	4.043E + 01
3.059E − 03	6.516E + 05	1.997E + 04	1.808E + 03	7.665E + 01
1.290E − 03	1.994E + 06	6.543E + 04	2.325E + 03	1.181E + 02
5.441E − 04	7.023E + 06	1.884E + 05	3.463E + 03	1.603E + 02
2.294E − 04	2.754E + 07	5.407E + 05	5.758E + 03	2.107E + 02
9.674E − 05	1.136E + 08	1.606E + 06	1.008E + 04	2.780E + 02
4.079E − 05	4.774E + 08	5.013E + 06	1.794E + 04	3.718E + 02

Table A5.11. $kT/Z^{4/3} = 5.0E + 03$

ZV	$P/Z^{10/3}$	$-DP/Z^{8/3}$	$E/Z^{7/3}$	$-DE/Z^{5/3}$
6.861E + 04	1.167E − 01	1.439E − 14	7.520E + 03	− 5.068E + 00
2.893E + 04	2.769E − 01	6.836E − 10	7.520E + 03	− 5.582E + 00
1.220E + 04	6.566E − 01	1.354E − 08	7.520E + 03	− 6.025E + 00
5.146E + 03	1.537E + 00	9.017E − 08	7.528E + 03	− 6.078E + 00
2.170E + 03	3.691E + 00	2.090E − 07	7.519E + 03	− 6.596E + 00
9.149E + 02	8.754E + 00	4.095E − 07	7.518E + 03	− 6.782E + 00
3.853E + 02	2.076E + 01	1.129E − 06	7.518E + 03	− 6.925E + 00
1.627E + 02	4.921E + 01	3.982E − 06	7.517E + 03	− 7.037E + 00
6.831E + 01	1.167E + 02	1.815E − 05	7.515E + 03	− 7.107E + 00
2.895E + 01	2.767E + 02	1.033E − 04	7.514E + 03	− 7.146E + 00
1.220E + 01	6.562E + 02	5.135E − 04	7.511E + 03	− 7.225E + 00
5.146E + 00	1.595E + 03	2.833E − 03	7.508E + 03	− 7.205E + 00
2.170E + 00	3.660E + 03	1.586E − 02	7.504E + 03	− 7.225E + 00
3.858E − 01	2.073E + 04	4.964E − 01	7.495E + 03	− 7.054E + 00
2.841E − 01	2.814E + 04	9.143E − 01	7.490E + 03	− 6.974E + 00
1.627E − 01	4.912E + 04	2.782E + 00	7.485E + 03	− 6.738E + 00
6.361E − 02	1.165E + 05	1.560E + 01	7.478E + 03	− 5.973E + 00
1.220E − 02	6.570E + 05	4.852E + 02	7.479E + 03	9.336E − 02
5.146E − 03	1.569E + 06	2.646E + 03	7.517E + 03	9.798E + 00
2.170E − 03	3.791E + 06	1.396E + 04	7.640E + 03	3.109E + 01
9.149E − 04	9.411E + 06	6.868E + 94	7.971E + 03	7.576E + 01
3.838E − 04	2.472E + 07	2.947E + 05	8.800E + 03	1.449E + 02
1.627E − 04	7.178E + 07	1.045E + 06	1.076E + 04	2.357E + 02
6.861E − 05	2.390E + 08	3.117E + 06	1.511E + 04	3.289E + 02
2.893E − 05	9.013E + 08	8.846E + 06	2.409E + 04	4.313E + 02
1.220E − 05	3.651E + 09	2.610E + 07	4.127E + 04	5.664E + 02
5.146E − 06	1.521E + 10	7.979E + 07	7.269E + 04	7.502E + 02

Table A5.12. $kT/Z^{4/3} = 1.0E + 04$

ZV	$P/Z^{10/3}$	$-DP/Z^{8/3}$	$E/Z^{7/3}$	$-DE/Z^{5/3}$
1.720E + 04	9.312E − 01	2.619E − 12	1.502E + 04	− 4.597E + 00
7.254E + 03	2.208E + 00	2.237E − 10	1.502E + 04	− 5.131E + 00
3.060E + 03	5.239E + 00	1.292E − 08	1.507E + 04	− 4.656E + 00
1.290E + 03	1.242E + 01	1.190E − 07	1.502E + 04	− 6.095E + 00
5.440E + 02	2.945E + 01	5.933E − 07	1.502E + 04	− 6.406E + 00
2.294E + 02	6.985E + 01	1.630E − 06	1.502E + 04	− 6.639E + 00
9.675E + 01	1.656E + 02	5.479E − 06	1.502E + 04	− 6.814E + 00
4.090E + 01	3.926E + 02	2.714E − 05	1.501E + 04	− 6.976E + 00
1.720E + 01	9.310E + 02	1.368E − 04	1.501E + 04	− 7.040E + 00
7.254E + 00	2.208E + 03	7.141E − 04	1.501E + 04	− 7.125E + 00
3.060E + 00	5.234E + 03	4.017E − 03	1.501E + 04	− 7.151E + 00
1.290E + 00	1.241E + 04	2.225E − 02	1.500E + 04	− 7.187E + 00
2.294E − 01	6.977E + 04	7.006E − 01	1.499E + 04	− 7.070E + 00
1.689E − 01	9.475E + 04	1.291E + 00	1.499E + 04	− 7.005E + 00
9.675E − 02	1.654E + 05	5.932E + 00	1.498E + 04	− 6.812E + 00
4.080E − 02	3.922E + 05	2.208E + 01	1.497E + 04	− 6.174E + 00
7.254E − 03	2.209E + 06	6.913E + 02	1.497E + 04	− 1.045E + 00
3.060E − 03	5.261E + 06	3.821E + 03	1.502E + 04	7.330E + 00
1.290E − 03	1.262E + 07	2.060E + 04	1.517E + 04	2.618E + 01
5.440E − 04	3.077E + 07	1.065E + 05	1.556E + 04	6.649E + 01
2.294E − 04	7.783E + 07	5.003E + 05	1.656E + 04	1.425E + 02
9.675E − 05	2.116E + 08	1.997E + 06	1.896E + 04	2.572E + 02
4.080E − 05	6.477E + 08	6.550E + 06	2.445E + 04	3.889E + 02
1.720E − 05	2.276E + 09	1.886E + 07	3.626E + 04	5.223E + 02
7.254E − 06	8.881E + 09	5.410E + 07	5.979E + 04	6.819E + 02
5.060E − 06	3.647E + 10	1.607E + 08	1.038E + 05	8.949E + 02

Table A5.13. $kT/Z^{4/3} = 5.0E + 04$

ZV	$P/Z^{10/3}$	$-DP/Z^{8/3}$	$E/Z^{7/3}$	$-DE/Z^{5/3}$
2.166E + 03	5.692E + 01	4.922E − 09	7.502E + 04	− 4.025E − 02
1.627E + 02	4.925E + 02	5.478E − 07	7.502E + 04	− 4.224E + 00
6.580E + 01	1.108E + 03	9.321E − 07	7.502E + 04	− 5.004E + 00
2.894E + 01	2.768E + 03	8.232E − 06	7.501E + 04	− 5.588E + 00
1.220E + 01	6.565E + 03	4.798E − 05	7.501E + 04	− 5.904E + 00
5.145E + 00	1.557E + 04	2.627E − 04	7.501E + 04	− 6.354E + 00
2.169E + 00	3.692E + 04	1.585E − 03	7.500E + 04	− 6.597E + 00
9.150E − 01	8.755E + 04	8.855E − 03	7.500E + 04	− 6.777E + 00
3.853E − 01	2.076E + 05	4.949E − 02	7.499E + 04	− 6.903E + 00
6.860E − 02	1.167E + 06	1.564E + 00	7.497E + 04	− 6.975E + 00
5.052E − 02	1.585E + 06	2.885E + 00	7.497E + 04	− 6.947E + 00
2.894E − 02	2.767E + 06	8.788E + 00	7.496E + 04	− 6.850E + 00
1.220E − 02	6.563E + 06	4.941E + 01	7.495E + 04	− 6.458E + 00
2.169E − 03	3.693E + 07	1.558E + 03	7.495E + 04	− 5.046E + 00
9.150E − 04	8.766E + 07	8.714E + 03	7.502E + 04	2.679E + 00
3.638E − 04	2.086E + 08	4.846E + 04	7.524E + 04	1.610E + 01
1.627E − 04	4.990E + 08	2.643E + 05	7.584E + 04	4.674E + 01
6.860E − 05	1.208E + 09	1.595E + 06	7.756E + 04	1.140E + 02
2.894E − 05	3.004E + 09	6.859E + 06	8.105E + 04	2.486E + 02
1.220E − 05	7.903E + 09	2.946E + 07	8.985E + 04	4.755E + 02
5.145E − 06	2.298E + 10	1.044E + 08	1.101E + 05	7.610E + 02
2.169E − 06	7.647E + 10	3.118E + 08	1.545E + 05	1.056E + 03
9.150E − 07	2.877E + 11	8.846E + 08	2.454E + 05	1.380E + 03

Table A5.14. $kT/Z^{4/3} = 1.0E + 05$

ZV	$P/Z^{10/3}$	$-DP/Z^{8/3}$	$E/Z^{7/3}$	$-DE/Z^{5/3}$
4.079E + 01	3.927E + 03	2.964E − 06	1.500E + 05	− 3.409E + 00
1.721E + 01	9.311E + 03	8.939E − 06	1.500E + 05	− 4.393E + 00
7.253E + 00	2.208E + 04	7.279E − 05	1.500E + 05	− 5.271E + 00
3.059E + 00	5.236E + 04	5.236E + 04	1.505E + 05	− 5.682E + 00
1.290E + 00	1.242E + 05	2.267E − 03	1.500E + 05	− 6.097E + 00
5.441E − 01	2.944E + 05	1.249E − 02	1.505E + 03	− 6.400E + 00
2.294E − 01	6.982E + 05	6.985E − 02	1.500E + 05	− 6.622E + 00
4.079E − 02	3.927E + 06	2.212E + 00	1.500E + 05	− 6.837E + 00
3.004E − 02	5.332E + 06	4.077E + 00	1.500E + 05	− 6.834E + 00
1.721E − 02	9.309E + 06	1.243E + 01	1.499E + 05	− 6.789E + 00
7.255E − 03	2.209E + 07	6.990E + 01	1.499E + 05	− 6.489E + 00
1.290E − 03	1.242E + 08	2.207E + 02	1.499E + 05	− 3.662E + 00
5.441E − 04	2.947E + 08	1.237E + 04	1.500E + 05	1.157E + 00
2.294E − 04	7.003E + 08	6.098E + 04	1.503E + 05	1.254E + 01
9.674E − 05	1.669E + 09	3.820E + 05	1.510E + 05	3.901E + 01
4.079E − 05	4.008E + 09	2.059E + 06	1.528E + 05	9.858E + 01
1.721E − 05	9.784E + 09	1.064E + 07	1.572E + 05	2.259E + 02
7.255E − 06	2.478E + 10	5.000E + 07	1.678E + 05	4.660E + 02
3.059E − 06	6.748E + 10	1.998E + 08	1.926E + 05	8.290E + 02
1.290E − 06	2.066E + 11	6.552E + 08	2.485E + 05	1.246E + 03
5.441E − 07	7.249E + 11	1.886E + 09	3.680E + 05	1.667E + 03

Some mathematical relations for Chapter 13

$$P_S = \sum_{k=1}^{N} a_k{}^S (\sigma - 1)^k, \quad \sigma \equiv \frac{\rho}{\rho_0} = \frac{V_0}{V}$$

$$\frac{\mathrm{d}P_S}{\mathrm{d}V} = - \sum_{k=1}^{N} k a_k{}^S \frac{V_0}{V^2} (\sigma - 1)^{k-1}$$

$$\frac{\mathrm{d}^2 P_S}{\mathrm{d}V^2} = \sum_{k=1}^{N} k a_k{}^S \left\{ (k-1)(\sigma - 1)^{k-2} \frac{V_0{}^2}{V^4} + (\sigma - 1)^{k-1} \frac{2V_0}{V^3} \right\}$$

$$\frac{\mathrm{d}^3 P_S}{\mathrm{d}V^3} = - \sum_{k=1}^{N} k a_k{}^S \left\{ (k-2)(k-1)(\sigma - 1)^{k-3} \frac{V_0{}^3}{V^6} \right.$$

$$\left. + 6(k-1)(\sigma - 1)^{k-2} \frac{V_0{}^2}{V^5} + 6(\sigma - 1)^{k-1} \frac{V_0}{V^4} \right\}$$

Same for P_H with $a_k{}^S \to A_k{}^H$

$$\left(\frac{\mathrm{d}P_S}{\mathrm{d}V} \right)_{V_0} = - \frac{a_1{}^S}{V_0}$$

$$\left(\frac{\mathrm{d}^2 P_S}{\mathrm{d}V^2} \right)_{V_0} = \frac{2!}{V_0{}^2} (a_1{}^S + a_2{}^S)$$

$$\left(\frac{\mathrm{d}^3 P_S}{\mathrm{d}V^3} \right)_{V_0} = - \frac{3!}{V_0{}^3} (a_1{}^S + a_2{}^S + a_3{}^S), \quad \text{etc.}$$

$$E_S - E_0 = - \int_{V_0}^{V} P_S \mathrm{d}V = - \int_{V_0}^{V} \sum_{k=1}^{N} a_k{}^S \left(\frac{V_0}{V} - 1 \right)^k \mathrm{d}V$$

For $V \rightarrow V_0$

$$\int_{V_0}^{V} \left(\frac{V_0}{V} - 1 \right)^k dV \approx \frac{1}{V_0^k} \int_{V_0}^{V} (V_0 - V)^k dV = - \frac{(V_0 - V)^{k+1}}{(k+1)V_0^k}$$

$$E_S - E_0 = \sum_{k=1}^{N} \frac{a_k^S (V_0 - V)^{k+1}}{(k+1)V_0^k}$$

$$P_S - P_H = \sum_{k=1}^{N} \frac{(a_k^S - A_k^H)(V_0 - V)^k}{V_0^k}$$

$$E_S - E_H = \sum \frac{(a_k^S - \frac{1}{2}A_k^H)(V_0 - V)^{k+1}}{V_0^k}$$

A note on the Lawson criterion

The fusion energy is given by

$$E_{\text{fusion}} = \tfrac{4}{3}\pi r^3 n_D n_T < \sigma v > w\tau \quad [\text{erg}] \tag{A7.1}$$

where $w = 17.6\,\text{MeV}$ for D–T, and assuming $n_D = n_T = n/2$ and plasma neutrality give $n_e = n$, one has ($\tau =$ is the nuclear reaction time)

$$E_{\text{fusion}} = \frac{\pi}{3} r^3 n^2 \langle \sigma v \rangle w\tau \tag{A7.2}$$

$$E_{\text{thermal}} = \tfrac{4}{3}\pi r^3 \cdot \tfrac{3}{2} nk(T_e + T_i) \tag{A7.3}$$

assuming $T_e = T_i = T$

$$E_{\text{thermal}} = 4\pi r^3 nkT \tag{A7.4}$$

From (A7.2) and (A7.4) in order to get energy gain

$$E_{\text{fusion}} \geqslant E_{\text{thermal}} \tag{A7.5}$$

so that break-even is defined by

$$E_{\text{fusion}} = E_{\text{thermal}} \tag{A7.6}$$

Now using (A7.2) and (A7.4) in (A7.6) one has

$$n\tau = \frac{12kT}{w\langle \sigma v \rangle} \tag{A7.7}$$

And for D–T process (A7.7) yields for $kT = 10\,\text{keV}$

$$n\tau = 10^{14}\,\text{cm}^{-3}\text{s} \quad \textit{Lawson criterion} \tag{A7.8}$$

Derivation of the equation describing hydrostatic equilibrium for a completely degenerate gas

The starting point is (15.234)

$$\frac{A}{B}\frac{1}{r^2}\frac{d}{dr}\left[\frac{r^2}{x^3}\frac{df(x)}{dr}\right] = -4\pi GB[x(r)]^3 \tag{A8.1}$$

where

$$f(x) = (2x^3 - 3x)(1 + x^2)^{1/2} + 3\sinh^{-1}x \tag{A8.2}$$

$$x = x(r) = \frac{\hbar}{m_c}[3\pi^2 n_e(r)]^{1/3} \tag{A8.3}$$

and $n_e(r)$ represents the electron density; other symbols have been defined in Section 15.11. Thus

$$\frac{df(x)}{dr} = \frac{df(x)}{dx}\frac{dx}{dr}$$

$$= \frac{8x^4}{(1 + x^2)^{1/2}}\frac{dx}{dr} \tag{A8.4}$$

or

$$\frac{r^2}{x^3}\frac{df(x)}{dr} = r^2\frac{8x}{(1 + x^2)^{1/2}}\frac{dx}{dr} = 8r^2\frac{d}{dr}(1 + x^2)^{1/2}$$

$$= 8r^2\frac{dy(r)}{dr} \tag{A8.5}$$

where

$$y(r) \equiv [1 + x^2(r)]^{1/2} \tag{A8.6}$$

Thus (A8.1) can be written in the form

$$\frac{1}{r^2}\frac{d}{dr}\left[r^2\frac{dy(r)}{dr}\right] = -\frac{\pi GB^2}{2A}[y^2(r) - 1]^{3/2} \tag{A8.7}$$

which is the same as (15.235).

REFERENCES

Chapter 1

Brush, S.G. (1967). *Prog. High Temp. Phys. Chem.*, ed. C.A. Rouse, 1, 1.
Bushman, A.V. & Fortov, V.E. (1983). *Sov. Phys. Usp.* 26, 465.
Dirac, P.A.M. (1926). *Proc. Roy. Soc.* A112, 661.
Euler, H. (1936). *Ann. Physik* 26, 398.
Fermi, E. (1928). *Z. Phys.* 48, 73.
Golden, K.I. (1983). *Laser Part. Beams* 1, 1.
Görlich, P., Hora, H. & Macke, W. (1957). *Exp. Tech. Phys.* 5, 217.
Görlich, P. & Hora, H. (1958). *Optik* 15, 116.
Griem, H.R. (1981). *Laser Interaction and Related Plasma Phenomena*, eds. H. Schwarz *et al.* Vol. 5, p. 811. New York: Plenum.
Hansen, J.P. (1973). *Phys. Rev.* A8, 3096.
Harrison, E.R. (1964). *Proc. Phys. Soc.* 84, 213.
Harrison, E.R. (1965). *Astrophys. J.* 142, 1643.
Heisenberg, W. (1934). *Z. Phys.* 90, 209.
Henry, B.I. (1983). *Laser Part. Beams* 1, 11.
Henry, B.I. & Hora, H. (1983). *Opt. Comm.* 44, 218.
Hora, H. & Müller, H. (1961). *Z.Phys.* 164, 360.
Hora, H. (1978). *Lettre Nuovo Cim.* 22, 55.
Hora, H. (1981). *Physics of Laser Driven Plasmas.* New York: John Wiley.
Hora, H. (1981a). *Nuovo Cim.* 64B, 1.
Hora, H. & Romatka, R. (1982). *Naturwissensch*, 69, 399.
Hora, H., Lalousis, P. & Eliezer, S. (1984). *Phys. Rev. Lett.* 53, 1650.
Ichimaru, S. Tanaka, S. & Iyetomi, H. (1984). *Phys. Rev.* 29A, 2033.
Inglis, D.R. & Teller, E. (1939). *Astrophys. J.* 90, 439.
Joos, G. (1976). *Theoretische Physik.* p. 537. Wiesbaden: AVG.
Kippenhahn, R. (1983). *One Hundred Billion Suns*, New York: Basic Book Inc.
Madelung, E. (1918). *Z. Phys.* 19, 524.
Niu, K. (1981). *Proc. Topical Seminar Part. Beam Appl. Fusion Res.*, Nagoya Univ. Inst. Plasma Phys.
Pauli, W. (1957), *Nuovo Cim.* 6, 204.
Pines, D. & Noziéres (1966). *Theory of Quantum Fluids.* New York: Benjamin.
Scheid, W., Müller, H. & Greiner, W. (1974). *Phys. Rev. Lett.* 32, 741.
Schottky, W. (1920). *Ann. Physik* 64, 142.
Stock, R. (1984). *Phys. Bl.* 40, 278.
Stöcker, H., Greiner, W. & Scheid, W. (1978) *Z. Phys.* A286, 121.
Stöcker, H. (1986). *Phys. Today*, 39, 5–49.
Traving, G. (1959). *Mitteil. Astron. Ges.*, No. 1, Hamburg.

Wulfe, H. (1976). *Nuovo Cim.* **31B**, 92.
Zeldovich, Y.B. & Novikov, I.D. (1965). *Sov. Phys.* Usp. **8**, 522.

Chapter 2

Becker, R. (1955). *Theorie der Wärme.* Heidlberg: Springer.
Einstein, A. (1916). *Mitt. Phys. Ges.* No. 18, Zürich.; Phys. Z. **18**, 121.
Hora, H. (1981). *Physics of Laser Driven Plasmas.* New York: John Wiley.
Joos, G. (1949). *Theoretical Physics*, p. 554. London: Blackie & Son.
Landau, L.D. & Lifshitz, E.M. (1966). *Theoretical Physics*, vol. 5. Oxford: Pergamon.
Planck, M. (1900). *Ann. Physik* (4) **1**, 719.

Chapter 4

Ghatak, A.K. & Lokanathan, S. (1984). *Quantum Mechanics.* New Delhi: Macmillan India.
Gopal, E.S.R. (1974). *Statistical Mechanics and Properties of Matter.* Chichester: Ellis Horwood.
Hill, T.L. (1960). *Introduction to Statistical Thermodynamics.* Reading, Mass: Addison Wesley.
Mayer, J.E. & Mayer, M.G. (1940). *Statistical Mechanics.* New York: John Wiley.
Pathria, R.K. (1972). *Statistical Mechanics.* Oxford: Pergamon.
Saha, M.N. & Srivastava, B.M. (1958). *A Treatise on Heat.* Allahabad: The Indian Press.

Chapter 5

Gopal, E.S.R. (1974). *Statistical Mechanics and Properties of Matter.* Chichester: Ellis Horwood.
Hill, T.L. (1960). *An Introduction to Statistical Thermodynamics.* Reading, Mass: Addison Wesley.
Landau, L.D. & Lifshitz, E.M. (1958). *Statistical Physics.* Oxford: Pergamon.
Landsberg, P.T. (1954). On Bose–Einstein condensation, *Proc. Camb. Phil. Soc.*, **50**, 65.
Landsberg, P.T. (1961). *Thermodynamics with Quantum Statistical Illustrations.* New York: Interscience Publishers.
Lee, T.D., Huang, K. & Yang, C.N. (1957). *Phys. Rev.* **106**, 1135.
Pathria, R.K. (1972). *Statistical Mechanics.* Oxford: Pergamon Press.
Ter Haar, D. (1962). Elements of quantum statistics. In *Quantum Theory II: Aggregates of Particles*, ed. D.R. Bates. New York: Academic Press.
Thyagarajan, K. & Ghatak, A.K. (1981). *Lasers: Theory and Applications.* New York: Plenum Press.

Chapter 6

Abramowitz, & Stegun (1965). *Handbook of Mathematical Functions.* New York: Dover.
Brush, S.G. (1967). *Prog. High Temp. Phys. Chem.* C.A. Rouse **1**, 1.
Chandrasekhar, S. (1939). *An Introduction to the Study of Stellar Structure.* Chicago: University of Chicago Press. Now available in Dover edition.
Chandrasekhar, S. (1939). *An Introduction to the Study of Stellar Structure.* Chicago:
Ghatak, A.K. & Kothari, L.S. (1972). *An Introduction to Lattice Dynamics.* Reading, Mass.: Addison Wesley.
Kittel, C. (1956). *Introduction to Solid State Physics.* New York: John Wiley.
Pathria, R.K. (1972). *Statistical Mechanics.* Oxford: Pergamon.

Chapter 7

Brush, S.G. (1967). *Prog. High Temp. Phys. Chemi.* ed: C.A. Rouse, **1**, 1.
Landau, L.D. & Lifshitz, E.M. (1969). *Statistical Physics.* Oxford: Pergamon.
Saha, M.N. & Srivastava, B.N. (1965). *A Treatise on Heat.* Allahabad: The Indian Press.
Zemansky, M.W. (1957). *Heat and Thermodynamics.* New York: McGraw Hill.

Chapter 8

Brush, S.G. (1967). *Prog. High Temp. Phys. Chem.* ed. C.A. Rouse **1**, 1.
Debye, P. & Hückel, E. (1923a). *Physik Z.* **24**, 185.
Debye, P. & Hückel, E. (1923b). *Physik Z.* **24**, 305.
Henry, B.I. & Hora, H. (1982). *Opt. Comm.* **44**, 195.
Henry, B.I. (1983). *Lazer and Particle Beams.* **1**, 11.
McQuarrie, D.A. (1976). *Statistical Mechanics,* New York: Harper & Row.
Milner, S.R. (1913). *Phil. Mag.* **25**, 660.
Milner, S.R. (1913a). *Phil. Mag.* **25**, 742.

Chapter 9

Al'Tshuler, L.V. Bakanova, A.A. & Trunin, R.F. (1962). *Sov. Phys., JETP,* **15**, 65.
Beer, A.C., Chase, M.N. & Choquard, P.F. (1955). *Helv. Phys. Acta,* **28**, 529.
Bloch, F. (1929). *Zeits. f. Phys.,* **57**, 545.
Brillouin, L. (1934). *Actualites Sci. Industr.,* No. 160, Paris: Hermann.
Brush, S.G. (1967). *Prog. High Temp. Phys. Chem.,* ed. C.A. Rouse, **1**, 1.
Chandrasekhar, S. (1939). *Introduction to the Study of Stellar Structure.* Chicago: University of Chicago Press. Now available in Dover edition.
Cowan, R.D. & Ashkin, J. (1957). *Phys. Rev.,* **105**, 144.
Fermi, E. (1928). *Zeits. f. Phys.,* **48**, 73.
Feynman, R.P., Metropolis, N. & Teller, E. (1949). *Phys. Rev.,* **75**, 1561.
Ghatak, A.K. & Lokanathan, S. (1984). *Quantum Mechanics,* New Delhi: Macmillan.
Gilvarry, J.J. (1954). *Phys. Rev.,* **96**, 934.
Goldstein, H. (1950). *Classical Mechanics.* Reading, Mass: Addison Wesley.
Gombas, P. (1949). *Die statistische theorie des atoms und ihre anwendungen.* Wien: Springer–Verlag.
Hora, H. (1981). *Physics of Laser Driven Plasmas.* New York: John Wiley.
Jensen, H. (1938). *Zeits. f. Physik,* **111**, 373.
Jensen, H., Meyer–Gossler, G. & Rohde, H. (1938). *Zeits. f. Physik,* **110**, 277.
Kirzhnits, D.A. (1957). *Sov. Phys. JETP,* **5**, 64.
Kirzhnits, D.A. (1959). *Sov. Phys. JETP,* **8**, 1081.
Kompaneets, A.S. & Pavlovskii, E.S. (1957). *Sov. Phys. JETP,* **4**, 328.
Landau, L.D. & Lifshitz, E.M. (1965). *Quantum Mechanics: Non Relativistic Theory.* Oxford: Pergamon.
Latter, R. (1955). *Phys. Rev.,* **99**, 1854.
March, N.H. (1957). *Adv. Phys.,* **6**, 1.
McCarthy, S.L. (1965). *The Kirzhnits Corrections to the Thomas-Fermi Equation of State.* Lawrence Radiation Laboratory Report UCRL–14364.
McDougall, J. & Stoner, E.C. (1938). *Trans. Roy. Soc.* A237, 67.
Molière, G. (1947). *Zeits. f. Naturforsch.,* **2A**, 133.
Moore, R.M. (1981). Atomic Physics in Inertial Confinement Fusion. Lawrence Livermore Laboratory Report UCRL–84991, Parts I and II.
Pathria, R.K. (1972). *Statistical Mechanics.* Oxford: Pergamon.
Plaskett, J.S. (1953). *Proc. Phys. Soc.,* A66. 178.
Saha, M.N. & Srivastava, B.N. (1958). *A Treatise on Heat.* Allahabad: The Indian Press.

Schwinger, J. (1980), *Phys. Rev.*, A22, 1827; see also Schwinger, J. (1981). *Phys. Rev.*, A24, 2353; De Raad, L.L., & Schwinger, J. (1982). *Phys. Rev.* A25, 2399.
Slater, J.C. & Krutter, H.M., (1935). *Phys. Rev.*, 47, 559.
Sommerfeld, A. (1932). *Zeits. f. Physik*, 78, 283.
Thomas, L.H. (1927). *Proc. Camb. Phil. Soc.*, 23, 542.
Torrens, I.M. (1972). *Interatomic Potentials*. New York: Academic Press.
Umeda, K. & Tomishima, Y. (1953). *J. Phys. Soc. Japan*, 8, 360.

Chapter 10

Altshuler, L.V. (1965). *Sov. Phys. Usp.* 8, 52.
Born, M. & Huang, K. (1954). *Dynamical Theory of Crystal Lattices*. Oxford: Clarendon Press.
Brush, S.G. (1967). *Prog. High Temp. Phys. Chem.*, vol. 1, 1.
Debye, P. (1912). *Ann. Physik*, 39, 789.
Dugdale, J.S. & McDonald, D.K.C. (1953). *Phys. Rev.* 89, 832.
Einstein, A. (1907). *Ann. Physik*, 22, 180
Einstein, A. (1911). *Ann. Physik*, 34, 170
Grüneisen, E. (1926). *Handbuch der Physik*, ed. H. Greiger & K. Scheel, 10, 1–59. Berlin: Springer.
Kittel, C. (1956). *Introduction to Solid State Physics*. New York: John Wiley, pp. 153–5.
Romain, J.P. Migault, A. & Jacquesson, J. (1979). *High Pressure Science and Technology*, vol. 1, p. 99, eds. K.D. Timmerhaus & M.S. Barber, New York: Plenum.
SESAME library, Bennett, B.I. Johnson, J.D. Kerley, G.I. & Rood, G.T. (1978). Los Alamos Scientific Laboratory, LA-7130.
Slater, J.C. (1939). *Introduction to Chemical Physics*. New York: McGraw Hill.

Chapter 11

Zeldovich, Y.B. & Raizer, Y.P. (1966). *Physics of Shock Waves and High-Temperature Hydrodynamic Phenomena*, vol. I. New York: Academic Press.
Zeldovich, Y.B. & Raizer, Y.P. (1967). *Physics of Shock Waves and High-Temperature Hydrodynamic Phenomena*, vol. II. New York: Academic Press.

Chapter 12

Blatt, J.M. (1959). *Prog. Theor. Phys.* 22, 745.
Blatt, J.M. & Opie, A.H. (1974). *J. Phys.* A7, 1895.
Denisse, J.F. & Delcroix, J.L. (1963) *Plasma Waves*, New York: John Wiley.
Eliezer, S. & Ludmirsky, A. (1983). *Laser Part. Beams*, 1, 251.
Hansen, J. (1980). *Laser Plasma Interactions*, p. 433, ed. R.A. Cairns, St. Andrews: Scottish University.
Hora, H. Lalousis. P. & Jones, D.A. (1983). *Phys. Lett.* 99A, 89.
Lalousis, P. & Hora, H. (1983). *Laser Part. Beams*, 1, 287.
Ray, P.S. & Hora, H. (1977). *Zeits. f. Naturforsch.* 32A, 538.
Valeo, E.J. & Kruer, W.L. (1974). *Phys. Rev. Lett.* 33, 750.

Chapter 13

Anisimov, S.I., Prokhorov, A.M. & Fortov, V.E. (1984). *Sov. Phys.* Usp. 27, 181.
Altshuler, L.V. (1965). *Sov. Phys.* Usp. 8, 52.
Altshuler, L.V., Kormer, S.B., Bakanova, A.A. & Trunin, R.F. (1960). *Sov. Phys. JETP*, 11, 573.

Altshuler, L.V., Drupnikov, K.K., Ledenev, B.N., Zhudukin, V.I. & Brazhnik, M.I. (1958). *Sov. Phys. JETP,* **34**, 606.
Davison, L. & Graham, R.A. (1979). *Phys. Rev.* C**55**, 255.
Dugdale, I.S. & McDonald, D. (1953). *Phys. Rev.* **89**, 832.
Kormer, S.B. & Urlin, V.D. (1960). *Sov. Phys.* Doklady, **5**, 317.
Krupnikov, K.K., Bakanova, A.A., Brezhnik, M.I. & Trunin, R.F. (1963). *Sov. Phys.* Doklady **8**, 205.
McQueen, R.G. & Marsh, S.P. (1960). *J. Appl. Phys.* **31**, 1253.
McQueen, R.G., Marsh, S.P., Taylor, J.W., Fritz, J.N. & Carter, W.J. (1970). The equation of state of solids from shock wave studies. In *High Velocity Impact Phenomena*, ed. R. Kinslow, pp. 293–417, 515–68. New York: Academic Press.
Ragan, C.E., III (1980). *Phys. Rev.* A**21**, 458.
Rice, M.H., McQueen, R.G. & Walsh, J.M. (1958). Compression of solids by strong shock waves. In *Solid state Physics*, vol. 6 eds. F. Seitz & D. Turnbull, pp. 1–63. New York: Academic Press.
Salzmann, D., Eliezer, S., Krumbein, A.D. & Gitter, L. (1983). *Phys. Rev.* A**28**, 1738.
Zeldovich, Y.B. & Raizer, Y.P. (1967). *Physics of Shock Waves and High Temperature Hydrodynamic Phenomena*, vols. I and II. New York: Academic Press.

Chapter 14

Ahlstrom, H.G. (1982). *Physics of Laser Fusion*, vol. II, p. 331. Lawrence Livermore Laboratory.
Alexandrov, V.V., Kovalski, N.G., Pergament, M.I. & Rubenchik (1984). *Laser Part. Beams* **2**, 211.
Christiansen, J.P., Asby, D.E.T.F. & Roberts, K.V. (1974) *Comp. Phys. Comm.* **7**, 271.
Cicchitelli, L., Dragila, R., Golden, K.I., Hora, H., Jones, D.A., Kemeny, L.G., Kentwell, G.W., Lalousis, P., Goldsworthy, M.P. & Tubbenhauer, G. (1984). *Laser Interaction and Related Plasma Phenomena*, vol. 6, p. 437. New York: Plenum.
Clark, R.G., Hora, H., Ray, P.S. & Titterton, E.W. (1978). *Phys. Rev.* C**18**, 1127.
Cooper, N.G. (ed.) (1979). *An Invitation to LASL-EOS Library*. Los Alamos, LASL-79-62.
Cooper, N.G. (ed.) (1983). *Equation of State and Opacity Group SESAME 83*. Los Alamos, EOS-Library, LALP-83-4.
Destler, W.W., O'Shea, P.G. & Reiser, M. (1984). *Phys. Rev. Lett.* **52**, 1978.
Destler, W.W., O'Shea, P.G. & Reiser, M. (1984a). *Phys. Fluid.* **27**, 1897.
DeWitt, H.E. (1984). *Equation of State of Strongly Coupled Plasma Mixtures*, LLNL Report UCRL-90357, *J. Qu. Spectr. Rad. Transf.* (submitted).
Eliezer, S. & Ludmirsky, A. (1983). *Laser Part. Beams* **2**, 251.
Emmett, J.H., Nuckolls, J.H. & Wood, L. (1974). *Sci. Am.* **230**, 24.
Glass, A.J. (1983). *Laser Focus*, No. 2.
Hora, H. (1964). *Inst. Plasmaphysik, Garching, Rept.* 6/23, U.S. Govt. Res. Rept. Nat. Res. Council – Tech Translat. – 11931.
Hora, H. (1981). *Physics of Laser Driven Plasmas*, New York: Wiley.
Hora, H. & Pfirsch, D. (1972). *Laser Interaction and Related Plasma Phenomena*, H. Schwarz *et al.* eds. vol. 2, p. 515. New York: Plenum Press.
Hora, H., Castillo, R., Clark, R.G., Kane, E.L., Lawrence, V.F., Miller, D.C., Nicholson-Florence, M.N., Novak, M.M., Ray, P.S., Shepanski, J.R. & Tsivinsky, A.I. (1979). *Plasma Physics and Controlled Nuclear Fusion Research 1978*, Vol. 3, p. 237. Vienna: IAEA.
Hora, H., Lalousis, P. & Jones, D.A. (1983). *Phys. Lett.* **99A**, 87.
Hora, H. & Miley, G.H. (1984). *Laser Focus* **20**, No. 2, 59.
Hora, H., Lalousis, P. & Eliezer, S. (1984a). *Phys. Rev. Lett.* **53**, 650.

Hora, H. & Ghatak, A.K. (1985). *Phys. Rev.* **31A**, 3473.

Kalitkin, N.N. (1960). *Sov. Phys. JETP* **8**, 1106.

Kormer, S.B., Funtikov, A.I., Vrlin, V.D. & Kolesnikova, A.N. (1962). *Sov. Phys. JETP* **15**, 477.

Kidder, R.E. (1974). *Nucl. Fusion* **14**, 63.

Kirshnits, D.A. (1959). *Sov. Phys. JETP* **8**, 1081.

Lawson, J.D. (1957). *Proc. Phys. Soc.* **B70**, 6.

Linhart, J.G. (1984). *Laser Part. Beams* **2**, 87.

McCarthy, S.L. (1965). *The Kirshnits Correction Thomas–Fermi Equation of State*, UCRL-14364.

Meyer, F.J., Osborn, R.K., Daniels, D.W. & McGrath, J.F. (1978). *Phys. Rev. Lett.* **40**, 30.

Meyer-ter-Vehn, F. *et al.* (1984). *Equation of State Effects on Laser Accelerated Foils*, Report Max-Planck-Institute for Quantum Optics, Garching.

Meldner, H.W. (1981). In *Encyclopedia of Physics*, p. 497, eds. R.G. Lerner & G.L. Trigg Reading, Mass: Addison-Wesley.

More, R.M. (1981). In *Laser Interaction and Related Plasma Phenomena*, eds. H. Schwarz, H. Hora *et al.* Vol 5. p. 255, New York: Plenum.

Nuckolls, J.L., Wood, L., Thiessen, A. & Zimmerman, G. (1972). *Nature* **239**, 139.

Oliphant, M.L.E., Harteck, P. & Lord Rutherford (1934). *Proc. Roy. Soc.* **A144**, 692.

Slater, D.C. (1977). *Phys. Rev. Lett.* **31**, 196.

Spitzer, L. Jr. (1962). *Physics of Fully Joined Gases*. New York: John Wiley.

Szichman, H., Krumbein, A.D. & Eliezer, S. (1983). *Two-Temperature Equations of State for Aluminium*. SOREQ, Yavne, IA-1390.

Tahir, N.A. & Long, K.A. (1984). *Nuclear Fusion*.

Thompson, S.L. & Lanson, H.S. (1972). *Improvements in the CHART-D Code III: Revised Analytical EOS*. Albuquerque: Scandia, SC-RR-710714.

Trainor, R.J., Graboske, H.C., Long, K.S. & Shaner, J.W. (1978). *Application of High Power Lasers to EOS Research at Ultralight Pressures*, Los Alamos Res. Lab. Rept. 52567.

Yamanaka, C. (1981). *European Conference on Plasma Physics & Fusion Energy*, Sept., Moscow.

Yamanaka, C. (1984). *Review on Inertial Confinement Fusion in Japan* (*Heavy Ions Fusion Conference*, Jan. Tokyo), see *Laser Part. Beams* **2**, No. 4.

Yamanaka, C. *et al.* (1986). *Laser Interaction and Related Plasma Phenomena*, vol. 1. H. Hora & G.H. Miley Eds. New York, Plenum.

Yonas, G. (1984). *Nuclear Fusion*, **24**, 505.

Chapter 15

Aller, H. (1953). *Astrophysics: The Atmospheres of the Sun and Stars*. New York: Ronald Press.

Aller, H. (1954). *Astrophysics: Nuclear Transformations, Stellar Interiors and Nebulae*. New York: Ronald Press.

Aller, H. & McLaughlin, D.B. (eds.) (1965). *Stellar Structure*. Chicago: University of Chicago Press.

Chandrasekhar, S. (1935). *Monthly Notices of the Royal Astronomical oc.* **95**, 201.

Chandrasekhar, S. (1939). *An Introduction to the Study of Stellar Structure*. Chicago: University of Chicago Press. Reprinted by Dover Publications, New York, 1957.

Cox, J.P. & Giuli, R.T. (1968a). *Principles of Stellar Structure, Vol. 1: Physical Principles*. New York: Gordon and Breach.

Cox, J.P. & Giuli, R.T. (1968b). *Principles of Stellar Structure, Vol. 2: Applications to Stars*. New York: Gordon and Breach.

Giacconi, R. & Ruffini, R., eds. (1978). *Proceedings of the International School of Physics, Course LXV: Physics and Astrophysics of Neutron Stars and Black Holes.* Amsterdam: North Holland Publishing.

Goldstein, H. (1950). *Classical Mechanics.* Reading, Mass.: Addison Wesley.

Hayashi, C., Hoshi, R. & Sugimoto, D. (1962). *Prog. Theor. Phys. Suppl.* No. 22.

Jeans, J. (1928). *Astronomy and Cosmogony.* Cambridge University Press.

Menzel, D., Bhatnagar, P.L. & Sen, H.K. (1963). *Stellar Interiors.* London: Chapman & Hall.

Mestel, H. (1965). The Theory of White Dwarfs. In *Stellar Structure*, eds. Aller and McLaughlin, Chicago: University of Chicago Press.

Riez, A. (1949). *Astrophys. J.*, **109**, 303.

Schatzman, E. (1958). *White Dwarfs.* Amsterdam: North Holland Publishing.

Schwarzschild, M. (1958). *Structure and Evolution of the Stars.* Princeton, N.J.: Princeton University Press.

Stein, R.F. & Cameron, A.G.W., eds. (1966). *Stellar Evolution.* New York: Plenum Press.

Stodolkiewicz, J.S. (1973). *General Astrophysics.* New York: American Elsevier Publishing.

Zeldovich, Y.B. & Novikov, I.D. (1971). *Relativistic Astrophysics, Vol. 1: Stars and Relativity.* Chicago: University of Chicago Press.

Chapter 16

Chew, E.G. (1961). *S-Matrix Theory of Strong Interactions.* New York: Benjamin.

Chew, E.G. & Frautschi, S.C. (1961). *Phys. Rev. Letts.*, **7**, 394.

Chew, E.G. & Mandelstam, S. (1961). *Nuovo Cimento* **19**, 752.

Dolan, L. & Jackiw, R. (1974). *Phys. Rev.* D**9**, 3320.

Eliezer, S. & Weiner, R. (1974). *Phys. Letts.* **50B**. 4631.

Eliezer, S. (1975). *Phys. Rev.* D**12**, 1739.

Eliezer, S. & Weiner, R. (1976). *Phys. Rev.* D**13**, 87.

Eliezer, S. & Weiner, R. (1978). *Phys. Rev.* D**18**, 879.

Fermi, E. (1950). *Prog. Theor. Phys.* **5**, 570.

Goldstone, J. (1961). *Nuovo Cimento* **19**, 154.

Hagedorn, R. (1971). *Thermodynamics of strong interactions.* CERN 71–12.

Imaeda, K. (1967). *Nuovo Cimento* **48A**, 482.

Kirzhnits, D.A. & Linde, A.D. (1972). *Phys. Letts.* **42B**, 471.

Koppe, H. (1948). *Zeits. f. Naturforsch*, **3A**, 251.

Koppe, H. (1949). *Phys. Rev.* **76**, 688.

Nambu, Y. & Jona-Lasinio, G. (1961). *Phys. Rev.* **122**, 345.

Salam, A. (1969). Weak and electromagnetic interactions. In *Elementary Particle Theory: Relativistic Groups and Analyticity*, Nobel Symposium No. 8, ed. N. Svartholm, p. 367. New York: John Wiley.

Weinberg, S. (1967). *Phys. Rev. Lett.* **19**, 1264.

Weinberg, S. (1974). *Phys. Rev.* D**9**, 3357.